The Use of Satellite Data
in Rainfall Monitoring

The Use of Satellite Data in Rainfall Monitoring

ERIC C. BARRETT

Department of Geography
University of Bristol, U.K.

DAVID W. MARTIN

Space Science and Engineering Centre,
University of Madison, Wisconsin, U.S.A.

1981

ACADEMIC PRESS
A Subsidiary of Harcourt Brace Jovanovich, Publishers
London   New York   Toronto   Sydney   San Francisco

ACADEMIC PRESS INC. (LONDON) LTD
24–28 Oval Road,
London NW1

U.S. Edition published by
ACADEMIC PRESS INC.
111 Fifth Avenue,
New York, New York 10003

British Library Cataloguing in Publication Data

Barrett, E. C.
   The use of satellite data in rainfall monitoring
   1. Rain and Rainfall – Measurement
   2. Meteorological satellites
   I. Title   II. Martin, D. W.
   571.57′781   QC925
   ISBN 0-12-079680-5

   LCCCN 81-66677

Printed in Great Britain by
W & G Baird Ltd

Preface

The invitation to prepare a book concerned with the use of satellites in monitoring rainfall came unexpectedly. The proposal, however, was of immediate interest because of our long involvement in this field of enquiry, our complementary concerns with different methods and types of applications, and our judgment that an overview of progress in monitoring rainfall from satellite observations is needed.

In writing this volume we have attempted to summarize present methods designed to augment conventional rainfall data using satellites to provide information on rainfall occurrence, rate, amount, and distribution; to denote primary uses of such methods in meteorology, climatology, hydrology, and allied environmental sciences; and to examine the prospects for obtaining still further improved rainfall data in the future, especially through new technological developments, improvements in inter-relations between different types of observational networks, and better data processing and analysis.

It is our hope that this book will be of interest and value to many scientists—both pure and applied—whose subjects involve in some direct or indirect way questions in the realm of rainfall hydrometeorology. In preparing it, we have tried less to be exhaustive in our review of the literature than to identify the more significant work which has been undertaken, and reported mostly in the refereed literature; we have tried to provide the readers with that critical synthesis of the field which we believe is necessary at this time. In order to ensure that the book will be of interest and help to many for whom satellite methods might be of real practical significance, but for whom satellite meteorology is a relatively unfamiliar field, we have set our critical review of methods and applications of satellites in rainfall monitoring within a framework of basic facts and principles. Thus, it is our hope that, as a textbook of science, this book will be seen by all to be suitably self-explanatory and self-contained; we also hope that, as a report and discussion of research and research results, it will be seen by all to be fair to the work which has been done in its field, and that it may go on to be seminal to some of what is yet to come.

In the preparation of this book we have depended much on the help and advice of friends and colleagues in many parts of the world. Our thanks are due to all of them, but especially to Drs Jan-Hwa Chu, Roger Davies, Frederick R. Mosher, and Edward J. Zipser, who read parts of the manuscript, and to Professor Verner E. Suomi and Mr Thomas O. Haig, who provided essential support to one of us (DWM) during the course of this work. Thanks also are due to Angela Crowell, Sandra Bero and Gillian Barrett for skill and patience in typing various drafts of the manuscript.

Finally, we recall the comment Dr Will Kellogg made fifteen years ago in opening a workshop on Satellite Data in Meteorological Research: "But just keep in mind that anybody who can measure the precipitation pattern of the atmosphere from a satellite gets some kind of fur-lined teapot." It seems we have not one, but many candidates for the teapot award.

August 1981   ERIC C. BARRETT
                                                       DAVID W. MARTIN

Contents

Preface   v

PART I: Present Problems and the Scope for their Solution

CHAPTER 1
Monitoring Rainfall: The Nature of the Problem   3

I.   Characteristics of rainfall   4
     1. The variability of rainfall   4
     2. The processes of rainfall   9
     3. The types of rainfall   11
II. The measurement of rainfall by conventional means   12
III. The need for rainfall monitoring by satellites   14

CHAPTER 2
Satellite Remote Sensing Systems of Use in
Rainfall Monitoring   17

I.   The use of satellite remote sensing   17
II. Satellite remote sensing systems   18
III. The utility of environmental satellites in weather
     monitoring   20
IV. Existing satellite systems for rainfall monitoring   22

PART II: Satellite Rainfall Monitoring Methods

CHAPTER 3
Preliminary Studies   31

I.   Cloud brightness and rainfall   31
     1. Snapshot views   31
     2. Time lapse views   37

II. Cloud top temperature and rainfall   39
III. Cold, bright clouds   40
IV. Summary   41

CHAPTER 4
Cloud-indexing Methods   43

I.   Introduction   43
II. Bristol methods   44
III. Ness methods   52
IV. Related methods   59

CHAPTER 5
Life-history Methods   65

I.   Stout, Martin and Sikdar   65
II. Griffith and Woodley   69
III. Scofield and Oliver   75
IV. Summary   81

CHAPTER 6
Bi-spectral and Cloud Model Methods   82

I.   Bi-spectral schemes   82
     1. An example of bi-spectral thresholding   83
     2. Early work   83
     3. Statistical population discrimination   87
     4. Assessment   94
II. Convective model methods   95
     1. Parameterization of deep convection   95
     2. A one dimensional cloud model   97

CHAPTER 7
Rainfall from Visible and Infrared Images:
A Physical Explanation   102

I.   Infrared images   103
II. Visible images   105
     1. Cloud thickness and precipitation   106
     2. Brightness and thickness   108
III. The factors governing cloud brightness   110

CHAPTER 8
Properties of Microwave Radiation
in the Atmosphere   124

I.   Brightness temperature and polarization   125
II. Sources and sinks of microwave energy   127
     1. Surface emission and reflection   127

2. Absorption by the gaseous atmosphere   127
3. Absorption and scattering by cloud particles   129
III. A complete description—radiative transfer
equation for microwaves   134
1. Plane parallel clouds   137
2. Finite clouds   142

CHAPTER 9
Passive Microwave Methods   144

I.   The observing problem   144
II. History   145
III. ESMR-5   147
1. Characteristics   147
2. Rain mapping—the 1972 Wilhejt model   147
3. Tropical storms— the Gaut and
Reifenstein Model   152
4. The 1977 model of Wilheit, Chang, Rao,
Rodgers and Theon   153
5. Non-beam filling and other problems   159
IV. ESMR-6   164
1. Characteristics   164
2. Evolution of models   165
VI. The future   177

PART III: Satellite Rainfall Monitoring Applications

CHAPTER 10
Rainfall Inventories   181

I.   Introduction to Part III   181
II. Macroscale inventories of rainfall   181
III. Macroscale rainfall mapping over the oceans   183
1. The distribution of major rainfall areas over
the world's oceans   195
2. The sub-structure of major rainfall areas   197
3. Interannual variations of rainfall   199
IV. Macroscale rainfall mapping over the continents   199
V.   Mesoscale rainfall mapping over the oceans   201
1. The concept of moisture budgets   201
2. Applications of satellite rainfall techniques
to moisture budgets   202
3. Additional applications and outlook   205

CHAPTER 11
Applied Hydrology   206

I.   Introduction   206
II. Water resource evaluation   206

III. Catchment monitoring   212
IV. River monitoring and control   216
V. Conclusions   220

CHAPTER 12
Floods, Droughts and Plagues   221

I.   Satellite remote sensing of environmental hazards
     and disasters   221
II. Flooding from intense rains   223
     1. Flash floods   223
     2. Tropical storm floods   236
III. Monitoring longer-term rainfall anomalies
     and extremes   238
IV. Pests and plagues   241

CHAPTER 13
Crop Growth and Production   251

I.   The rise of interest in agrometeorology
     and agroclimatology   251
II. The assessment of conditions affecting vegetation
     and crop growth   252
     1. Crop growth   252
     2. Biomass estimation   256
III. Crop prediction programmes   258
IV. Towards future integrated crop prediction
     and food information systems   264

PART IV: Future Prospects and Possibilities

CHAPTER 14
Active Microwave Systems   269

I.   Radar rainfall physics   270
     1. The reflectivity of raindrops   270
     2. The reflectivity of the Earth   273
     3. Attenuation   275
     4. Polarization   280
     5. Summary   280
II. Early proposals for satellite weather radar   281
     1. A visionary view   281
     2. The pragmatic view   286
III. Recent proposals for satellite weather radars   287
     1. Present needs   287
     2. Proposed satellite radar systems   288

CHAPTER 15
Integrated and International Programmes for
Rainfall Monitoring   293

I.   Introduction   293
II. Potential users of satellite-assisted rainfall
     monitoring methods   294
     1. The global atmospheric research programme
        (GARP) of WMO   296
     2. The United States Climate Program   297
III. The present choice of satellite-assisted rainfall
     monitoring methods   301
     1. Single sensor approach   301
     2. Multi-sensor approach   302
IV. The future of satellite-assisted rainfall
     monitoring methods   306

References   312

Index   331

To our wives, Gillian and Linda,

and our children, Andrew and Stella, Stacy and Graeme,

for unfailing patience, understanding and encouragement

during the three-year gestation period of this book

PART I

Present Problems
and the Scope for their Solution

CHAPTER 1

Monitoring Rainfall: The Nature of the Problem

Water is the most ubiquitous mineral in the world. It is also, perhaps, the most vital for human life and activity. Unfortunately its availability to man is restricted by factors of local supply, natural purity, and its unique ability to be present in gaseous, liquid, and solid forms within the common range of environmental conditions found at or near the surface of the Earth.

Ultimately the source of all water in its most desirable state is rainfall, whose natural purity is also generally high. Small wonder, then, that so much time and effort has been, and is being, spent in the evaluation of rainfall through both time and space.

Rainfall studies are of great scientific significance also: precipitation is a vital exchange process within the hydrological cycle, and represents the net of heating from condensation in the atmosphere. The importance of precipitation monitoring is growing, too, within the context of current trends in environmental research. It was E. N. Lorenz, a leading atmospheric modeler, who said in 1969 that "The next generation of modellers [of the global circulation of the atmosphere] will be talking about the dynamics of water systems."

But rainfall studies are growing in significance also because of the steep upward curve in the demand of human society for water supplies as populations, and living standards and expectations have risen in many parts of the world. This demand is certain to go on growing, a fact which prompted a representation of the US Department of Commerce to make this arresting statement at the opening of the World Water Conference in 1976:

3

"If, today, we are suffering from problems associated with the provision of a suitable supply of oil at a price we can afford, so, by 2000 AD, we shall be suffering from comparable problems associated with the provision of water."

This was echoed by a 3-year study, prepared for the White House by the President's Council on Environmental Quality and the Department of State, with the aid of eleven other American Government agencies. This spoke of "an increasing severity of regional water shortages because of forest distruction and increased demand", and concluded that time was running out to prevent "alarming global problems".

There seems little doubt that such predictions will be fulfilled in many, if not all, countries of the world. There are some countries, notably in the arid zone of the Old World, in which water is already more scarce and expensive than oil; and recent weather anomalies—not only in more marginal countries of Africa but even in traditionally well-endowed parts of north-west Europe and North America—have underlined how sensitive economies may be to the balancing of water demand and water supply, even when nations at very different levels of development are considered.

I. Characteristics of Rainfall

It is unfortunate, especially in view of man's need for accurate observations of precipitation amounts and distributions, that rainfall is one of the most variable elements of weather. Precipitation varies notably with respect to its frequency, duration, intensity and spatial pattern, not to mention its propensity to fall as rain, hail, sleet or snow. As the physical state of snow is so different from rain, resulting in very different problems of monitoring and possible solutions involving satellites as aids to ground data collection systems, this book will focus upon rainfall to the exclusion of hail, sleet and snow except insofar as these may be expected to fall from clouds from which rain is the dominant hydrometeor: problems of snow and snowfall monitoring comprise a theme worthy of treatment in a separate volume.

1. The variability of rainfall

Expressed most conveniently in "rain days" the climatic average frequency of rainfall ranges from over 180 p.a. in some humid coastal areas (e.g. Olympic Peninsula, State of Washington) to less than 1 p.a. in the arid zone. Topographic influences can be strong, resulting in steep gradients of rain days, for example in trade wind regions in which volcanic islands may have penetrated the trade inversion prompting very rainy climates on the windward mountain flanks. Frequency may undergo marked seasonal changes, which are largest in areas dominated by monsoon circulations.

Ordinarily rainfall events last up to a few hours. They tend to be shorter where rainfall is influenced by the diurnal cycle of heating and cooling, and longer in topographic situations exposed to moist airstreams and in regions affected by large cyclonic disturbances, e.g. mid-latitude lows and tropical cyclones, especially when these are slow-moving.

There is a wide range of intensities of rainrate, ranging from practically zero, to above 100 mm h^{-1}. Intensity is an inverse function of duration: the highest intensities are recorded over the smallest intervals. World record rainfall intensity, for example, is approximately proportional to the inverse square root of duration (Chow, 1964). Maximum radar rainfall rates observed during the GARP Atlantic Tropical Experiment on the other hand, were approximately proportional to the inverse cube root of duration (Fig. 1.1), because, as Hudlow and Patterson (1979) point out, precipitation during GATE came mainly from convective, transient weather systems. The variation of intensity with duration can be large from region to region as well as from one storm to the next (Miller, 1977); however, each region tends to have a characteristic family of curves relating intensity to duration for storms of different severity (Fig. 1.2). Rain is most intense in the tropics. There

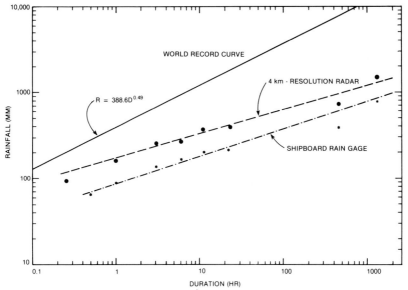

FIG. 1.1   Plots of maximum point rainfall accumulations observed during GATE for various durations and an analogous curve for variable durations based on world record events. The GATE radars covered a circle of radius 204 km centred at 08°30'N, 23°30'W. After Hudlow and Patterson, 1979.

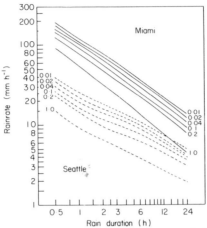

FIG. 1.2   Duration–intensity curves of rain for return periods of 1, 5, 10, 25, 50 and 100 years (frequencies of 1·0, 0·2, 0·1, 0·04, 0·02, 0·01 per year) and durations of ½ to 24 h. Solid lines are for Miami, Florida; dashed lines are for Seattle, Washington. These curves indicate that only once in ten years at Seattle is it expected (on the basis of a long series of observations) that rainfall averaged over 12 h will reach or exceed 5·5 mm h^{-1}. This is about one-third the intensity expected for a ten year, 12 h storm at Miami. From data in Hershfield, 1961.

50% of rainfall occurs during 10% of rain interval (Riehl, 1954; Garstang, 1972; Woodley et al., 1974).

The gross features of the global distribution of rainfall are closely related to the General Circulation. Rainfall is relatively abundant where uplift is encouraged and enhanced in the great convergence zones of the equatorial trough of low pressure and polar frontal zones of middle latitudes; it is scarce in areas of surface divergence, such as the trade wind anticyclones and the anticyclones which dominate cool and cold land surfaces in middle latitudes in winter and high latitudes throughout the year.

Patterns of rainfall organization and distribution prompted by the overriding influence of the General Circulation of the atmosphere are greatly enriched and confused by global geography. The sizes, surface configurations, and dispositions of the land masses influence their own precipitation climatologies; so do the geometries and water circulation patterns of the oceans and seas. Land and water bodies also have varying influences upon the precipitation climatologies of each other.

The overall results of such factors and influences on rainfall are distribution patterns of great complexity not only on meteorological time-scales but in climatological perspectives also. Figure 1.3 is a map of average annual rainfall around the world. Comparable statements of rainfall on much shorter time scales (e.g. 24 h) are nowhere prepared on an operational

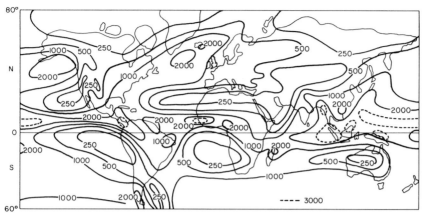

FIG. 1.3   Global mean annual precipitation (mm). By R. Geiger; from NASA, 1976.

basis, and there is no conventional counterpart for rainfall to compare with maps of instantaneous patterns of other important meteorological elements such as atmospheric pressure. To make matters worse for the consumer of rainfall data, rainfall is more susceptible than many atmospheric parameters to short- to medium-term fluctuations of spatial distribution: climatic statements of rainfall compiled from variable periods in the past may not be representative of current climatic conditions, whilst monthly and seasonal rainfall is, in some regions, highly variable, percentage variability levels of up to 40–50% being by no means uncommon.

On smaller time scales variability increases. An instantaneous distribution of rainfall measured by ship radar over the eastern Atlantic Ocean is shown in Fig. 1.4, from Hudlow and Patterson (1979). Gradients of rainfall approach 100 mm 10 km^{-1}. Spatial correlation of rainfall decreases rapidly with increasing distance between points. In Fig. 1.4 rainfall at points separated by distances as small as 50 km is practically unrelated. Spatial and temporal variabilities of rain over the north Pacific Ocean were assessed through correlation and autocorrelation by Dennis (1963) from the radar observations of picket ships. The correlation of rain over 20 × 20 km blocks dropped to 0·8 at distances ranging from 45 to 140 km, and over intervals of 30 to 60 min. Spatial correlations decreased most rapidly in the north, where rain was most frequent. These results are probably typical of marine rainfall in temperate latitudes.

Often, variability in the spatial distribution of rainfall is expressed in area–depth relations: that is, for particular storms or storm types the area covered by rainfall of a given depth or greater. Figure 1.5 is an example, from Miller (1977), of area–depth relations for two very large storms. It is found that depth decreases in exponential fashion with increasing area

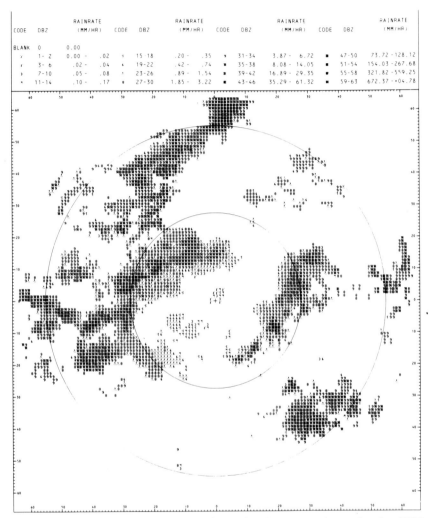

CODE	DBZ	RAINRATE (MM/HR)	CODE	DBZ	RAINRATE (MM/HR)	CODE	DBZ	RAINRATE (MM/HR)	CODE	DBZ	RAINRATE (MM/HR)
BLANK	0	0.00									
.	1- 2	0.00 - .02	.	15-18	.20 - .35	.	31-34	3.87 - 6.72	■	47-50	73.72 -128.12
.	3- 6	.02 - .04	.	19-22	.42 - .74	■	35-38	8.08 - 14.05	■	51-54	154.03 -267.68
.	7-10	.05 - .08	.	23-26	.89 - 1.54	■	39-42	16.89 - 29.35	■	55-58	321.82 -559.25
.	11-14	.10 - .17	.	27-30	1.85 - 3.22	■	43-46	35.29 - 61.32	■	59-63	672.37 -*04.78

FIG. 1.4   Instantaneous rainrate map from the *Oceanographer* radar for 1730 GMT September 2. The ship was located at 7°46′N, 22°11′W. The interval between range marks is 100 km and the grid spacing is 4 × 4 km. From Hudlow and Patterson, 1979.

(Fig. 1.5). Thus one standard model of storm rainfall assumes a bivariate normal (Gaussian) distribution of depth (Miller, 1977). The full range of variability of storm rainfall is expressed most concisely in tables of depth–areà–duration.

The increasing variability of rainfall on smaller scales of time and space can be expressed in another way. It is considered that, on average 50% of the

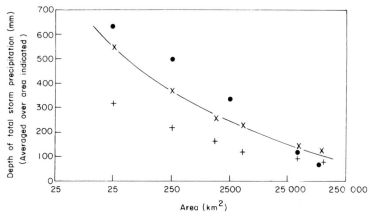

FIG. 1.5   Area–depth relations in the storm of 20–23 September 1941 in New Mexico over durations of 24 h (+) and total storm (78 h, x), and in hurricane Camille, 19–20 August 1969 in Virgina (·) for 17 h. From Miller, 1977.

Earth's surface is covered by cloud (Henderson-Sellers, 1979; Berlyand and Strokina, 1975). Even in thunderstorms, of which there might be 2000 over the world at a time (Brooks, 1925), the ratio of echo to cloud area is seldom larger than 0·1 (Griffith *et al.*, 1978). Considering that many clouds do not rain at all, the fraction of the Earth actually receiving rain at an instant in time could not be much more than one half of 1%. In temporal terms, the distribution is equally uneven. Because of the particular combinations of circumstances required for precipitation to occur, the small fraction of clouds which delivers the Earth's rain is active over but a small part of the cloud lifetime. As Lilly (1979) remarked, even stations most frequented by thunderstorms are under their active influence only a few percent of the time.

Let us explore these considerations a little further in order to clarify the relationship between clouds and precipitation. We may do so most conveniently under two headings, namely, the processes of rainfall and the types of rainfall.

2. The processes of rainfall

Before moisture can be removed by precipitation, a parcel of air must undergo a cycle which has become known as the "precipitation process". Following Mason (1971), to whom the reader is referred for details, the factors governing the occurrence of rainfall are the motion of cloud air and its aerosol properties, for these determine the concentration, initial size distribution, and nature of cloud particles. Once a cloud has formed, condensation and aggregation of droplets deform the spectrum, increasing the average radius and the range of sizes. Whether particles grow to precipitation size then is largely determined by air motion, through its control of macroscale properties of the cloud—dimensions, water content, and lifetime. These govern the roles of microscale processes and the length of time over which they operate, and therefore the maximum size which a cloud particle can attain. In general the different states of water in the atmosphere, and the important range of presentations of water in its liquid state, can be expressed in terms of size: the water molecule in the vapour phase approximating to 10^{-4} μm in diameter, whilst cloud droplets range upwards to 100 μm, drizzle to 500 μm, and raindrops to 6000 μm. Snow, ice and hail attain still larger sizes. Size is critical to precipitation, for two reasons: terminal fall velocity is roughly proportional to the second power of radius, and the distance a drop falls in unsaturated air before evaporating is proportional to almost the fourth power of radius. Terminal velocity must be greater than updraft velocity for a droplet to fall out of a cloud. Evaporation must be less than water substance for the droplet to reach the surface. According to calculations of Findeison (1939), for air of 90% relative humidity, only droplets larger than 100 μm in radius will fall more than 150 m before evaporating. As a practical matter, therefore, precipitation is composed of cloud droplets of diameter larger than 200 μm.

Because droplet concentrations are large (~ 100 cm^{-3}), condensation is ineffective in producing cloud droplets of precipitation size. If, however, the cloud is cooled below freezing, liquid and ice phases may coexist. At temperatures between -10 and $-20°C$ water vapour equilibrium with liquid will be 10 to 21% supersaturated with respect to ice. The ice crystals therefore will grow rapidly by sublimation, at the expense of the water drops. This is commonly called the Wegener-Bergeron, or Bergeron-Findeisen process. Precipitation occurs even in clouds not cooled to freezing, through the process of coalescence. Coalescence depends on a range of terminal fall velocities: hence, a range of droplet sizes. Droplets collide and coalesce, further increasing the range of size and fall velocity, and the likelihood of more collisions. Coalescence is sufficient to explain heavy showers which are observed in some warm cumulus clouds. Its efficiency increases as the spread of droplet size increases, becoming significant when larger "collector" drops grow to a diameter of $\geqslant 40$ μm. When droplets reach this size, coalescence (and accretion of supercooled droplets by ice crystals) is important to production of precipitation in partly supercooled clouds as well as in warm clouds. Droplets which grow beyond a diameter of 6 mm are liable to break apart. If the depth of the cloud is large and updrafts are strong, a chain reaction may be started, which multiplies the number of relatively large droplets (Langmuir, 1948; Simpson, 1975).

According to Mason (1971), the relative importance of condensation, Wegener-Bergeron, coalescence, accretion, and Langmuir chain reaction processes is largely a function of three parameters: cloud thickness, base temperature, and top temperature. Together with updraft velocity, which in turn is a function of the thermodynamic stability of air masses and the scale and magnitude of low level convergence (Betts, 1978; Lilly, 1978; Palmén and Newton, 1969), these largely determine precipitation intensity. Other factors such as base height, subcloud relative humidity, low level winds, vertical shear of the horizontal wind and downdrafts, may be important locally.

3. The types of rainfall

The most common approach to the classification of rainfall is based on the mechanisms which prompt cooling, condensation, and associated rain. The following are of greatest significance (see Palmén and Newton, 1969; Mason, 1971):

(a) Stable upglide precipitation, which is relatively steady and moderate in intensity and amounts. This type is prompted dominantly by slow, widespread, dynamically forced uplift of moist air over large areas in association with low atmospheric pressure and airstream convergence along baroclinic zones in middle latitudes.

(b) Convectional precipitation, which is generally of shorter duration, but of higher or even much higher intensities. This type is linked mainly with moist thermal convection, which, in turn, is dependent on heating and moistening from below. Convectional precipitation occurs in cells, which often congregate in clusters or rows. Rarely (e.g. in tropical storms and tropical cyclones), such precipitation falls from subsynoptic-scale cloud masses with which extensive severe weather is associated.

Palmén and Newton (1969, p. 353) also distinguish a second stable precipitation type, which results from cooling of moist marine airstreams being advected over colder surfaces. Such conditions are most often found ahead of cyclones over the polar oceans. Rarely is this precipitation more than a drizzle.

Precipitation of either main type may be augmented by orographic lifting, resulting from the passage of moist airstreams across high ground. Typically, cyclonic-type rainfall is enhanced in duration and/or intensity as moist airstreams approach the crest lines of hill or mountain ranges, the patterns of rainfall being complicated locally by the differing effects of spurs, valleys, and other terrain features (Elliot, 1977).

It is not surprising that, in a fluid continuum such as the Earth's atmosphere, intermediate types of precipitation frequently occur. The most common is a mixture of convectional and cyclonic types, especially in occluded frontal situations, where local uplift is often strong enough to stimulate convective cells to grow within and through belts of more steady, less heavy cyclonic rain. Investigations employing radar and dense raingauge networks show a surprising degree of cellular and mesoscale band organization of precipitation in extratropical cyclones (Mason, 1971; Cunningham, 1952; Browning and Harold, 1969), especially in the organization of heavier precipitation (Hobbs *et al.*, 1979).

It is of great significance to much work in the monitoring of rainfall by satellites that different types of rainfall are closely associated with different types of clouds—and that different cloud types usually have characteristic appearances in satellite images. This point will be enlarged in other chapters, and discussed and exemplified in considerable detail in various parts of this book. Suffice it to say for the moment that cyclonic and orographic types of precipitation fall mainly from sheet (stratiform) cloud types, whilst convectional and forced convectional rains fall mainly from tower (convective) cloud types; mixed cyclonic/convectional rain falls from cloud masses where both sheet and tower cloud types are present.

II. The Measurement of Rainfall by Conventional Means

In view of the wide range of uses of rainfall data and the physical questions outlined above, it is not surprising that monitoring rainfall with the desired detail and accuracy is often difficult. Here we summarize problems arising from measurement of rainfall by conventional means. "Conventional" refers to those means which rely upon well-established types of instruments: not only traditional *in situ* devices (raingauges), but also ground-based remote sensing systems (weather radar), which in certain regions have been used for rainfall monitoring, including measurement of amounts, for more than 20 years.

Measurement of rainfall by gauges is affected in particular by the interrelated factors of topography, site, wind, and gauge design. The gauge catch may be representative of a small or large area depending on slope, aspect, elevation, and location in relation to hills and ridges. Taking an extreme example, raingauges located on volcanic islands in the tropics are not likely to give readings representative of rainfall over the ocean in their vicinity. On a very small scale, the gauge catch is influenced by the nature of the local surface and the presence of nearby objects and structures: these may act as shields, decreasing the catch from what would have been measured in their absence (Weiss and Wilson, 1958). Exposure becomes acute for measurement of forest rainfall (Miller, 1977): it is especially complicated on ships, which do not provide a stable platform. Wind is the single factor contributing most significantly to errors in gauge measurements. Even if a raingauge is well sited and exposed, owing to deflection of raindrops in the disturbed airstream around the orifice of the gauge, monthly average winds as light as 5 m s^{-1} may result in errors of underestimation as large as 12% in regimes where half the monthly rainfall comes at rates less than $1\cdot8$ mm h^{-1} (Struzer et al., 1965). Raingauge designs are intended to reduce wind effects: for example, raingauges may be mounted flush with the ground, or provided with shields. However, flush mounting increases the possibility of error from splash, makes the gauge more vulnerable to shielding by natural objects, and is relatively expensive; and, according to a study of Weiss and Wilson (1958), shields attached to gauges are not entirely effective.

Woodley et al. (1974) found relatively small differences—5 to 10%—in the measurement of heavy convective rainfall by pairs of co-located, unshielded raingauges. Systematic differences among three types of gauges (wedge, 8 inch dipstick, and tipping bucket) were no more than 2%. However, Miller (1977) assesses rain measurement errors in "storms that are not too windy" at up to 20 to 30%.

To those problems must be added a plethora of practical and/or organizational difficulties, some of which are more significant in certain types of regions than others. These difficulties commonly include:

(i) accessibility of desired raingauge locations;
(ii) the availability of suitable personnel for reading and servicing the gauges;
(iii) a suitable power supply for some types of continuous-recording raingauges;
(iv) preserving the rainfall catch in accumulating raingauges especially when read infrequently;
(v) security of the rainfall station from vandalism and other damage;
(vi) access to a suitable communication link so that the data can be sent quickly to a central facility for processing and archiving; and
(vii) transcription and transmission errors, which tend to introduce positive errors in reported rainfall (R. W. Burpee, 1979, personal communication).

In the case of weather radar for rainfall monitoring there is a comparable range of problems which circumscribe the use of such a system. It is sufficient for present purposes to say that there are difficult problems as yet not completely solved relating to the proper relationship of backscattered microwave energy to drop size spectrum, partial filling of the radar beam, attenuation of the radar beam by intervening drops, absorption and reflection by the ground (anomalous propagation), and signal calibration. Discussions of these points may be found in Battan (1973), Mason (1971), Hudlow *et al.* (1979) and in the context of satellite radar, in Chapter 14 of this book. In practice the calibration of radars used for measurement of precipitation includes an adjustment to match gauge measurements of rainfall within the radar scan.

Radar has the singular advantage over gauges of providing a spatially continuous view. It is employed extensively in support of raingauge networks in some more advanced nations of North America, Europe and Asia, but elsewhere, because it is costly and requires sophisticated technical and engineering support, the operational use of radar for rainfall monitoring is extremely limited. Despite the enormous local value of radars, most of the world's information on rainfall continues to come from raingauges.

III. The Need for Rainfall Monitoring by Satellites

The World Meteorological Organization (WMO) has set advisory standards for "optimum" and "minimum" gauge networks. The minimum network is represented in Table 1.1. Optimum networks are denser, depending on the type of region involved—and the chief user's need for rainfall data.

Unfortunately, the present network for rainfall monitoring around the world by conventional means is deficient in many areas, the more so if rainfall data are required very quickly (in "near real-time"), for many raingauges and raingauge networks provide data weekly, monthly, or even

TABLE 1.1 Minimum density of precipitation network, set by the World Meteorological Organization (1965) (from Wiesner, 1970).

Type of region	Normal tolerance (area for 1 station)	Tolerance under difficulties (area for 1 station)
Flat areas of temperate, mediterranean, and tropical zones	600–900 km² 230–350 square miles	900–3000 km² 350–1160 square miles
Mountainous regions as above	100–250 km² 39–100 square miles	250–2000 km² 100– 770 square miles
Small mountainous islands with irregular precipitation	25 km² 10 square miles	
Arid and polar zones	1500–10 000 km² 580– 3860 square miles	

less frequently, and with time-lags which further restrict their utility to that section of the user community which is satisfied with essentially "historical" data. In general terms the regions of greatest deficiency include most of the desert and semi-desert regions, most major mountain regions, and extensive humid regions in the tropics, to which must be added the world's oceans, over almost all of which rainfall observation by raingauges is impractical. It is difficult to assess the extent of the rainfall data-deficient regions *in toto*, but they must constitute between 80–90% of the surface of Earth. Indeed, there is evidence to suggest that the rainfall-observing networks are actually deteriorating from a peak probably reached in the 1930s—contrary to casual expectation, which is that man must continually improve the monitoring of his home planet (Barrett, 1980a).

Although no special comprehensive study of the decline in raingauge operation is thought to have been made, some indication and exemplification of the nature of this deterioration may be given, in relation to short-, medium- and long-term studies and applications:

(a) Short-term. Data from meteorological stations in the international network, disseminated via the Global Telecommunication System (GTS) of the World Weather Watch (WWW), are now frequently below, and sometimes from certain regions, far below their intended levels. Table 1.2 illustrates this fact by reference to a block of countries in north-west Africa in which the first author has recently worked.

(b) Medium-term. Here the uses of rainfall data often depend as much, if not more, upon weather and rainfall station networks which are supplementary to the principal stations reporting via the GTS: data from such stations are usually less quickly available for analysis nationally or internationally. Work in several tropical and subtropical areas has revealed that supplementary rainfall stations have become fewer—often much fewer—and less dependable since World War II.

TABLE 1.2   GTS rainfall data from Morocco, Algeria, Libya and Tunisia, 22–28 March 1977.

Date	22		23		24		25		26		27		28	
Time	06	18	06	18	06	18	06	18	06	18	06	18	06	18
Reports on GTS	60	61	61	60	63	62	60	64	50	71	63	69	62	68
Nil Reports on GTS	24	43	23	24	21	22	24	20	34	13	21	15	22	16

(c) Long-term. Numerous meteorological stations whose records have been ideal for long-term data analyses have been, and are being, lost to the scientific community, especially as a result of ever-changing patterns of international relations. Examples of long-term records disrupted recently at key locations on account of such effects are Masirah Island at the entrance to the Gulf of Oman, and Gan in the North Indian Ocean.

Such developments are extremely serious, for, as we approach the end of the twentieth century there are ever-increasing demands for more and better rainfall data stemming from a growing body of specialist consumers: traditional consumers of such data (e.g. meteorologists, climatologists, and hydrologists) are spawning an ever-broadening range of applications, whilst being joined by new consumers especially in agriculture, industry and commerce. Today, important short-term (near real-time) uses of rainfall data are found in weather forecasting, river management and flood control; important medium-term uses are found in water supply prediction, crop forecasting, humanitarian aid programmes such as famine and flood relief, and in commodity market activities; and long-term uses of particular significance include agricultural and hydrological planning, civil engineering design activities, design of microwave communication systems, and the modelling of atmospheric cycles and selected key parameters.

Thus, many factors—particularly cultural, political, and economic—have combined to prompt an unexpected, and probably unprecedented, situation in which a growing demand for rainfall data is faced with diminishing data returns. This book is not the proper place to comment further on the factors thought responsible for this situation; but it is the place to emphasize that new tools are available in the form of satellite systems to help reverse the trend: the satellite affords the best hope we have for substantial improvements of our knowledge and understanding of global rainfall in the foreseeable future. Before we turn, in Part II, to survey the methods by which this hope is being realized, we shall, in the next chapter, introduce and discuss the satellite systems and sensors which have opened up such possibilities.

CHAPTER 2

Satellite Remote Sensing Systems of Use in Rainfall Monitoring

I. The Use of Satellite Remote Sensing

Remote sensing has been defined as "The observation of a target by a device separated from it by some distance" (Barrett and Curtis, 1976). It is, therefore, clearly different from long-established ("conventional") observation of the environment of Man for this has relied upon *in situ* sensors or sensor systems: systems in which the observing instruments are either touching, or even immersed in, the object or medium of interest. As a scientific approach remote sensing was effectively born in the mid-nineteenth century with the invention of photography. As a tool in rainfall monitoring it began with the development of meteorological radar after World War II. As a recognized discipline its rise has been both recent and meteoric: the term "remote sensing" was coined as recently as 1960. It was the birth of space exploration by satellites which prompted the new discipline to take off in the late 1960s as one of the most rapidly expanding fields of scientific technological endeavours. Today the variety and flexibility of satellites both as communication links and platforms for Earth observation systems rivals that of their conventional predecessors. This chapter briefly summarizes the satellite systems and sensors available for rainfall monitoring research and operations especially for those readers whose previous experience has not been in the satellite field: some appreciation of the characteristics of satellites, satellite systems, sensors, data types, and the past, present, and future availability of such data types, is necessary if the methodologies of rainfall

17

monitoring from satellites are to be adequately understood and properly evaluated for use in any given situation.

II. Satellite Remote Sensing Systems

It has become fashionable and convenient to differentiate between two groups of Earth observation satellites, namely "Earth resource" and "environmental" satellites. Comprehensive technical details of all such satellites, including their payloads and principal investigators, may be found in Vette and Vostreys (1978). In more general terms, the differences between the two groups arise from design capabilities for the frequency, and spatial resolution on the ground, of the data they provide. These capabilities are inversely related to each other largely because of data-flow constraints. Therefore, for example, the high (80 m) resolution of the multispectral scanners (MSS) on Landsat 1, 2, and 3 has been accompanied by a low (18-day) imaging frequency of any one area, whilst the lower (1 km) resolution of the high resolution picture taking (HRPT) system of the US National Oceanic and Atmospheric (NOAA) satellites has been accompanied by the much higher imaging frequency of 12 h.

It is obvious that, for precipitation studies, Earth resource satellites have relatively little to offer, for rainfall is so highly variable through time. It is fortunate that, although no dedicated hydrological or hydrometeorological satellite has yet been numbered amongst the environmental satellites (but see Section IV, Chapter 15), many of these have been equipped with sensors yielding data of value in rainfall monitoring. Consequently it is upon these environmental satellites that our attention must henceforth solely rest.

Environmental satellites have commonly occupied two types of orbits, and may be grouped accordingly into:

(a) polar-orbiting, and
(b) geostationary satellite families.

In the first case above, the satellites occupy relatively low-level orbits (usually between 500–1500 km above the surface of Earth), crossing the equator at high angles so that each orbit takes such a satellite close to the North and South Poles. These satellites are usually *sun-synchronous*, i.e. their orbits are so organized that, as the Earth rotates on its polar axis within the orbital ellipses, each new orbit results in the presentation of a new strip of the global target to the satellite in such a way that the relationship between Earth surface and sun angle is kept relatively constant. Characteristic orbital periods are *c.* 100 min, 14–15 orbits being required for each satellite to investigate the entire globe under daylight conditions every 24 h.

Of course, since such satellites spend one-half of each orbit over the night-time side of the globe, each unit area on the surface is thus viewed twice every 24 h, once in daylight and once at night. Crossing times at the equator differ from system to system, but commonly NOAA satellites have crossed at about 0900 and 2100 h local time.

Other presently noteworthy families of American polar-orbiting environmental satellites have included Tiros and ESSA, their Russian counterparts including Molniya, Meteor, and a few of the huge and varied Cosmos series. Present American families include the civilian Tiros-N (and succeeding NOAA operational satellites), Defense Meteorological Satellite Program (DMSP) Block 5D satellites, and the ongoing research and development Nimbus family.

In the second case above, the satellites are placed into orbit at approximately 35 400 km in the plane of the equator, and advance in the same direction as the rotation of the Earth. This type of orbit is *geosynchronous*, the satellite keeping pace with the rotation of the Earth on its polar axis, and *geostationary* in that it ensures that any satellite occupying it appears to be fixed or stationary above a given point on the Earth's surface. Such a satellite is able to record the same geographic field of view very frequently throughout the diurnal cycle, commonly at intervals of 30 min.

The earliest geostationary satellites were the experimental Applications Technology Satellites (ATS) of the early 1970s. They were followed by the Synchronous Meteorological Satellites (SMS). In 1979, five geostationary satellites were intended to provide a complete circumglobal coverage between latitudes of effectively about 50°N and S of the equator. These included two American Geostationary Operational Environmental Satellites (GOES-1 and 2), one covering the eastern Americas and the western Atlantic, and the other the western Americas and eastern Pacific; a European Space Agency (ESA) satellite, Meteosat, located over West Africa; an American satellite over the Indian Ocean controlled by ESA, and named GOES-I0; and a Japanese satellite, GMS-1, located over Borneo.

Since data from non-Russian weather satellites are much more readily available in western-orientated and Third World countries our account will, henceforth, deal chiefly with those. The first "operational" weather satellite system was inaugurated in early 1966 by two polar-orbiting ESSA satellites. One of these was a recording satellite which stored imagery for the whole globe and played this back over the USA for archiving. The other was an Automatic Picture Transmission (APT), or direct read-out satellite, which transmitted imagery locally around the world for near real-time use in weather analysis and forecasting. This system has continued to the present time with few breaks in coverage; NOAA, and very recently, Tiros-N satellites have succeeded the initial operational family. For the First Garp Global Experiment (FGGE) of the Global Atmospheric Research Programme (GARP), special arrangements were made for 1979, so that the five geostationary satellites listed earlier might supplement the existing polar orbiters. During this period effective circumglobal coverage of low and middle latitudes was obtained from geostationary satellites for the first time. However, the geostationary satellite coverage was more restricted and fragmentary than had been hoped, because of malfunctions in the SMS satellites and GOES-I0. Bearing in mind, too, the fact that GOES-I0 was only temporarily moved to a location at 60°E to make good the non-appearance of the planned Russian geostationary satellite over the Indian Ocean during the FGGE year, it must be stressed that it is not known when a fully operational satellite network of the FGGE type may be available. Coverage from DMSP satellites is subject to time and space restrictions related to their military sponsorship, and data from Nimbus satellites cannot be used for operational programs because of the research and development nature of this family, and delays in data availability related to this status.

III. The Utility of Environmental Satellites in Weather Monitoring

Satellites have contributed enormously to the related sciences of meteorology and climatology through the last two decades. The special attributes of satellite systems which have made this possible include the following:

(a) Satellite systems can provide a complete global coverage of data, thereby greatly extending our appreciation of the atmospheric environment into previously data-remote regions.

(b) Satellite imaging systems yield spatially continuous data, contrasting strongly with those obtained from the commonly open, irregular, networks of surface weather observatories.

(c) Satellites can investigate the distributions of selected elements much more homogeneously than *in situ* observing networks which deploy large numbers of instrument packages.

(d) Some satellites can give a much higher temporal frequency of information than is commonly obtained from surface and upper air weather stations.

(e) Satellites provide a new view of the atmosphere, observing it from above, rather than from within. Thus satellite observing systems, unlike *in situ* systems, do not modify the parameters which are being measured, and integrations of parameters (e.g. radiation fluxes) along lines, over areas, or through volumes of the atmosphere can be more readily obtained than from *in situ* point observation stations.

(f) Data from weather satellites, being remote sensing data, have different physical meanings from conventional (*in situ*) data.

(g) Satellite data can be obtained for broad areas, by suitable facilities, in real or near real-time.

Clearly satellite data are of great realized and potential utility in atmospheric studies because they complement conventional data in space and time, and through their physical implications. However, the general usefulness of satellite data in meteorology and climatology is often limited in practice by a number of problems and associated difficulties. These include:

(a) The data from satellites are mostly non-synoptic. Thus there are problems of correcting and/or calibrating the data for a number of factors which may have influences upon them.

(b) Elaborate transformation procedures are required to convert satellite-observed radiances into information which can be integrated with that provided by conventional networks. Remote soundings of the atmosphere, and evaluations of cloud drift winds and rainfall are relevant examples. The outstanding exception is in the detection of clouds—but even so it is only very recently that hydrodynamic models have begun to accept satellite cloud information for initialization purposes.

(c) The user of satellite data is distant from his data source. He may have little or no apparent influence on sensor, platform, or operation design, and/or on system calibration and the format of satellite data products. By choosing to process raw data himself he gains a measure of control over format and quality, but must have his own facility to effect this. By merely accepting data products provided from elsewhere an important part of the processing operation may remain obscure to him.

(d) For many users meteorological satellite data have a low information content: the data of interest may be a tiny fraction of the total provided. The associated problems of information extraction may be technically difficult for a particular use or user.

(e) Detailed methods for the analysis, interpretation, and use of satellite data for specified operational applications have been slow to be developed, largely because of a lack of funding beyond the limits of basic academic research.

(f) The types and resolutions of available data are less than optimal for many classes of potential users, who, finding that the data they feel they need are not presently available, are often unaware of the opportunities which are available for influencing future satellite policy and design.

(g) Existing operational procedures require adjustment if new data types and dependent practices are to be deployed: operational inertia, often manifest through a worker's resistance to change, can be very strong.

B

For further information on the role of satellites as remote sensing systems for studies of the Earth's atmosphere the reader is directed to the broad reviews by Suomi (1970) and Houghton (1978); applications in applied meteorology have been summarized by Anderson and Veltischev (1973), and in climatology by Barrett (1974). It will become apparent through this book that the problem of rainfall monitoring by satellites is, in many ways, a microcosm of the entire picture of satellite meterology and climatology, not least in terms of the satellite attributes and disattributes which it entails.

IV. Existing Satellite Systems for Rainfall Monitoring

Satellite evaluations of rainfall have depended largely on data provided by a small number of different types of sensors. These are:

(a) Scanning radiometers, e.g. those flown on polar-orbiting operational satellites. These monitor radiation from the target via a revolving mirror which scans the target across the sub-satellite track. As the satellite advances along its orbit each new revolution of the mirror yields a new scan line adjacent to the one before. The radiation collected by the mirror is passed through a beam splitter and spectral filter(s) to give the desired wavelength separation. The wavelengths most commonly used in rainfall studies have included:

(i) visible (VIS), observing radiation with wavelengths generally between about 0·5 and 0·7 μm;

(ii) infrared (IR), i.e. thermal infrared radiation, escaping from the Earth's atmosphere through one of its most important "window wavebands", namely between about 3·5 and 4·2 μm or 10·5 and 12·5 μm;

(iii) microwave (MW), involving naturally emitted radiation at radar wavelengths, particularly at 0·81 or 1·55 cm. Microwave instruments employ antennas or antenna reflectors to achieve scan.

Some scanning radiometers have had only bispectral capabilities; the present trend is towards multispectral systems, e.g. the four channel Advanced Very High Resolution Radiometer (AVHRR) on Tiros-N (including visible, infrared, near infrared and water vapour (WV) channels), and the Scanning Multispectral Microwave Radiometer (SMMR) on Nimbus-7, observing microwave radiation at five different wavelengths.

(b) Spin-scan radiometers, flown on geostationary satellites. Incoming radiation is received by a fixed mirror angled at 45° to the optical axis of the system, which is aligned parallel to the axis about which the satellite itself spins. The spinning motion of each satellite provides a west-to-east scan motion across the target when the spin axis is parallel to the polar axis of the Earth. Latitudinal scan is achieved by sequentially tilting the scanning mirror north to south at the end of each rotation. Thus an image of the entire visible disc of the Earth is built up over a period of about 20 min, after which the mirror is returned to its initial inclination ready for the imaging cycle to recommence. Commonly these systems have viewed in the visible and infrared, with Meteosat viewing additionally in the water vapour emission region (5·7–7·1 μm).

(c) Vertical profile radiometers. These "snapshot" radiometers flown on polar orbiters have fixed fields of view, within which they integrate and record emitted radiation at a number of closely spaced intervals within atmospheric window and/or carbon dioxide absorption wavebands (for vertical temperature profiling), and/or a water vapour rotational waveband (for vertical water vapour profiling): every observation in one of these bands may be related to the middle of a radiation-emitting layer in the sub-satellite atmospheric column. The Tiros Operational Vertical Sounder (TOVS) is a 20-channel instrument. This yields moisture soundings of low spatial resolution and somewhat doubtful accuracy. It is conceivable that useful measurements of rainfall might be derived from areas of precipitable water, but this has yet to be confirmed. Consequently we will not be concerned further with vertical profiling radiometers.

Parts II and III of this book are vitally concerned with the analysis of data from such sensing systems, especially two-dimensional data in the visible, infrared and microwave regions of the spectrum. It is appropriate, therefore, to summarize in qualitative terms at this stage the key characteristics of image data in these regions so that the possibilities and problems of cloud analysis and evaluation in terms of associated rainfall in Chapters 3–5 will be properly appreciated. In Chapters 7 and 8 there will be more advanced discussions of such data, partly stemming from the research in cloud/rainfall relationships itself.

(a) Visible data. Although these are influenced to some extent by the imaging system, and methods of data preprocessing, processing, and presentation, visible data relate most strongly to the albedo (percentage reflectivity) of the target: highly reflective surfaces (including ice, snow, desert sand,

TABLE 2.1 Characteristics of clouds portrayed by satellite visible images (from Barrett and Curtis, 1976).

Cloud type	Size	Shape (organization)	Shadow	Tone (brightness)	Texture
Cirriform	Large sheets, or bands, hundreds of km long, tens of km wide.	Banded, streaky or amorphous with indistinct edges.	May cast linear shadows esp. on underlying cloud.	Light grey to white, sometimes translucent.	Uniform or fibrous
Stratiform	Variable, from small to very large (thousands of square km).	Variable, may be vertical, banded, amorphous, or conforms to topography.	Rarely discernible except along fronts.	White or grey depending on sun angle and cloud thickness.	Uniform or very uniform.
Strato-cumuliform	Bands up to thousands of km long; bands or sheets with cells 3–15 km across.	Streets, bands, or patches with well-defined margins.	May show striations along the wind.	Often grey over land, white over oceans, due to contrast in reflectivity.	Often irregular, with open or cellular variations.
Cumuliform	From lower limit of photo-resolution to cloud groups, 5–15 km across.	Linear streets regular cells, or chaotic appearance.	Towering clouds may cast shadows downsun side.	Variable from broken dark grey to white depending mainly on degrees of development.	Non-uniform alternating patterns of white, grey and dark grey.
Cumulo-nimbus	Individual clouds tens of km across. Patches up to hundreds of km in diameter through merging of anvils.	Nearly circular and well-defined, or distorted, with one clear edge and one diffuse.	Usually present where clouds are well-developed.	Characteristically very white.	Uniform, though cirrus anvil extensions are often quite diffuse beyond main cells.

and clouds) are relatively bright in visible images. It is well established that, under suitable viewing conditions, many different cloud types can be identified therein. Table 2.1 provides a basic guide to correct cloud identification.

(b) Infrared data. Here the dominant influence is the radiation temperature of the target. Since cloud tops at different levels possess different brightness temperatures in the infrared, such image data may be interpreted rather broadly in terms of cloud top height. Thus infrared images have a three-dimensional quality lacking in the visible. Dependent operational products include photo maps of cloud top height, and satellite wind fields deduced from the positions of identified cloud elements in successive images for low, middle, and high levels in the troposphere (see Dismachek, 1977).

(c) Visible and infrared data. Taken together, these often successfully resolve ambiguities in cloud type recognition based on either type of data alone. Figure 2.1 illustrates relationships between cloud type and brightness in the visible and infrared.

(d) Multispectral data. With the advent of multichannel radiometers on operational satellites, e.g. the Advanced Very High Resolution Radiometer (AVHRR) on Tiros-N, cloud recognition schemes like that included here as Fig. 2.2 may be expected to become increasingly valuable in the 1980s. However, the use of a multispectral system for such a purpose carries with it inevitable implications of higher reception and processing facility costs.

(e) Microwave data. Satellite-borne microwave radiometers recording target radiation in the 0·81 cm and 1·55 cm bands have been shown to reveal not clouds, as in (a)–(d) above, but rain areas embedded in the clouds, with the qualification that these are most obvious over sea areas, being often obscured over land by the stronger "background" radiances from such surfaces. It is in the microwave region, therefore, that rain has been most directly evidenced from satellite data available until now; all rainfall monitoring schemes involving visible and/or infrared data depend on some less physically direct relationships between clouds and rain.

In an essentially prophetic status report on the applications of space technology (WMO, 1967) three possible types of methods were proposed for rainfall monitoring from satellite altitudes. These were:

(a) "Passive" methods, by which precipitating clouds could be distinguished from non-precipitating clouds through differences in emitted radiation from raindrop clouds and their backgrounds and/or environments. It was recognized that a careful wavelength selection should be made to permit the penetration of clouds, and that wavelengths near to, or longer than, 1 cm should prove suitable for such purposes. Visible and infrared wavebands were considered unsuitable.

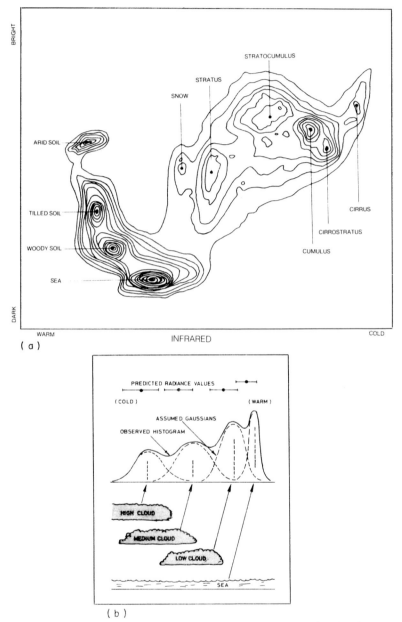

FIG. 2.1   A two-dimensional histogram prepared automatically for simultaneous visible and infrared data through a cluster procedure. The resulting peaks are identified with the aid of an interpretation table. Various types of clouds, and Earth surfaces between the clouds, are differentiated thereby. From Fusco *et al.*, 1980.

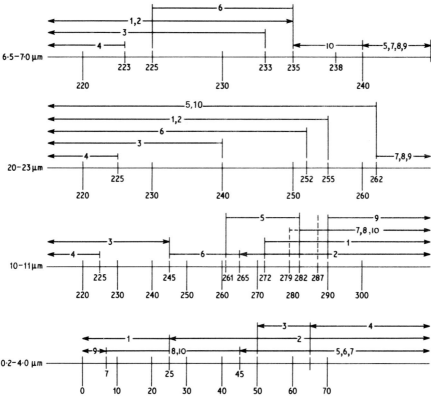

FIG. 2.2   A basis for automatic cloud-type recognition and nephanalysis using multispectral cloud observations. The "Cloud Type Decision Matrix" shown here was constructed for data from the Nimbus-3 Medium Resolution Infrared Radiometer. Horizontal graduations indicate equivalent blackbody temperature (K, top three decision lines) or spectral albedo (%, bottom decision line). Arrow heads indicate < or >; no arrow heads indicate ≤ or ≥. Broken lines show thresholds of nadir angles >40°. After Shenk *et al*., 1973. Key to cloud types: 1. Cirrus. 2. Cirrus with lower clouds. 3. Cumulonimbus and/or cirrostratus. 4. Cumulonimbus. 5. Middle clouds. 6. Middle clouds with cirrus above. 7. Stratus or stratocumulus. 8. Cumulus. 9. Clear. 10. Cumulus or stratocumulus with chance of cirrus.

(b) "Active" methods. Here the suggestion was that precipitating clouds should be identifiable through their special reflection characteristics, although difficult technical and interpretational problems were foreseen.

(c) "Delayed response" methods. Here, it was argued that infrared data might permit the identification of surfaces recently exposed to precipitation through reduced surface temperatures over both land and sea. However, the need for post-rainfall sky clearance, and difficulty of inferring intensity, seemed likely to limit the value of such approaches.

In the chapters which follow it will be seen that progress has been slower than had been envisaged in 1967. Type (b) methods, physically the most attractive, still lie in the future; type (a) methods, whilst tested with some considerable, but geographically restricted success, have not yet been put on an operational footing. All operational programmes, and much research, involving the use of satellites in rainfall monitoring have, by necessity, so far been based on those visible and infrared data considered "unsuitable" for such purposes in the WMO report. However, the children of necessity in this respect include pragmatism and practicality, and until the problem of operational rainfall monitoring by satellites can be approached in a more physically direct way visible and infrared methods will continue to dominate.

In any method, however, whether physically direct or indirect, there may be recognized the following seven elements:

(a) a need, in terms of area, period, resolution in time and space, and accuracy;
(b) a set of conventional (raingauge and/or radar) data for reference, calibration, and verification;
(c) a satellite data base;
(d) a measurement technique;
(e) a rainfall algorithm;
(f) a procedure for organizing and displaying results;
(g) a means of evaluating method performance.

These elements may all be recognized in the methods outlined and discussed in Part II. Differences between the methods result from differences in one or more of the method elements as set out above.

PART II

Satellite Rainfall Monitoring Methods

CHAPTER 3

Preliminary Studies

Three years before the launch of Sputnik 1, at a Symposium on Space Travel held in New York City in 1954, Harry Wexler posed the question, "What value might a satellite have for observing the weather?" "Inestimable" was the answer, for satellites could image the Earth and its cloud cover; locate thunderstorms; and measure solar radiation, Earth albedo, meteoritic dust, and the temperature of the surface and of the atmosphere. Using radar, they could detect precipitation areas. Just months before Sputnik, Widger and Touart (1957) published a design for a satellite system which had as one of its objectives the measurement of precipitation by means of radar.

In the papers of Wexler and of Widger and Touart we find the first concrete expression of the idea that rainfall might be monitored from satellites in orbit around the Earth. Although these proposals were found to be impractical (see Chapter 14, in which satellite radar is discussed), the need which impelled them has actually grown. This chapter describes the work, mostly from the 1960s, that first gave life and substance to the idea of monitoring rainfall from satellites. It is this work—involving clouds in visible and infrared wavelength pictures and the precipitation associated with them—which forms the observational basis of most contemporary schemes for monitoring rainfall using satellites.

I. Cloud Brightness and Rainfall

1. Snapshot views

Tiros I was the first meteorological satellite designed to provide images of the Earth. (Tiros is short for Television and Infrared Observation Satellite.

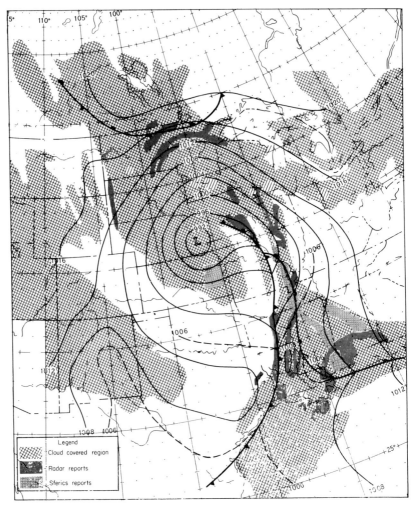

FIG. 3.1   Sea level pressure chart, 2100 GMT, April 1, 1960, with satellite cloud cover and radar and spherics reports. From Bristor and Ruzecki, 1960.

The early history of Tiros is covered by Vaeth, 1965.) In the months following its launch on 1 April 1960, there was a flurry of papers built around pictures showing dramatic patterns of clouds. The centerpiece in most of these papers was a large vortex, and a main point was to show how satellite cloud patterns were related to familiar patterns of weather—lows and highs, wind, cloud and precipitation. Of special interest is the 1960 paper of Bristor and Ruzecki, for upon their map of surface pressure and satellite clouds are superimposed echoes observed by weather radars

(Fig. 3.1). Over the next year Jones, Timchalk and Hubert, and Fritz published papers showing Tiros pictures marked with surface weather observations. Each noted an association of precipitation with clouds of relatively high brightness.

These studies of opportunity verified the idea that clouds in pictures from space could be typed. They also fathered a corollary idea that precipitation occurs with clouds of high brightness.

(a) *The classification of clouds*

From studies of individual satellite cloud systems, it was only a small step to classification of clouds. Systematic treatments of clouds by type appeared as early as 1961, when Erickson and Hubert published a comparison of 5 days of Tiros pictures with thousands of synoptic cloud reports. Soon after came the landmark study of Conover (1962, 1963). Conover compared satellite clouds with ground observations, station reports, and radar reports, and in addition with clouds in photo-reconnaissance pictures taken by U-2 aircraft programmed to Tiros overflights.

The characteristics of clouds important to interpretation were thought by Erickson and Hubert (1961) to be size, contrast and brightness. The criteria Conover proposed were form, pattern, texture, brightness and structure, and dimensions. Brightness, a "very important" characteristic, depends on illumination, scene geometry (incident, viewing and relative angles) and cloud reflectivity. Reflectivity, in turn, is related to "thickness, droplet size distribution, liquid water content, solar elevation, and probably the character of [the cloud's] upper surface. It varies directly as the thickness and liquid water content and inversely with the drop size". Citing measurements by Neiburger (1949) and calculations by Fritz (1954), Erickson and Hubert (1961) warned against indexing brightness to thickness: "Brightness itself is so complex that only qualitative inferences may be made, and measurements are meaningful only in the comparative sense." But Conover (1962) took a more sanguine view, for "Increases of brightness are generally associated with increases in total cloud thickness, especially in cyclonic cloud systems, and it appears that a fair guess can be made as to whether precipitation is occurring or not, depending on the brightness."

(b) *Synoptic systems*

The nephanalysis of Bristor and Ruzecki demonstrates the power of radar for exploring relationships between clouds and precipitation. This was the approach taken by Nagle and Serebreny (1962) in a study aimed at " . . . establishing principles for deducing precipitation patterns in cases where

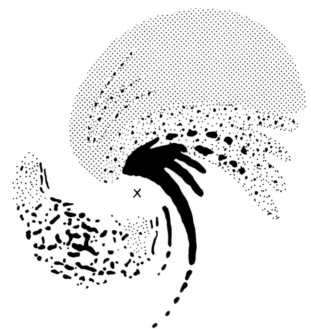

FIG. 3.2   Schematic model of the radar precipitation echo distribution around an occluded maritime cyclone. Light stippled area denotes continuous stratiform type precipitation; checked area, ragged patches of stratiform type precipitation; solid area, convective type precipitation; x, pressure vortex center. From Nagle and Serebreny, 1962.

only the satellite cloud observations may be available". Their study, like Conover's, used special observations—here, an airborne radar— synchronized with satellite passes. Nagle and Serebreny expressed their findings as a model, shown in Fig. 3.2, of precipitation in a mature marine cyclone. Echoes occurred with clouds and in their organization marked the major elements of cloud structure. But Nagle and Serebreny also found, as can be seen in Fig. 3.1, that " . . . at any . . . instant of time a very low percentage of clouds precipitate".

Repeating Nagle and Serebreny's study with better satellite data several years later, Blackmer and Serebreny (1968) came to the more encouraging result that the strength of the relationship of echoes to clouds was a function of synoptic regime. Coincidence was best for post-cold frontal cellular clouds, especially toward the centre of a deep trough. There "the distribution of precipitation echoes is nearly comparable to the distribution of the largest, brightest cloud elements, especially when they are arranged in polygonal cells . . .". Echoes were banded and cyclonically curved within the cyclone vortex. They became extensive near the broader warm frontal cloud shield, then linear further outward along the cold frontal band, and progressively smaller. Precipitation was most likely within the brighter areas within these features, and with the more distinct vortex patterns. Stratiform cloud associated with blocking anticyclones, even though extensive, did not contain precipitation.

Between these studies (as well as before and after) there was pessimism in good measure. In a synthesis of interpretations of extratropical vortex patterns, Widger (1964) expressed doubts that any simple relationship between cloud brightness and rainfall would be found. Acknowledging a good correspondence of precipitation with bright clouds, he stressed the poor association of bright clouds with precipitation, and concluded that criteria had not been found which would distinguish precipitating clouds and non-precipitating clouds. Nowhere in the dry prose of an early report on *Synoptic Use of Meteorological Satellite Data and Prospects for the Future*, prepared by the US National Weather Satellite Center for the World Meteorological Organization (Fritz *et al.*, 1965), is there mention of precipitation mapping or monitoring. But the study of Blackmer and Serebreny (1968) had suggested that associations with precipitation might be stronger in clouds where convective processes were important. In other words, it is not enough to know there are clouds. One should also ask, are the clouds the kind that produce rain?

(c) *Convective clouds*

Conover noted in 1962 that with similar illumination water clouds of a given thickness are brighter in satellite pictures than ice clouds. He reported in 1965 that cloud albedo measured in Tiros pictures was highest—about 90%—for cumulonimbi. Indeed, even before Conover, Jones (1961) had associated bright areas in large cloud masses with convective clouds embedded in stratus. In the same year Whitney and Fritz published a study of a square or rhomboid-like cloud found in a picture of Texas and Oklahoma taken by Tiros-I on 19 May 1960. This cloud was distinctive for being small, isolated and very bright. Surface stations recorded towering cumulus and cumulonimbi, then thunderstorms, and later, to the north-east, hail and tornadoes. Research aircraft reported a squall line with tops to 17 000 m (55 000 ft) in the area of the bright cloud one-half hour before the Tiros picture. Radar showed a line of rapidly growing echoes, 110–150 km long, co-located with the bright cloud. Whitney and Fritz concluded that the "bright cloud represented [a] concentrated area of heavy convective activity . . ." and indeed " rain was falling at the surface in at least one area beneath the bright cloud at picture time."

The ideas appearing in these results—that bright clouds often are convective, and the very brightest clouds are cumulonimbi—were tested and refined in a long series of studies continuing to the present day. Hizer *et al.* (1963) reported on comparisons of Tiros pictures of the Florida area with radar plan position indicator (p.p.i.) displays. Very bright clouds were identified as cumulonimbi, and "echoes generally were contained within the brighter cloud masses . . .". However, the area of echoes was much less than the area of cloud masses—often, according to Golden (1967), by more than 50%—and some of the brighter cloud masses contained no echoes. Erickson (1964) and Fritz (1964) also associated very bright cells and masses of cloud with cumulonimbi.

An exhaustive study of tropical marine convection was reported in two parts by Zipser and LaSeur in 1964 and 1965. In a series of case studies, Tiros cloud pictures were compared with time-lapse photographs taken from the ground and from an instrumented hurricane reconnaissance airplane. All available observations were brought to bear on the question of what can be learned about clouds over the tropical oceans from Tiros pictures. In the 1964 report LaSeur and Zipser presented cloud cross

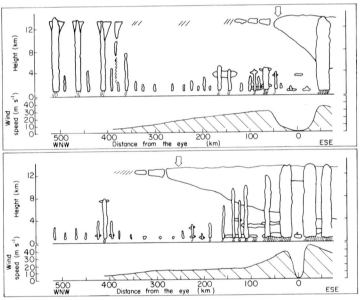

FIG. 3.3   Cloud and precipitation cross section and wind speed profile through hurricane Edith constructed from time-lapse motion pictures and observer's notes. The downward pointing arrow is the approximate location of the edge of the bright cloud shield observed by Tiros VI. Top: 1100–1300 GMT 24 September 1963. Bottom: 1100–1300 GMT 25 September 1963. From LaSeur and Zipser, 1964.

sections and cloud and echo maps for a hurricane (Edith, 1963). The cloud cross sections (Fig. 3.3), which were based on aircraft observations, show precipitation falling only from the deeper clouds. They imply that the area of precipitation was small compared with the area of clouds. This was confirmed in a plan view comparing radar-measured precipitation with the Tiros-measured hurricane cloud shield. In the 1965 report Zipser and LaSeur were concerned with patterns and systems of convective clouds. They found that convective clouds had a distinctive appearance in Tiros pictures, being small, discrete and isolated. Extensive layer cloudiness was sometimes left in the wake of widespread convection as it dissipated. Thus, quoting from the 1964 report, "Each mesoscale cloud system, as each single cumulonimbus, has a life cycle during which it undergoes profound changes. To adequately interpret a Tiros photograph of such a group, it should be properly placed in this cycle . . . ." Zipser and LaSeur also found a high correlation between the intensity of convection deduced from aircraft observations and the brightness of corresponding clouds in Tiros, and so came to a remarkable conclusion: "the depth of the convective layer can be usefully estimated by careful qualitative evaluation of cloud unit brightness, combined with size . . . Four classes of convective regime . . . can be distinguished."

2. Time lapse views

In their 1964 report LaSeur and Zipser observed,

> "In this as in many satellite meteorology studies, the more coordinated observational facilities that are concentrated in a given observational region, the more the value of each is enhanced. To existing observational systems, the advantage of adding an approximation to time-lapse photography from a geosynchronous satellite would be tremendous, for time changes are to be studied, and time changes are frequently the observational missing link."

This wish was realized in 1967, with the launch of the first Applications Technology Satellite, ATS-1. Again there was a round of papers reporting results of studies of the new data. Looking at tornado and hail producing thunderstorms over Texas and Oklahoma, Vonder Haar (1969) found "a very good correspondence between the brightest reflected-radiance regions and the location of areas of heavy precipitation". Some relationships of tropical clouds to precipitation were demonstrated by Martin and Suomi (1972) through superposition of radar images on concurrent satellite images which had been remapped to radar scale and projection. Large echoes were associated with very bright clouds, though not all bright clouds had echoes and echo area was very much less than cloud area.

Explosive growth was added to the criteria distinguishing cumulonimbus

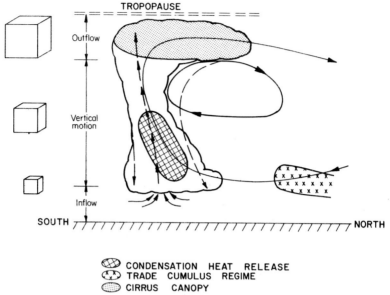

CONDENSATION HEAT RELEASE
TRADE CUMULUS REGIME
CIRRUS CANOPY

FIG. 3.4 Meridional section through a large tropical cumulonimbus or cumulonimbus ensemble. The boxes on the left represent a unit mass of air rising through the cloud. After Sikdar and Suomi, 1971.

clouds. From ATS movies D. N. Sikdar (personal communication, 1968) surmised that the upward transports of cumulonimbus mass, energy, and moisture could be inferred from measurements of anvil expansion. The idea, which is sketched in Fig. 3.4, was tested by Sikdar et al. (1970), who examined two April 1968 outbreaks of severe thunderstorms over the Southern Great Plains and Ohio River Valley. Transports of mass and latent heat were found to be ten to one hundred times larger than in ordinary, air mass thunderstorms. Sikdar and his co-workers were encouraged to suggest that with the anvil expansion model, precipitation from thunderstorms might also be estimated; we shall come back to this idea in Chapters 5 and 6.

A systematic examination of the association of bright ATS clouds with rainfall was made by Woodley and Sancho (1971). They hypothesized that "the brightest cloud masses on the satellite photographs . . . correspond to clouds that [have] penetrated a significant depth of the troposphere and consequently [are] those most likely to contain areas of precipitation." ATS-1 pictures of south Florida were colour enhanced to isolate bright clouds and compared with simultaneous radar pictures. Some aircraft photographs also were available. Eight-seven per cent of radar echoes (322) coincided with bright clouds; 70% of bright clouds (87) had echoes. But the relation between bright clouds and precipitation was not constant. The best correspondence of echo and bright cloud area was for small, growing cumulonimbi. Relatively small echoes, or none at all, were found with the large bright anvils left as cumulonimbus clouds decayed.

Sikdar (1972) examined radar and raingauge observations of precipitation occurring with the April 1968 severe thunderstorms. Echoes were small compared with cumulonimbus anvils (Fig. 3.5a), and most were located within bright anvil clouds; however, not all anvils had echoes. The larger gauge rainfall amounts were measured within or close to cumulonimbus anvils (Fig. 3.5b), but, again, precipitation area was small compared with anvil cloud area, and not observed at all for a few anvils.

II. Cloud Top Temperature and Rainfall

Although all these studies were based on satellite images of reflected shortwave energy, through most of this period (beginning with the launch of Tiros-II late in 1960) images of emitted longwave radiation were also available.

Wexler (1954) and Widger and Touart (1957) suggested that surface temperatures could be inferred from satellite measurements of infrared radiation. If the surface being viewed is a cloud, the temperature measured is that of the top, and with an appropriate sounding the satellite provides a measure of cloud top altitude. This idea is implicit in papers by Weinstein and Suomi (1961), Nordberg et al. (1962), and Wark et al. (1962). It was tested by Fritz and Winston (1962), who compared window infrared radiances from Tiros-II with cloud distributions determined from Tiros visible pictures and ground and aircraft observations. Fritz and Winston reported "useful quantitative estimates of tops of cloud systems" and a clear portrayal of large scale systems composed of thick middle and high clouds. Visible and infrared data were used together to measure the heights of low clouds.

Comparisons of infrared temperature maps and radar echo maps were made by Hawkins in 1964. The coldest temperatures corresponded with a wide band of echoes; otherwise, temperatures and echoes were not well matched. Radok (1966) and Rainbird (1969) also explored the relation of cloud top temperature to rainfall, using window radiation measurements of Tiros-III and gauge measurements of rainfall from Indochina. The correspondence of low temperatures (cold, high cloud tops) to rainfall was best a few hours after heavy rainfalls, but even then could only be described as slight. Cherkirda and Yakovleva (1967) reported some success in associating precipitation with cold cloud top temperatures. For the most part, however, results of comparisons of infrared temperature and precipitation were disappointing.

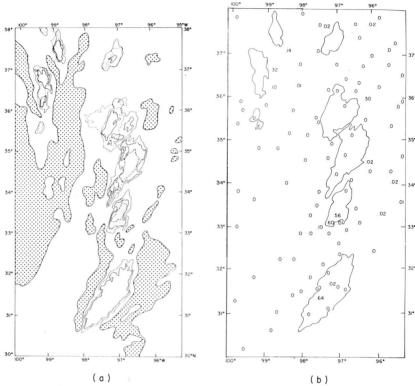

(a)   (b)

FIG. 3.5   Contoured ATS-3 cloud brightness and precipitation for 19 April 1968. From Sikdar, 1972. (a) 2223 GMT. Clear areas are stippled. Cloud outlines and shaded centers represent, respectively, the upper 40 and 30% of the total brightness range of the picture. Echoes, outlined by broken lines, are for 2217 GMT. (b) Gauge rainfall (in inches) for the hour ending at 2300 GMT. Contours are for the upper 40% of brightness at 2233 GMT.

III. Cold, Bright Clouds

In the history of rain measurement by satellites the 1967 paper of Leth- bridge marks a milestone. Using the relatively dense network of stations over the United States for measurements of precipitation, Lethbridge related Tiros-IV window infrared radiation to probability of precipitation for a late spring period. As expected the probability of precipitation occurring during the 3 h following satellite passage increased as the infrared tempera- ture decreased, but only as high as 50% even for $T < 249$ K. For reflected shortwave radiation, 3 h precipitation probability was somewhat higher,

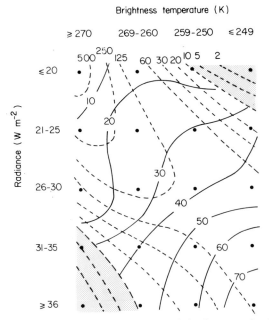

FIG. 3.6   Probability of late May–early June precipitation over land as a function of scene (brightness) temperature and radiance. Dashed lines show the number of temperature–radiance pairs, with areas less than ten shaded; solid lines represent percentage probability of precipitation within 3 h of a satellite observation. From tables of Lethbridge, 1967.

reaching 65% for radiance exceeding 35 W m^{-2}. But the strongest relationships were found for window and shortwave radiation together. Three hour probabilities increased from 5% for dark, warm scenes to 75% for bright, cold scenes (Fig. 3.6). For 12 h periods the probability of precipitation was as high as 85%. A March–April period gave similar results. Thus, Lethbridge concluded, "cloud brightness clearly matters," because it distinguishes between low cloud and no cloud for high radiation temperatures and thin cloud and thick cloud for low temperatures.

IV. Summary

A thread running through these early studies is the search for a key to detecting and mapping precipitation. To some that key seemed to be brightness. But not all bright clouds precipitate, and in those that do the area of precipitation is usually much less than the area of cloud. Others added a requirement that clouds be convective. The association of brightness with precipitation was much stronger; nevertheless, the appearance of cumulonimbus clouds was found to change greatly as they evolved, and so also did the relation of precipitation area to cloud area. Some workers, considering infrared temperature as an alternative to visible brightness, found a strong correspondence of temperature with cloud top height, but also found many high clouds which never rained. Perhaps, then, the best approach involves multispectral combinations of data: rain is most certain in clouds that are both cold and bright. We shall return to this possibility in Chapters 6 and 7.

Three years into Tiros, both the failure and the hope of researchers in satellite meteorology was so expressed by Sigmund Fritz: "Hydrodynamical and thermodynamical processes produce the observed cloud forms when an adequate supply of water vapor is present. The problem is to deduce the nature of these processes and the state of the atmosphere." Although today we have gone far beyond instruments which merely image clouds, the sentiment is apt, especially to measurement of rainfall by satellite. Despite formidable problems, by no means solved, several modest steps had been taken toward realizing the vision of Wexler, Widger, and Touart: a foundation was laid in the 1960s for development of methods to measure precipitation by satellites. The methods of the 1970s are described in the chapters which follow.

CHAPTER 4

Cloud-indexing Methods

I. Introduction

After our discussion of the need for and background to the monitoring of rainfall by satellites we turn, in this and succeeding chapters of Part II, to review the more important groups of methods developed more recently for such purposes. In every case three relevant questions may be posed:

(a) What are the needs addressed by each method?
(b) How is rainfall derived from satellite evidence?
(c) How well does the method perform?

Of the different groups of methods to be reviewed in Part II, those which have become known as "cloud-indexing" methods may be considered first. These have been applied to the widest range of climatic regions (defined by latitudinal zones and/or maritime or continental characteristics). Furthermore, it is logical to consider cloud-indexing methods first because they are generally the least dependent on sophisticated software and hardware systems.

The centres of origin of the two most widely used cloud-indexing methods (or families of methods) are the Applied Climatology Laboratory of the Department of Geography, University of Bristol, UK, and the Applications Group of the National Environmental Satellite Service (NESS) of NOAA, Camp Springs, Md, USA. Related work has been undertaken in a number of other laboratories and institutions. Although most cloud-indexing methods have grown from a common root (Barrett, 1970), and much cross-fertilization has taken place both within and between the families of cloud-indexing methods, it is convenient to subdivide this chapter into three corresponding sections.

One particularly important trend in cloud-indexing philosophy and practice is worthy of special note. In the early days of rainfall monitoring by cloud-indexing methods it seemed to be generally assumed that the satellite was a *competitor* to the raingauge and/or ground-based radar, although in some cases this was justifiable because the interest was in assessing the level of the information content of the satellite data rather than in the preparation of the best possible rainfall inventories. Consequently, many early studies invoked conventional data only in calibration and verification of satellite rainfall estimates, and not as inputs to final improved data sets (see e.g. Barrett, 1970; Follansbee and Oliver, 1975). More recently it has become more usual for the satellite system to be seen as a *complement* to the conventional observing network. Today, satellite data are often used in the inference of rainfall patterns and amounts between the positions of ground observing stations: available conventional data are invoked as "ground truth" in the fullest sense: not only for satellite data calibration and verification, but also as the references around and between which improved rainfall patterns are derived from satellite evidence.

Results of cloud-indexing methods today are of particular value to the applied climatology community where the chief need is for the best possible rainfall data over large land areas for near real-time operations. This will be illustrated by and attested in Part III.

II. Bristol Methods

The initial cloud-indexing method from Bristol was formulated in an attempt to use satellite data to homogenize rainfall mapping for unit periods of time (initially one month) for a large region designated the "Tropical Far East" (Barrett, 1970, 1971). In response to the recognition that any evaluation of the global hydrological cycle depends heavily upon assessments of rainfall over the oceans, it was decided to test the utility of satellite data (initially in the form of NOAA nephanalyses drawn from ESSA visible image mosaics) in rainfall mapping from the inferences of the imaged clouds. This work, first described in an oral presentation at the Adelaide sysposium of the Australia and New Zealand Association for the Advancement of Science in 1969, seems to have been seminal to much of the subsequent work in cloud-indexing in Bristol and elsewhere.

Viewed very generally, it is assumed in all the Bristol methods that rainfall accumulated over a period may be represented as:

$$R = f(c, i(A)) \tag{4.1}$$

R is rainfall accumulated over a period at grid intersections or (more commonly in the recent work) in grid squares ("cells") of selected size; c is cloud area, i is cloud type, and A is altitude (usually above sea-level, or the lowest calibration station in the study area). Since the Far East studies it has become usual to express i as a single (cloud type) index, evaluated differently for ranges of chosen cloud types, and for different regions. In each exercise a suitable rank of cloud type is accorded a set of values on a dimensionless scale. At first (Table 4.1) the ranks and values were based on "meteorological expectations". More recently, it has been possible to draw also on accumulated experience and local evidence.

TABLE 4.1   Rainfall probabilities and intensities as related to satellite-observed states of the sky. From Barrett, 1970.

1 States of the sky (nephanalysis cloud categories)	2 Assigned probabilities of rainfall (relative scale range 0–1·00)	3 Assigned intensities of rainfall (relative scale range 0–1·00)
Cumulonimbus	0·90	0·80
Stratiform	0·50	0·50
Cumuliform	0·10	0·20
Stratocumuliform	0·10	0·01
Cirriform	0·10	0·01
Clear skies	—	—

In the early Bristol work a synoptic weighting factor, S_w, appeared instead of A. This was to account for the more intense rains which often fall in the tropics from the better-organized ("synoptically-significant") weather systems. More recently, as attention has turned to meteorological, not climatological rainfall estimation, the function of S_w has been fulfilled more flexibly by a procedure to "float" rainfall estimates in keeping with the evident intensity of each separate rain-cloud system (see Barrett, 1980a).

There are two basic preparatory stages in the Bristol methods. These involve:

(a) the construction of an equation to provide, for any selected point or area through a chosen period of time, a rainfall coefficient or cloud index to be evaluated from features of the satellite-imaged cloud field (visible and/or infrared); and

(b) the compilation of a suitable regression diagram relating cloud indices to raingauge observations of rain.

In the first cloud-indexing study (for the tropical Far East) the necessary regression (Fig. 4.1) was constructed from satellite cloud indices and

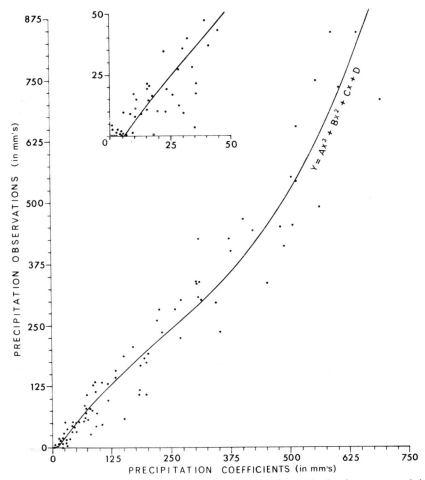

FIG. 4.1   Regression of "precipitation coefficients" (cloud indices) versus precipitation observations for 29 stations in the Tropical Far East, March–June 1966. From Barrett, 1971.

monthly rainfall totals for selected stations through 12 months preceding the year from which two high-season months (July 1966 and January 1967) were chosen for the satellite rainfall mapping exercise. For these two months rainfall estimates were obtained for 2½° grid intersections by evaluating cloud indices from the satellite nephanalyses, and translating these into estimates of rainfall through the computed regression line. Isohyetal maps were constructed from the point estimates, as illustrated and discussed in Chapter 10. Contingency tables for estimated versus observed rainfall indicated that the satellite-derived patterns and amounts conformed well enough with climatological expectations to justify further development of the technique.

Attention in Bristol turned next to the possibility that short-period rainfall *forecasts* might be improved by satellite methods, especially in windward coastal zones (Barrett, 1973). Correct 24 h forecast categories were achieved for a single station (Valentia, Ireland), and a small catchment (the Lower Parrett basin, Somerset, England) on 74% and 70% respectively of all days through periods of 6 months in either case (Barrett, 1975a). However, this promise has not been built on since: the demand and scope for post-event rainfall estimation have been much greater than for rainfall prediction.

In the meantime, two developments elsewhere were to exercise considerable influence on the future Bristol work. First, early analyses of data from the Nimbus-5 Electrically Scanning Microwave Radiometer (ESMR: see Chapters 8 and 9) had shown that that sensor was not likely to yield the quality of rainfall data over land which had been anticipated prior to the launch of Nimbus-5. Secondly, adaptations of the initial Bristol Cloud-indexing method by workers in NESS (see Section III) for *daily* rainfall estimation were seen as something of a challenge by the British team. The first NESS method gave rainfall estimates daily rather than monthly through the expedient of averaging the rainfall performance of clouds over large areas rather than relatively long periods of time. Given the incentive of a study contract for a small area of northern Sumatra, a new Bristol method was drawn up to given short-period rainfall estimates for very small ($\frac{1}{6}$° square) grid cells (Barrett, 1975b). This approach has formed the foundation for all subsequent efforts in Bristol. These have had the following objectives:

(a) The design and development of a satellite-assisted rainfall mapping method to capitalize simultaneously on the better attributes of both satellite and conventional observations: the *relative accuracy* of rain-gauge measurements (which represent accumulations of rainfall through time, but only for the point locations of the raingauges), and the *more complete areal indications* of rain given by satellite images (which give better views of the spatial distribution of rain cloud, but only for separated points in time).

(b) The development of a satellite-improved rainfall monitoring method which could be implemented more or less immediately in any area of the world. Associated practical considerations at first necessitated that the method should be based on polar-orbiting satellite data, but, if possible, it should be easily adaptable for use with data from geostationary satellites.

(c) The development of an operational technique which could be implemented at modest cost. Questions of cost are of special significance for developing countries and/or agencies of limited jurisdiction. At present this objective can be met best by a method or methods not necessitating expensive data processing equipment.

(d) The testing of such a method or methods in contrasting areas, for a variety of applications. Tests have included northern Sumatra (for irrigation design and implementation: Barrett, 1975b, 1976), part of the Sultanate of Oman (for water resource evaluation: Barrett, 1977, 1979; see also Section II, Chapter 11), and north-west Africa (primarily for desert locust monitoring and control: see Hielkema and Howard, 1976; Barrett, 1977b,c, 1980a; and Barrett and Lounis, 1979; see also Section IV, Chapter 12).

(e) The implementation of a suitable method in a fully operational form, elevating such an approach from research status to that of a fully fledged procedure in routine use. This final objective has been realized first in north-west Africa (see Chapter 12).

Usually the "Bristol Method" in its modern form (Fig. 4.2) is applied to grid squares of selected size (the choice has ranged from $1/6°$ – 1° grid squares) for 12 h units of time (the periods for which rainfall data are most commonly forthcoming from synoptic stations in developing countries of the Old World through the GTS (Global Telecommunication System) of the World Weather Watch).

At first the grid squares are identified as "gauge cells" (containing GTS stations) and "satellite cells" (lacking GTS stations). Then the study area is subdivided morphoclimatically, and related cloud index/observed rainfall regression diagrams are prepared for an historic period. During daytime both visible and infrared images are used whenever possible in the all-important differentiation of cloud types: the joint use of both types of imagery aids good cloud-type recognition (see Chapter 3). During nighttime infrared images are used alone. Where a grid square contains more than one precipitable cloud type the two most significant precipitable cloud types are considered separately, and the resulting cloud index for the square is taken to be their sum. Operationally, GTS rainfall data are plotted for the gauge cells, and cloud index/observed rainfall regressions are used to establish appropriate rainfall estimates for intervening, conventional data-sparse, areas of satellite cells. In this way the satellite evidence is used to improve the shapes of the isohyetal patterns between the "ground truth" observations.

Two details of the present Bristol Method are worthy of special mention. First, it is recognized that the analysis of satellite images 12 h apart may

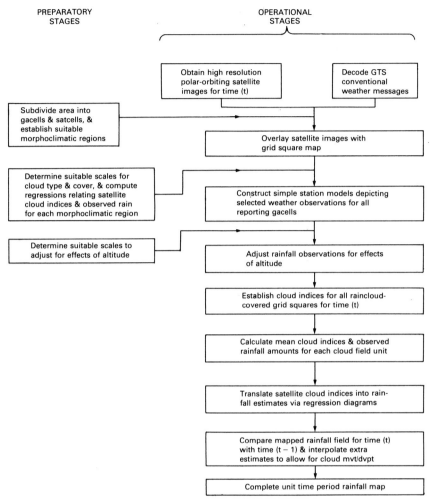

FIG. 4.2   Flow diagram for the "Bristol Method" in a recent form for use with polar-orbiting satellite imagery. From Barrett, 1980a.

result in a misleading picture of semi-diurnal rainfall, especially when rapid raincloud system movement and/or development occur. Consequently, initial satellite-improved rainfall maps ("set-piece analyses") are used only as first approximations on which final 12 h rainfall maps are based. In constructing these, the analyst may insert and/or adjust rainfall estimates in some squares to allow for likely raincloud performance which the satellite images did not reveal or adequately represent. If geostationary imagery is available this can be used to refine, confirm, or modify the results.

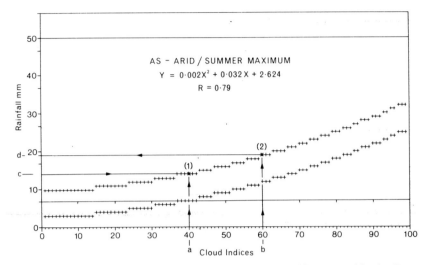

FIG. 4.3   A stepped regression for part of north-west Africa as used in the Desert Locust monitoring project (see p. 241) to facilitate the "floating" of rainfall estimates for satellite cells in relation to gauge cell evidence. In the example shown a gauge cell cloud index (a) and rainfall observation (c) are used to locate point (1) on the operative surface parallel to the computed regression. A duplicate transparent overlay is moved up so that the climatic (computed) regression intersects (1). For an associated satellite cell with cloud index (b), the appropriate rainfall estimate is (d), for it is considered that (2) should lie on the same surface as (1). The computed (climatic) regression alone is used to give satellite cell rainfall estimates where rain-cloud systems cannot be calibrated from gauge cell evidence; it is stepped to round rainfall estimates to the nearest mm. From Barrett, 1980a.

Secondly, as foreseen in Chapter 3, it is certain that clouds with similar appearances do not always precipitate equally. Consequently guidelines have been established for individual rainfall estimates to be "floated", upwards or downwards, guided by observed cloud index/rainfall relationships at gauge cells affected by the same rain-cloud system. This process is illustrated in Fig. 4.3. To be applied successfully, reasonable meteorological skill is required of the analyst.

To summarize, the present Bristol Method has been described as

"an extension of classical synoptic meteorology, involving as it does a strong element of qualitative judgement based on a variety of types of weather observations, and carried out within a framework of rules and practices evolved largely through practical experience." (Barrett, 1980a)

Verification of the Bristol Method has been attempted in various ways, with the following results (see Barrett, 1980a):

(a) Adjudged on a categoric basis, correct results have been indicated in approximately 75% of all cases tested in this way. In individual studies the accuracy of the rainfall estimates seem to have been affected particularly by the type(s) of climatic regions involved, and the densities of ground observation (calibration gauge) networks.

(b) Detailed tests in north-west Africa in 1977 revealed that, on a multi-categoric basis, twice-daily rainfall was evaluated correctly on 94% of all occasions, with rain estimated for rain cases on 76% of such occasions. The omnibus "skill score" (see Follansbee, 1976) of 0·80 was flatteringly high: it was boosted by a large number of zero rainfall estimates for dry days.

(c) Monthly rainfall estimates, aggregated from daily or twice-daily data, are generally within ±20% of ground observations. In one study in Oman (see Section II, Chapter 11) the mean annual rainfall estimation for a number of verification stations was 97% of observed rainfall.

(d) Reports of post-event vegetation flush in desert and semi-desert areas have confirmed rainfall events mapped in joint rain-gauge/satellite operations. Conversely, in quasi-operational projects no major vegetation developments have been reported which have gone unpredicted by satellite-assisted rainfall monitoring.

Most recently (Barrett, 1980c), guidelines have been drawn up for the supporting use of geostationary satellite data. Meteosat data will be of special value when rapid movement and/or development or dissipation of rain-cloud systems is evident (Barrett and Lounis, 1979). However, where raingauge data are provided no more frequently than every 12 or 24 h a much fuller use of geostationary imagery would seem to have little to recommend it unless the distribution of rainfall between gauge stations or between gauge times is important, as it might be in forecasting flash floods.

It is thought probable that a substantial degree of computer assistance could be built into the Bristol rainfall monitoring approach, especially if a suitable method of cloud-type identification could be devised. However, recent research has suggested that such identification procedures will not be established easily (Harris and Barrett, 1978). Multispectral data analysis seems potentially more promising than uni- or bispectral methods, but that, too, is in its early stages (see e.g. Shenk et al. 1976). Further research in multispectral cloud identification methods could prove very valuable, and pave the way for the design and construction of dedicated microprocessors for low-cost local agency use. Data from the new AVHRR radiometers on Tiros-N satellites are suitable for such work.

III. NESS Methods

It seems that active interest in satellite estimation of rainfall by NESS staff was born in 1971. The Director of Meteorology in Zambia, following disastrous floods in the Luangwa Valley, Eastern Province, Zambia in early March 1971, suggested to the Administrator of NOAA that the feasibility of devising a flood forecasting scheme using satellite data along the lines suggested by Barrett (1970, 1971) should be explored (Follansbee, personal communication, 1973). We have seen that, since the original work in Bristol was essentially climatological, not meteorological, modifications had to be made before such an approach could be put to daily rather than monthly use. The result was the first 24 h satellite rainfall estimation scheme (Follansbee, 1973). In this, statistical averaging of cloud/rainfall relationships was achieved areally, not temporally as in the first Bristol method.

In the first NESS method, rainfall estimates were based on afternoon images from NOAA polar-orbiting satellites. For each 24 h period, average rainfall across the broad study area (R) was calculated through a relationship which may be written:

$$R = (K_1 A_1 + K_2 A_2 + K_3 A_3)/A_0 \qquad (4.2)$$

A_0 is the area under study; A_1, A_2 and A_3 are areas of A_0 covered by the three most important types of rain-producing clouds (cumulonimbus, cumulocongestus, and nimbostratus); and K_1, K_2 and K_3 are empirical coefficients. Initially the values of these weights were very similar to the rainfall probability × rainrate cloud indices used by Barrett in his study of the tropical Far East. However, Follansbee used them as expressions of total daily rainfall, whereas Barrett had used them as dimensionless weighting factors: this obviated the need in the NESS method for continuous calibration against raingauge data.

Following the first applications of this method in eastern Zambia, further tests were made in several tropical and subtropical regions. These included peninsular Florida, Arkansas, Louisiana, Mississippi, the coastal basin of southern California, plus Zambia as a whole. The technique, being comparatively simple, could be taught easily. This was soon confirmed by experience in both the USA and south-eastern Asia (in the Mekong River Basin: see Section IV, Chapter 11). Some results were very good (Fig. 4.4a), but, not surprisingly in view of the use of fixed weights, some were not (Fig. 4.4b).

In a contingency-table test for the accuracy of aggregated results the class interval of estimated rain was that of the observed rain in 215 cases (44%) from 488; a further 172 (35%) were only one class astray (Follansbee, 1973).

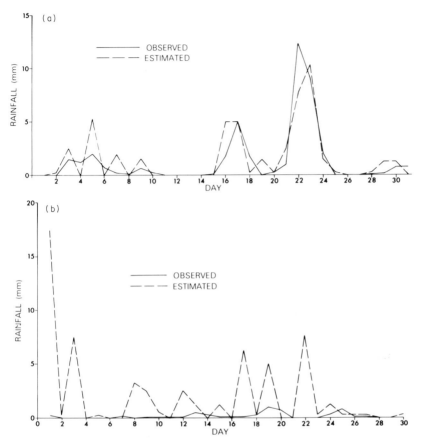

FIG. 4.4   Early graphs from NESS relating observed, and satellite-estimated rain using a simple fixed-weight cloud-indexing method over peninsular Florida south of 30°N: (a) May 1970; (b) April 1970. Such results (some good, some not) pointed to the need for variable weights to be used as meteorology, geography and altitude required. After Follansbee, 1973.

These trials and tests revealed three particular problems:

(a) The estimation of a single rainfall value for a large area often obscured important spatial elements of rainfall distributions, especially when convectional activity was dominant. For example, Follansbee (1973) refers to the case of 4 July 1971 in Louisiana. The state average from 87 rainfall stations was 2·5 mm—precisely the value proposed by the satellite image analyst. However, as much as 48·8 mm was reported from one station, whilst 60% of the stations reported zero rain.

(b) Absolute ceiling values were imposed on the rainfall estimates by the fixed cloud indices for different types of clouds. For example, the

C

maximum possible areal estimate for cumulonimbus was only 25·4 mm per day; for a continuous cover of nimbostratus the corresponding figure was 6·2 mm.

(c) The lack of some device to float rainfall estimates up or down according to synoptic weather conditions also led on some occasions to overestimates of rain. Figure 4.4(b) affords one such example: then atmospheric conditions were less conducive to rain than the clouds alone suggested.

A range of new subroutines was devised to reduce the adverse effects of these conclusions. First, it became common practice to reduce the areas covered by individual estimates so that spatial patterns of rainfall might be more accurately evaluated (see e.g. Follansbee, 1973).

Secondly, additional cloud weights were introduced to adjust for the particularly significant variations in rainfall from cumuliform clouds. For example, in applications of the early NESS method to the islands of the Caribbean, LeComte (personal communication, 1977, see also Section II, Chapter 13) allowed for important shower activity in winter through a five-fold increase in the index for cumulocongestus cloud.

Thirdly, special procedures were developed to increase the flexibility of the technique for use in areas subject to especially heavy rains (e.g. those affected by the south Asian summer monsoon) through comparison with other types of data. Follansbee (1973) listed a number of possibilities:

(a) Where necessary, the ratio $\bar{R}/<\bar{R}>$ could be used to increase the rainfall estimates. Here $\bar{R}$ represents the mean rainfall recorded at a single heavy rainfall station, and $<\bar{R}>$ that recorded over the wider region in which that station is located.

(b) $\bar{R}$ could be expressed as a fraction of $\bar{c}$, the average daily cloudiness over the wider area. Experience revealed that the relationship giving the best results for India is $\bar{R}/0\cdot605\bar{c}$.

(c) Automated mean cloud charts (e.g. those of Miller, 1971) could be used to evaluate station mean cloudiness $\bar{c}'$ at the heavy rainfall station for use in preference to area mean cloudiness $\bar{c}$.

Since $\bar{c}'$ is physically more closely related to the heavy rainfall station than $\bar{c}$, Follansbee considered that the best of these three options was the weighting factor in (b) above.

Attention in NESS subsequently advanced to a consideration of the significance for rainfall estimation procedures of the diurnal variability of convectional clouds in low latitudes. Follansbee and Oliver (1975) exemplified the advantages of sampling cloud fields in twice, not once, daily satellite images. They concluded that, whilst results of approximately equal accuracy were obtainable from either afternoon visible or nighttime infrared images alone, higher accuracies were possible for the "semi-arid" regions featured in their study if the means of the estimates from the two sets of images were calculated instead. Further, they confirmed a generally negative correlation between morning cloudiness and evening thunderstorms: this had been suggested earlier by Purdom (1973), and Purdom and Gurka (1974) using images from the first geostationary weather satellites ATS-I and ATS-III.

Diurnal variability of cloud and rain was considered so important for rainfall estimation in cumulonimbus-dominated low latitudes that Follansbee's original cloud index equation was simplified to

$$R = K_1 A_1 / A_0 \qquad (4.3)$$

for summer tests in Alabama, Georgia and South Carolina. In effect, estimated rain, in inches, was taken to be indicated simply by A_1/A_0, for the original value of $1 \cdot 0$ was retained for K_1. The most important implicaton of this is that, where geostationary satellite data are available to reveal the afternoon growth of convectional clouds over land, a cloud-indexing approach of extreme simplicity may yield data of sufficient accuracy for some applications.

Elsewhere, and in the absence of geostationary satellite coverage, there is always the need to use a different approach. This was recognized by Follansbee and Oliver (1975), who also pointed out that the effects of surface morphology, e.g. mountains and valleys, might sometimes require more detailed attention too. Indeed, as early as 1973, Follansbee had seen the need for the satellite approach to be "supported and modified by any tool available" including "all features of the synoptic situation that may affect the area under consideration, both local and large scale . . . radar scans, spot rainfall reports within and near the area, orographic effects and other pertinent terrain features, and persistence of various parameters". He went on to mention the likely utility of wind data for different levels, humidity values, and pressure patterns and changes, and concluded that the final method should be seen as a supplement to, rather than a replacement for existing procedures.

Both the more recent refinements of the Bristol Method, and the more recently developed life-history techniques of Scofield and Oliver (see Section III, Chapter 5) largely accord with, and, in different ways fulfill, the requirements of that philosophy. However, greater method flexibility and sophistication demand more, and more expert, image analysis, and/or more sophisticated supporting hardware and software. Not surprisingly, early NESS cloud-indexing methods are still in use today where quick, cheap, but very general assessments of rainfall meet a consumer's need. Elsewhere,

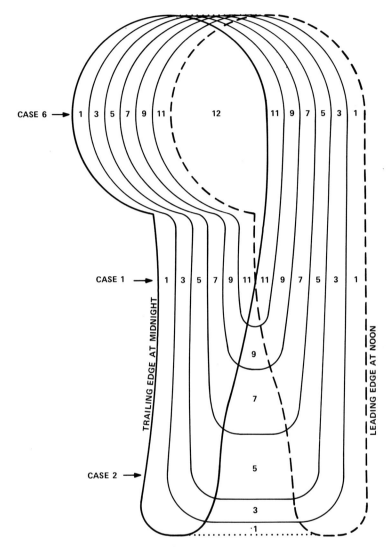

FIG. 4.5   A cloud motion model developed to summarize the duration of precipitation (in h) along the path of a precipitating cloud mass between succeeding 12 h satellite images. The results are used to weight the precipitation expected to fall (in this case) from an occluded mid-latitude low, identified as a suitable proportion of monthly normal rainfall. From Follansbee, 1976.

there have been opportunities to test more capital-intensive man-machine mix cloud-indexing methods. One such programme, designed to develop a cloud-indexing method for operational use involving a significant degree of automation, has been initiated in support of spring wheat harvest predictions (EarthSat Corporation, 1976). Here, modified versions of Follansbee's original equation have been used to evaluate rainfall in 25 n.mi. grid squares on a 6 h basis, employing GOES satellite data and a man-machine interactive system to identify and evaluate areas covered by clouds of different types. More recently the EarthSat method has been extended to provide assessments of rainfall over broader areas of the globe within the context of a commercial operation trade-marked "Cropcast" (see Chapter 13).

In verification tests for parts of the USA in 1975, differences between the satellite rainfall estimates and observations from stations in the US "Co-operative Observers" network were small (mostly 5–10%) for monthly aggregates, but often much larger for daily estimates. In absolute terms, over 20 000 comparisons were made between satellite estimates and co-operative station reports during the 1975 EarthSat exercise in the USA. Of this total 45% of the satellite estimates fell within 1 mm, 72% within 3 mm, and 95% within 13 mm of the co-operative station reports. Overall conclusions were that:

(a) spatial patterns of major rain areas are fairly well represented;
(b) major rain areas tend to be spread out by this method;
(c) low rainfall is often overestimated, and heavy rainfall underestimated;
(d) the daily spatial variability of rain becomes smoothed out; and
(e) maps for larger periods show better fits in terms of both spatial patterns and absolute rainfall amounts.

Such conclusions echo those of the Bristol and NESS groups. Later chapters in this Part will emphasize that they are prompted largely by the behaviour of rainfall itself, especially insofar as this is manifested through its noisy distributions in both space and time.

Cross-fertilization, which has already been so evident in this summary of satellite cloud-indexing activities, reasserts itself through the most recent cloud-indexing studies carried out in NESS. Once the particular problems posed by convective clouds in low latitudes had been clearly recognized and described by Follansbee and Oliver in 1975, attention in NESS was focussed onto two related problems. These involved:

(a) studies of area:brightness relationships in severe convectional storms (see Section III, Chapter 5); and
(b) studies of rainfall outside the tropics, where rain falls mostly from well-organized mid-latitude storms (Follansbee, 1976).

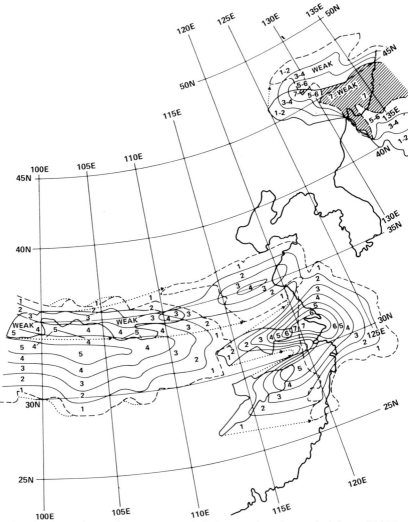

FIG 4.6 A workmap for east-central Asia covering the period from 0900 h to 2000 h, 10 March, 1975, showing isopleths of rainfall duration established through satellite image analysis and the use of models like that in Fig. 4.5. From Follansbee, 1976.

It is the second of these that we must examine here. This new work—leading to an essentially new type of rainfall estimation method—highlighted for the first time the importance of rain-cloud movement between (polar-orbiting) imaging times. It also prompted the subsequent attention to this question in the development of the present Bristol Method as explained in Section II, Chapter 4.

In one sense the new NESS method (see Follansbee, 1976) is not so much a cloud-indexing method as a "cloud-system indexing" or "rainfall climatology" technique. Its basis is the relationship between climatologically averaged rainfall and the long-term average contributions made to it by key synoptic weather systems. In the first areas studied these were mobile mid-latitude depressions. Cloud-type indexing is used secondarily to evidence spatial variations of rain within the systems of rain-bearing clouds. This approach was prompted by an increasing demand for improved, near real-time, rainfall data from extensive, globally significant, crop-growing regions in Asia and other continents (see Chapter 13). In practice the method involves:

(a) the delineation, by an analyst, of the area(s) covered by precipitation cloud in pairs of consecutive polar-orbiting satellite images 12 h apart;
(b) the invoking of cloud motion models to delineate time-envelopes of associated precipitation, assessed in hours;
(c) the use of an empirically derived relationship to evaluate likely rainfall totals for one-half day periods for each grid point in the study region. For rainfall estimation in China, the USSR, and about one-half of the USA, Follansbee adopted an association which can be written:

$$P_{\frac{1}{2}} = 0 \cdot 09 f_{\frac{1}{2}} E(P_{30}) \tag{4.4}$$

Here, f is the fractional period of precipitation, $E(P)$ is the climatic normal (expected) precipitation, and the subscripts refer to periods in days. The type of cloud motion model constructed for such purposes is exemplified by Fig. 4.5; Fig. 4.6 is a workmap for one precipitation system analysed through a single 12 h period. Sample precipitation maps are included in Chapter 13, Fig. 13.1.

Detailed verification tests with the results of rainfall estimation in the USA showed that, on a multicategoric basis, 75% of all daily-aggregated estimates were correct. As in earlier work, the most significant errors accompanied active thunderstorm situations. It would be interesting to test the performance of an approach of this nature developed to include rain-gauge data in the mapping stages, and to float rainfall estimates to suit the synoptic weather.

IV. Related Methods

Independently of the NESS efforts, Scherer and Hudlow (1971) developed a technique which nevertheless bears some similarities, especially in concept. Their scheme, a contribution to the 1969 Barbados Oceanographic and

Meteorological Experiment (BOMEX), was designed to extend the range of two 3 cm weather radars across the whole area of the BOMEX "budget box". It builds upon a relationship between echo length and precipitation established for BOMEX by Hudlow (1971). The distribution of echo lengths over an area beyond radar range is inferred from satellite infrared images which have been converted from temperature to equivalent pressure height using a nearby temperature sounding. The shape of the distribution of echo lengths is based on observed average distributions of echo length for each of three weather regimes—undisturbed, intermediate and disturbed. These regimes are characterized by progressively higher (and presumably deeper) clouds, with occurrence thresholds at 550 and 350 mb. Total echo area for a regime is one-quarter cloud amount at a second set of threshold pressures—800 mb for the undisturbed regime, and 500 mb for intermediate and disturbed regimes. From echo area and distribution, echo number is determined iteratively.

Martin and Scherer (1973) reported 30% accuracy in estimating echo area, and factor-of-two accuracy in estimating rainfall. They also noted that the scheme could be simplified and made less sensitive to instrumental biases by dropping the intermediate step involving echo area. Then cloud area for each regime would be related directly to precipitation, and one would have a three-part indexing scheme.

The problem of monitoring stream flow from remote catchments in the near-equatorial country of Surinam led Grosh et al. (1973) to a simplified version of Follansbee's scheme. Since their concern was the storms that produce the larger pulses in runoff, they treated only deep convective clouds. These were identified in visible wavelength images of the ATS-3 satellite by a threshold of brightness known to be associated with relatively high rates of rain. Area above this threshold was measured for a basin, and converted to a rain volume. Results for six storms show a linear relation between bright cloud area and the volume of runoff, determined by direct measurements of streamflow. This sample, however, is too small to be considered more than a demonstration of feasibility. We shall return to the question of hydrological applications of satellite rain monitoring methods in Chapter 11.

Further independent work was undertaken by a group at Stanford Research Institute. They developed a cloud-indexing scheme to estimate daily precipitation over river basins in the American north-west. Finding that basin average digital brightness was poorly related to precipitation, the Stanford group developed a nine category classification of cloudiness. Their approach is described in reports by Davis et al. (1971) and Davis and Weigman (1973); the cloud categories are given in Table 4.2.

The nine categories are assumed to cover the major precipitation systems

TABLE 4.2   Cloudiness categories and precipitation rates for the Flathead River Basin (after Davis and Weigman, 1973).

Category	Characteristic	Precipitation Rate (mm 12 h^{-1})	
		Fall-winter	Spring-summer
1	Bright shield, vortical	13·30	14·80
2	Bright band	7·11	7·95
3	Extensive convection	3·23	4·62
4	Shield fringe, dessicated shield	1·70	0·94
5	Broken, dissipating band	1·35	0·86
6	Limited convection	1·19	0·41
7	Multi-layered, disorganized	0·15	0·08
8	Single-layered, disorganized	0·03	0·03
9	Clear or mostly clear	0·00	0·00

affecting the basins of interest. Categories are assumed to be exclusive and identifiable in satellite pictures. The precipitation rate associated with each category is unique (cf. NESS methods of similar dates).

Cloud categories were determined for river basins over 12 h periods by interpolation from once-daily visible or infrared pictures from polar orbiting satellites. Coefficients of average precipitation rate were calculated separately for two 65 day periods, one autumn–winter, and one spring–summer. Precipitation was measured by gauges distributed through the Flathead River Basin, Montana. The relationship of 12 h cloudiness category to 12 h precipitation was determined by an iterative approximation procedure which assured equivalence between total observed and estimated basin precipitation for each calibration period. Coefficients were significantly larger for the spring–summer period, especially for the intermediate precipitation categories (Table 4.2).

Precipitation was estimated for the following year over the spring–summer period for the Flathead and two other basins. Compared with gauge measurements of basin precipitation, the method performed very well: it caught all major precipitation events, and cumulative satellite rainfall at the end of the 75 day period was within 16% of gauge rainfall (see Fig. 11.3). However, as might be expected, the satellite method did not catch differences in precipitation rate from basin to basin.

The Stanford scheme, like most other cloud-indexing schemes, hinges on the interpretative skill of the analyst. It is sensitive to misclassification, for there are large differences in precipitation rates between some adjacent categories. Errors in daily precipitation rates may be large, especially when precipitation regimes change quickly. Applications so far have been too limited to judge how much change might be expected for other areas and seasons. On the other hand, the Stanford technique applies to precipitation from all sources. It is straightforward, requires little equipment, and, at least for the basins of western Montana, produces precipitation estimates which are clearly useful on hydrological scales.

Lastly, we turn to a scheme almost elegant in its simplicity: that of Kilonsky and Ramage (1976). These authors sought measurements of rainfall over the oceans of the tropics, for use in studies of climate. They began with the premise that most of this rain comes from organized convective systems. These are large and bright in satellite visible imagery. If rates of rain are reasonably uniform, highly reflective cloud alone ought to be a useful index of ocean rainfall.

This hypothesis was tested by comparing the monthly frequency of highly reflective cloud over coral islands in the Pacific Ocean with monthly rainfall measured at these islands. Using NESS visible Mercator mosaics, highly reflective cloud was hand-measured daily, by fitting an ellipse of variable size and eccentricity to highly reflective cloud masses 2° latitude in radius or larger. Coded measurements of the size, eccentricity and orientation of the fitted ellipse and the location of its center were punched on cards, then monthly frequencies of highly reflective cloud at particular locations were calculated by computer. The results for 820 station-months of dependent data (Fig. 4.7) confirm a linear relation of monthly rainfall and highly reflective cloud. By least squares regression this was determined to be $R_{30} = 55\cdot3 + 39\cdot2N$, where R_{30} is given in millimetres and N is monthly frequency of highly reflective cloud. The correlation of R_{30} and N, $0\cdot75$, reflects the larger scatter in the observations and underlines the climatological framing of this technique.

Garcia (1981) discussed the particular problem of the finite rainfall which was predicted at zero frequency of highly reflective cloud: presumably this is due to the occurrence of highly reflective cloud over stations between satellite observations, and small or low rainclouds. It represents a floor for estimated rainfall, just as $N = 30$ represents a ceiling.

The scheme was tested by computing monthly rainfall over the Pacific between 20°N and 20°S at a grid interval of 1° by 1° for the period May 1971 through April 1973. The latitudinal average of annual rainfall is found to compare well with estimates of other authors based mainly on surface observations (see Fig. 10.3).

A second test was made by Garcia (1981), for the Atlantic Ocean during the period of the Global Atmospheric Research Programme Atlantic Tropical Experiment (GATE). Garcia compared his estimates of rain with those of Griffith et al. (1980). Because the Woodley maps are not for monthly periods, Garcia modified the regression relation of Kilonsky and Ramage.

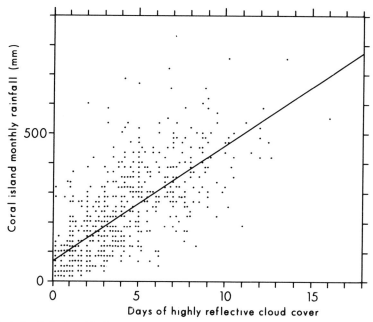

FIG. 4.7 Relationship between monthly observed rainfall totals on Pacific Ocean coral islands and the numbers of days with highly reflective cloud at the same locations and for the same months, as determined from NESS polar-orbiting satellite Mercator mosaics. The straight line is the least-squares best fit. From Kilonsky and Ramage, 1976.

For Phase estimates the y-intercept was reduced by the ratio of days-in-a-Phase to days-in-a-month. For GATE period estimates, the 85 days of GATE were divided into three segments, each almost a month in length. The y-intercept was reduced by the ratio of days-in-a-segment to days-in-a-month, then segment estimates were added to obtain the GATE period estimate. In addition, to avoid overestimating rainfall far from active zones, Garcia set Phase or segment rainfall to zero in those grid squares which reported no highly reflective cloud and shared no side with squares reporting highly reflective cloud.

Comparing his estimates with radar measurements for a 3° × 3° box centered on the ship hexagons, Garcia found differences of −6 to +15% for the three Phases of GATE. The large Phase 2 discrepancy reported by Woodley and his collaborators (see Section III, Chapter 10) did not appear. Comparisons with Woodley were made over a 1° × 1° grid. Agreement was better for ocean areas alone than for ocean and land. Over the ocean Garcia's estimates tended to be larger than Woodley's, by 20% for the GATE period; they were much less than Woodley's over land. The correlations, however, are generally high, ranging up to 0·81 for the entire GATE period, land and ocean, and 0·92, ocean only. The relationship is best seen in rain maps. That of Garcia for all of GATE is shown in Fig. 4.8 for comparison with the Woodley map, shown in Fig. 10.4. The agreement in large features is excellent. There is significant disagreement only along the coast of West Africa and over the interior. Garcia fails to show the very heavy rain found by Woodley, because convective rain there tends to fall at night.

The results, as Garcia concludes, proffer strong support for use of the Kilonsky-Ramage technique to estimate monthly rainfall over tropical oceans.

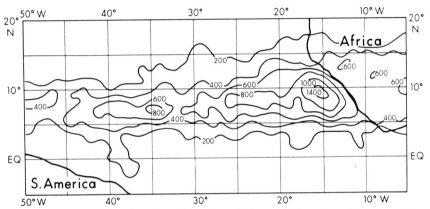

FIG. 4.8   Rainfall (mm) estimated by the technique of Kilonsky and Ramage. The estimate covers 80 days between 27 June and 20 September 1974. From Garcia, 1981.

CHAPTER 5

Life-history Methods

One class of satellite rainfall monitoring schemes is based on the twin prem- ises that significant precipitation comes mostly from convective clouds and convective clouds can be distinguished in satellite pictures from other clouds. Recognizing these convective clouds as well as estimating their rainfall depends in part on changes in the clouds. Thus each of the three schemes described here uses sequences of images from geostationary satel- lites. The interval between consecutive pictures must be short compared with the lifetimes of precipitating convective clouds, and there must be at least a relative calibration of the imaging sensors on the satellite.

I. Stout, Martin and Sikdar

The most straightforward scheme, that of Stout, Martin and Sikdar (1979), was conceived to provide maps and tables of convective scale rainfall for the Global Atmospheric Research Program Atlantic Tropical Experiment (GATE) outside the area covered by 5 cm shipboard radars. Here the amount of rainfall produced by a cumulonimbus cloud or cloud ensemble is estimated by the sum of area and area change terms:

$$R_v = a_0 A_c + a_1 dA_c/dt. \qquad (5.1)$$

R_v is volumetric rainrate (L^3/t) for a particular cloud, A_c is the area of the cloud at time t, dA_c/dt is cloud area change, and a_0 and a_1 are empirical coefficients. The basis of this scheme is the observation that plots of area and

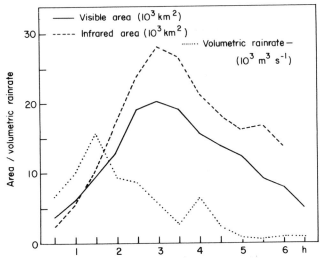

FIG. 5.1   The evolution of a typical cumulonimbus cloud ensemble. Cloud area in visible and infrared wavelength images, and volumetric rain rate (measured by radar) are plotted against time. From Stout *et al.*, 1979.

volumetric rainrate for particular clouds show similar shapes, but cloud area lags behind rainfall (Fig. 5.1). Implied here is the condition that there exists a threshold which defines an area closely related to production of rain.

During GATE the SMS-1 satellite provided visible and infrared wavelength images at an interval of 30 min or less. Ground measurements of rainfall were made by four calibrated 5 cm radars. Visible and infrared thresholds were selected to maximize the probability that a cloud would contain precipitation, and minimize the probability of precipitation in clouds below the threshold. In sequences of satellite pictures matched to simultaneous radar images the areas and volumetric rainrate of individual cumulonimbus clouds were measured throughout the cloud lifetimes. The coefficients a_0 and a_1 were calculated from the combined measurements by least squares regression of satellite cloud area on radar volumetric rainrate.

Area thresholds are 200 W m^{-2} (corresponding to a cloud albedo of 0·45, with overhead sun) and 245 K for visible and infrared data, respectively. Coefficients are

$$a_0 = 5\cdot2 \times 10^{-7}/5\cdot4 \times 10^{-7} \,[\text{m s}^{-1}] \text{ and}$$
$$a_1 = 2\cdot6 \times 10^{-3}/2\cdot8 \times 10^{-3} \,[\text{m}]$$

for visible (left of solidus) and infrared (right of solidus). In both visible and infrared, the area change term (treated as an absolute value) on the average was one-half the area term. The standard error of estimate was 60% of the volumetric rainrate for visible, 75% for infrared, "thus for one cloud for

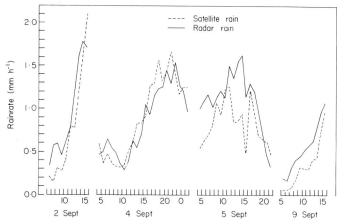

FIG. 5.2   Hourly satellite estimates of rain compared with radar estimates of rain. The area is a circle $1·31 \times 10^5$ km^2 in area. Hours in GMT are plotted on the abscissa. From Stout *et al.*, 1979.

one-half hour the expected error . . . is 60–75% of the rain estimate" (Stout *et al.*, 1979).

To make a rain estimate one needs a sequence of geostationary satellite pictures covering the area and period of interest. A cumulonimbus is identified. Its area and location are measured in the first picture, and in each picture in the sequence until the cloud disappears. This is repeated for all cumulonimbus clouds in the first picture. The analyst then looks for new cumulonimbi in the second and succeeding pictures, until the sequence is exhausted. Volumetric rainrate is calculated step by step for each cloud. It may be summed to get rain amount for any interval, or it may be mapped to get the distribution of rainfall.

This scheme was tested against calibrated radars for accuracy in estimating rain amount and distributions. The test covered 4 days (however, one-sixth of the test data overlapped measurements which determined the coefficients). Satellite and radar rainfall was estimated first at 1 h intervals for a circular area of radius 204 km (Fig. 5.2). On three occasions satellite rainfall significantly underestimated radar rainfall. Otherwise satellite and radar rainfall agreed well, through large changes in average rainfall rate. The correlation of satellite and radar rainfalls was 0·84 and the standard error of the hourly satellite estimate was 0·25 mm. Overall bias was insignificant.

Satellite and radar rainfalls were compared also through maps covering 24 h (Fig. 5.3). On both days the resolution of satellite and radar rain estimates was comparable and the magnitudes and orientation of the main centres were similar. The main discrepancy was a northward shift of the pattern of satellite compared to radar rainfall.

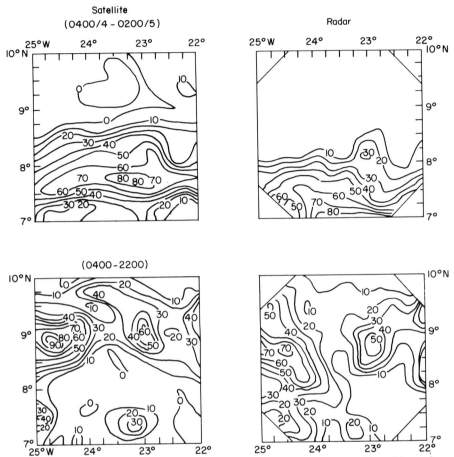

FIG. 5.3   Daily rainrates (mm day^{-1}) by satellite (Stout *et al.*, left) and by radar (courtesy of M. D. Hudlow; right). Radar estimates cover one calendar day. 4 September 1974 (top); 5 September 1974 (bottom). From Stout *et al.*, 1979.

The uses of the Stout *et al.* scheme have been largely confined to GATE. One hour rainfall was determined for 5 days over a synoptic-size area of the eastern Atlantic Ocean. Results are presented as tables and maps in Stout *et al.* (1978).

Wylie (1979) presents evidence of a need to adjust the coefficients if this scheme is used to estimate rainfall in higher latitudes. There is no provision accounting for rainfall from clouds which are not cumulonimbus, and each cumulonimbus cloud of a given size and growth rate produces an identical rainfall. One might wish for more flexibility in allocating rainfall, to reduce spatial bias of the kind shown in Fig. 5.3. The need to follow the evolution of individual cumulonimbus clouds and cloud ensembles makes the scheme demanding and tedious where convection is extensive and variable, unless spatial resolution can be sacrificed in the interest of simplicity.

Perhaps the outstanding advantage of this scheme is its conceptual simplicity: two pairs of straightforward measurements—cumulonimbus cloud area and location—are all that are needed to provide an estimate of rainfall. Stout *et al.* (1979) note that could be used for real-time estimations of flood rains.

II. Griffith and Woodley

The scheme of Griffith and Woodley (described in Griffith *et al.*, 1976, 1978; Woodley *et al.*, 1980) was intended to provide estimates of convective rainfall beyond the range of calibrated radars, over Florida and offshore waters as well as over the tropical Atlantic for GATE. Griffith and Woodley maintain that in the tropics precipitation occurs with bright (cold) clouds, and its intensity in cumulonimbus clouds is a function of stage of development. Cumulonimbus cloud area is shown to be coupled to echo area through a relation which changes as the cloud evolves. Echo area, in turn, is linearly related to volumetric rainfall rate. Thus from a sequence of measurements of cloud area, volumetric rainfall rate can be determined. Functionally, the relations are of the form $A_e = f[A_c(t)]$ and $R_v = RA_e$, therefore,

$$R_v = R \times f[A_c, A_m] \qquad (5.2)$$

Here A_e is echo area (at a threshold level of ~1 mm h^{-1}), A_m is maximum cloud area and R is a variable coefficient connecting echo area to volumetric rainrate. The value of R depends on whether echo area is increasing or decreasing. f is a higher order function, which is usually expressed in graphs or tables. Estimates may be made with either visible or infrared imagery. Functionally, this technique is like that of Stout *et al.* except in its involvement of echo area and in its implicit expression of area change (see Fig. 5.4).

The unknowns R and f in Eq. 5.2 were determined from measurements on Florida clouds viewed by a raingauge-calibrated 10 cm radar and by one of two geosynchronous satellites, SMS-1 and the third Applications Technology Satellite, ATS-3. Cloud thresholds are 253 K (infrared) and about 135 W m^{-2} (visible). R is $15\cdot6 \times 10^3$ m^{3}km^{-2}h^{-1} for growing echoes, $7\cdot92 \times 10^3$ m^{3}km^{-2}h^{-1} for shrinking echoes. Cloud-echo area relations are shown in Fig. 5.4.

The procedure followed to estimate rainfall is like Stout *et al.* up to the point of calculating rainfall from a series of cloud area measurements. Here for each cloud Griffith and Woodley divide cloud area by the maximum area for that cloud. They enter the cloud area–echo area graph to get echo areas, then use the appropriate echo area–volumetric rainrate relation to calculate

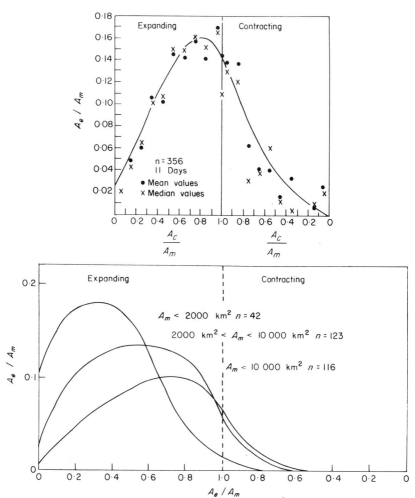

FIG. 5.4   Cloud and echo area relationships for Florida. Both cloud and echo area (A_c and A_e) are normalized to the relative maximum cloud area A_m. Data are averaged over intervals of 0·10 (A_c/A_m). Visible satellite data (top); infrared data (bottom). Infrared data have been stratified by maximum cloud area A_m. From Griffith et al., 1978.

R_v. R_v may then be summed over some interval and area to get total volumetric rainfall, or it may be mapped to get the distribution of rainfall.

The Griffith and Woodley scheme is unique in having been widely tested in various convective situations and different regions. The tests include thunderstorms in Florida, Montana, and Venezuela, hurricanes in the western Atlantic, and the thunderstorms and cloud clusters of GATE from West Africa across the Atlantic Ocean to South America. In Florida over an area of 10^4 km^2 the correlation of ATS visible image rainfall with gauge-calibrated radar rainfall was 0·80 for 53 cases of hourly rainfall, and 0·93 for 8 day-cases of accumulated 6–9 h rainfall. The correlations dropped somewhat for a larger area of 10^5 km^2: with 37 cases of 1 h rainfall the correlation was 0·68. Infrared imagery from SMS-1 gave comparable results: 49 estimates of hourly satellite rainfall for the larger area were correlated as 0·65 with radar measurements of rainfall; however, infrared data showed a tendency to overestimate rainfall. A nadir in tests of this scheme was reached in Venezuela. There, with poor quality satellite imagery and an inadequate network of raingauges, the correlation of satellite with gauge measurements was only 0·46. The scheme was modified to account for the presence of very bright or very cold interior clouds for testing on hurricanes (Griffith *et al.*, 1978; see Section II, Chapter 12). Satellite and gauge measurements of daily rainfall for three hurricanes (seven cases) correlated as 0·84, with satellite estimates tending to be high. This scheme has been used to estimate rainfall in flash flood storms also (Griffith *et al.*, 1978; see Section II, Chapter 12).

The outstanding application of the Griffith-Woodley method is estimation of rainfall for the GARP Atlantic Tropical Experiment. For this application, which is described in Woodley *et al.* (1980), several modifications were made, and the method was fully programmed to work automatically on a large digital computer (Fig. 5.5). In the GATE version the area conversion coefficient R assumes one of nine possible values, depending on the size of the ratio of (inferred) echo area to maximum echo area and its trend. In addition, the volumetric rainfall rate is adjusted upward by an amount proportional to those fractions of the cloud which are colder than two secondary temperature thresholds (233 and 213 K). Finally, rainfall is allocated to fixed grid boxes each 1/3° on a side. The allocation of rainfall from a cloud to a box depends on how much of the box is covered by cloud above each of the three threshold temperatures. This allocation was facilitated by a preprocessing step which converted the original infrared data to average temperature with a resolution of 1/3° in space but preserved for each grid box information on fractional coverage of cloud above each threshold. Rain is spread over the whole cloud (as defined by the 253 K threshold), but the colder boxes get a larger share.

One hourly digital infrared data from SMS-1 were processed on a large computer. Rainfall was estimated in 6 h intervals for the 85 day period from 27 June through 20 September 1974. Maps and statistics cover day, phase (21 days), and Experiment (85 days) periods, over an area extending from West Africa to South America. A few highlights of this effort appear in Section III, Chapter 10. Here we are concerned with the quality of the product.

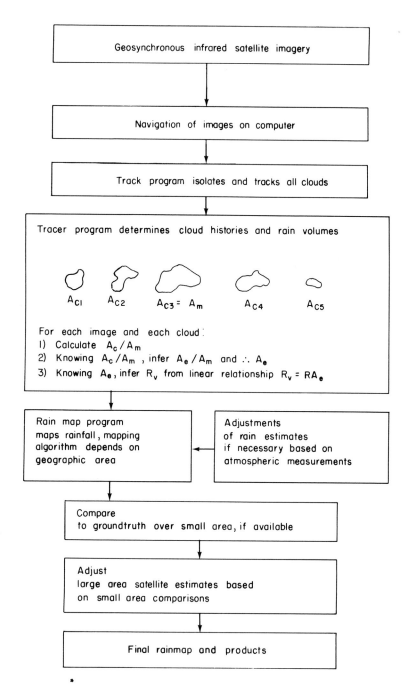

FIG. 5.5a Flow chart for the Griffith-Woodley technique in the computer auto-mated version used for GATE and after GATE. Figure courtesy of C. G. Griffith.

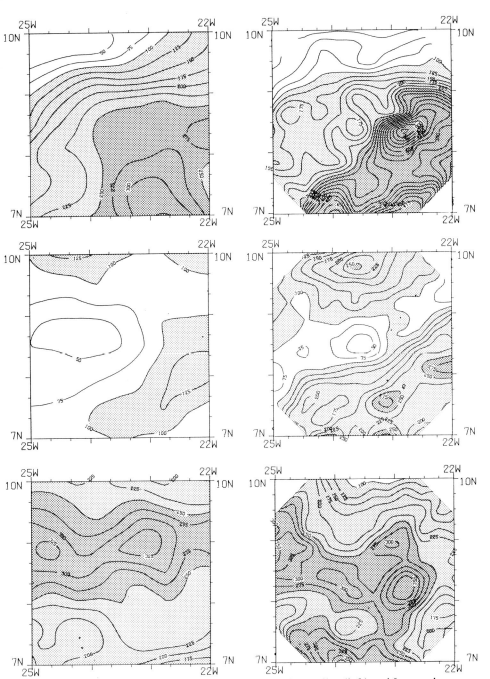

FIG. 5.5b   Phase total rainfall (mm) by Griffith and Woodley (left) and from radar (right). Phase 1 (top); Phase 2 (center); Phase 3 (bottom). From Woodley *et al.*, 1980.

This may be judged from comparisons of the Woodley/Griffith rainfall estimates with radar and gauge measurements of rainfall. Comparisons with radar were made for a 3° box centered at 08°30'N, 23°30'W. Looking at total satellite and radar rainfall for 21 day periods (Fig. 5.5), we see much less relief in the satellite map. Nevertheless, there is good agreement in the main features. Overall average rain volumes were within a few percent, except for the intermediate period, when satellite rainfall was 40% less than radar rainfall (Table 5.1). The loss of detail and damping of peaks and

TABLE 5.1   Radar rainfall and Woodley/Griffith satellite rainfall for a 3° Box[a]

	Rainfall ($m^3 \times 10^8$)		
	radar	satellite	Radar/satellite
27 June–16 July (Phase 1)	218·6	200·0	1·09
28 July–15 Aug. (Phase 2)	155·2	89·8	1·73
30 Aug.–19 Sept. (Phase 3)	228·9	221·1	1·04

[a]after Woodley et al. (1980).

valleys is due to coarser satellite resolution and smearing in the allocation of rain. The Phase 2 discrepancy is thought to be due to misallocation of rain and to the presence of many relatively warm rain-clouds (Augustine et al., 1981). It does not appear to extend to larger scales.

Comparisons also were made of satellite rainfall with gauge measured rainfall over West Africa (Fig. 5.6). Satellite rainfall over the three phases of GATE (63 days) was slightly higher than gauge rainfall equatorward of 11°N. However, in the drier precipitation regime closer to the Sahara, differences were larger, with satellite rainfall estimates as much as three times average gauge measurements. To cope with such environments Griffith et al. (1981) recently added a 1-D cloud model factor to their technique (see p. 98).

The main weaknesses of this life-history scheme compared to others are its formulation in graphical rather than analytic terms, its use of echo area as an intermediate step, and its complexity—three relationships govern the conversion of (infrared) cloud area to echo area, nine the conversion of echo area to rain rate, and there are two secondary cloud area thresholds. The gains these give in accuracy and flexibility tend to be offset by losses in simplicity and efficiency. The outstanding feature of this technique is the great effort invested in development. Automation of estimation procedures

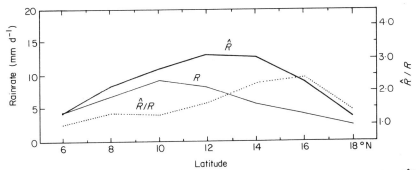

FIG 5.6   Rainrates over West Africa by the technique of Griffith and Woodley ($\hat{R}$) and as measured by gauges (R). $\hat{R}$ is the ratio of satellite to gauge rainrates. Rainrates are averaged over 2° strips of latitude. The area covered is the triangle with vertices at 5°N, 5°W; 20°N, 5°W; 20°N, 15°W. The period covered is the three phases of GATE (63 days). After Woodley *et al.,* 1980.

has enabled Griffith and Woodley to measure daily convective rainfall over the span of an ocean and the length of a season.

III. Scofield and Oliver

Rapid City, Big Thompson Canyon, and Johnstown are infamous in recent meteorological history. Together they represent 451 lives lost, some $335 000 000 damage to property, and misery beyond measure. But it was not the notorious demons of nature which brought disaster—not hurricanes, or tornadoes. Each of these places was the site of a *flash flood*. In each instance, despite radar, raingauges, and river gauges, forecasters were hand-icapped by lack of information on the intensity and magnitude of the storms producing the floods. It is this kind of information which the method of Scofield and Oliver (1977) attempts to provide.

Scofield and Oliver offer forecasters a way to measure convective rainfall in the enhanced infrared SMS/GOES satellite pictures which are now trans-mitted routinely to forecast offices in the United States. As originally configured the method produces estimates of convective rainfall rate at particular points or stations. Its premises are the following:

(a) Precipitation is favoured by high cloud brightness (low cloud top temperature).
(b) Heavy rainfall is favoured by cold cloud top temperatures, growth and merging; light rainfall is favoured by shrinking clouds and warming clouds.
(c) Rainfall is concentrated on the upshear side of an anvil.

A rainfall estimate is made by answering a series of questions—organized as a decision tree—for some particular location in a satellite picture sequence:

(1) Is the cloud convective?
(2) Is the cloud cold?
(3) Is the location (station) under the active part of the cloud, that is, that part of the cloud which is coldest, has the tightest gradient, contains towers, or shows texture?

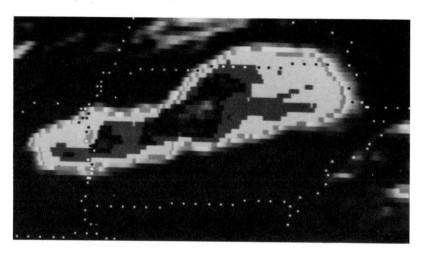

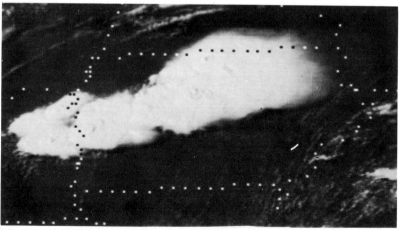

FIG. 5.7   GOES images of thunderstorms over Iowa, 2100 GMT 13 June 1976. Infrared image (top) has been (digitally) enhanced by repeating the grey scale. Visible image, bottom. From Scofield and Oliver, 1977.

(4) Is the cloud growing?

(5) Are there overshooting tops, merging cumulonimbi, or merging convective cloud lines?

Significant rainfall is predicted only if (1), (2) and (3) are answered affirmatively. Questions (4) and (5) then estimate this rainfall according to four additive terms: one involving both minimum cloud top temperature and growth of the coldest contour, and one term each for overshooting tops, merging thunderstorms, and merging convective cloud lines.

The assignment of particular rainfall rates to categories of cloud top temperature, growth, position, texture and relationship with other convective systems is based on standard gauge measurements of convective rainfall over the central United States for one summer season. These were adjusted by physical reasoning, and later, experience. The scheme is appropriate to summer conditions, with the tropopause between -60 and $-66°C$.

The 48 possible rain rates range from Trace to $17·8$ cm h^{-1} (7 in h^{-1}). The distribution is variable. Precision is highest just above 0, $2·54$, $5·08$ and $7·62$ cm h^{-1} (0, 1, 2, and 3 in h^{-1}). Half of the 48 possible rates fall in four intervals of $0·25$ cm h^{-1} ($0·1$ in h^{-1}) width—these begin at rates of Trace, $2·54$, $5·08$, and $7·62$ cm h^{-1}. Only three rates are greater than $10·16$ cm h^{-1} (4 in h^{-1}).

One of the first tests of this scheme involved a case of heavy thunderstorm rain over Iowa (Fig. 5.7). The gross features of the 6 h integrated rainfall pattern were captured very well (Fig. 5.8). Details, especially peaks, tended to be smoothed. A second test (Scofield and Oliver, 1977) involved rainfall over South and North Carolina as a tropical storm skirted the coast

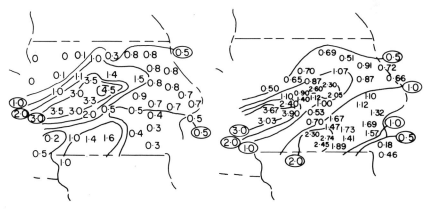

FIG. 5.8 Twenty-four hour rainfall over Iowa ending at 1200 GMT 14 June 1977 (inches) by the technique of Scofield and Oliver (left), gauge observations (right). From Scofield and Oliver, 1977.

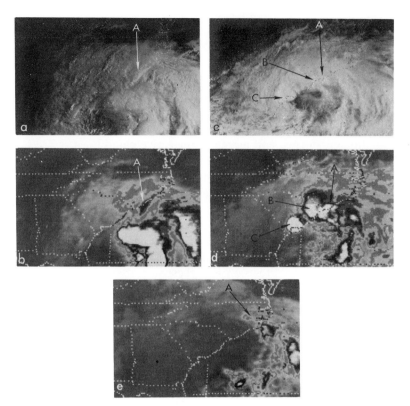

FIG. 5.9   GOES image sequence of tropical storm Dottie, over the Carolina coast. Lettered arrows point to major rain cells. From Scofield and Oliver, 1977. (a) Visible image, 1200 GMT 20 August 1976. (b) Enhanced infrared image, 1200 GMT 20 August. (c) Visible image, 2100 GMT 20 August. (d) Enhanced infrared image, 2100 GMT 20 August. (e) Enhanced infrared image, 1130 GMT 21 August.

(Fig. 5.9). The location, movement and strength of the main rain centre were very well reflected in 6 h totals of satellite rainfall (Fig. 5.10). Six hour average rainfall for an area $1 \cdot 5 \times 10^5$ km² agreed to within $0 \cdot 38$ cm ($0 \cdot 15$ in or 50%) of gauge measurements, and 24 h rainfall was just $0 \cdot 13$ cm ($0 \cdot 05$ in or 6%) away from gauge rainfall. Applications of this technique will be found in Chapter 11.

Others have applied the Scofield–Oliver scheme in different situations. Ingraham *et al.* (1977) describe its use in inferring the distribution of convective rainfall in Venezuela and Columbia. For this application the operation was speeded through use of a chart position digitizer to register grid points and a desktop computer to calculate and print grid point rain-rates. One of the more than 150 convective storms for which rain was

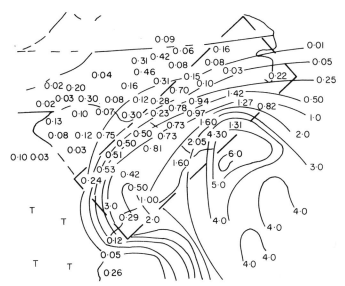

FIG. 5.10   Satellite rainfall contoured over gauge observations, for the 24 h ending at 1200 GMT 21 August 1976, in inches. The box marks the area for which rain was averaged. From Scofield and Oliver, 1977.

estimated is shown in Fig. 5.11. The distribution and magnitude of the satellite rainfall agrees very well with gauge rainfall. Rates, though not always in close agreement, indicate considerable skill.

To improve performance in estimating convective rainfall over a relatively dry region, the central High Plains of North America, Reynolds and Smith (1979) suggested a set of modified parameters for the Scofield–Oliver technique. Their parameters give rates as small as one-sixth of those from Scofield and Oliver. In addition, Reynolds and Smith reconfigured the Scofield–Oliver algorithm to produce rain volume using an electronic display and digital measurement system.

Most recently Moses (1980) has fitted empirical functions to the discrete values of cloud growth versus rainrate which had been produced by Scofield and Oliver, and has shown how a computer might be used to speed the process of making a rain estimate in flood forecast situations. In this interactive scheme the computer locates cold clouds in a sector of an infrared image, measures cloud area at a series of temperatures, assigns rainrates on a pixel by pixel basis and sums rainrates for each pixel over a series of infrared images. The meteorologist picks the sector, decides which temperature contours best represent both cloud growth and the pattern of rain and assigns such adjustment factors as might be needed. Two tests suggest that resulting estimates may be as good as those of the original method.

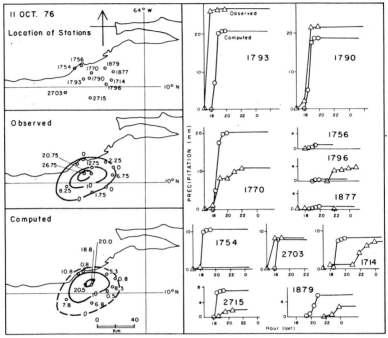

FIG. 5.11   Rainfall over coastal Venezuela on 11 October 1976, as measured by gauges and as estimated from infrared and visible imagery of the Geostationary Operational Environmental Satellite (GOES). Maps (left) show gauge locations, gauge rainfall, and satellite rainfall. Line plots (right) show cumulative gauge and satellite rainfall for the stations marked in the uppermost map. From Ingraham et al., 1977.

It may be expected that unevenness in the distribution of possible rainfall rates would lead to biases in estimates. Also, the configuration and history of this scheme invite additional factors to improve estimates or to extend the scheme to new situations. These include shifting rainfall to account for perspective and concentrating isohyets of heavy rainfall to 18·5 km or less (Scofield and Oliver, 1977). Areal estimates are provided by first locating minimum (0·1 cm/30 min) and maximum isohyets and then placing intermediate isohyets by interpolation (Scofield, 1978a). Higher rates are suggested for large, slow moving tropical storms. Each factor adds significantly to the complexity of the scheme, and moves it further from the realm of the field forecaster. (At present within NOAA the main responsibility for providing operational satellite estimates of potential flood rainfall lies with NESS, in Washington.) Most important, evaluation of this scheme has been difficult because in papers published so far little information is provided on how rainfall rates were selected, tests have been limited to events of exceptional rainfall, and there is no systematic summary of performance. The strength of this scheme is in its practicality: pared down, the scheme of Scofield and Oliver offers the forecaster a simple tool for monitoring storm rainfall with resources at hand.

IV. Summary

There is a core of empiricism in the life-history schemes, and in their exposition too little insight as to why they work as well as they do. Although each method applies to convective clouds, none attempts any explicit modelling of convective processes. Each method considers visible as well as infrared data, but none makes use of the two together. The strength of these schemes lies in their ability to provide quick, fairly accurate estimates of convective rainfall, on the scale of a thunderstorm, from ordinary visible and infrared geostationary satellite images. Methods which do combine visible and infrared imagery and/or model convective processes are described in the next chapter.

CHAPTER 6

Bi-spectral and Cloud Model Methods

One diverse group of satellite rainfall monitoring methods combines elements of simpler schemes, or incorporates models of convection. The first category is comprised of bi-spectral schemes, in which infrared and visible imagery are used together to map the extent and distribution of precipitation. Convective parameterization and cloud model techniques comprise the second category. In these, cloud- or cloud ensemble-scale processes of precipitation subject to observation by satellite are used to modulate estimates of rainfall. Cloud model methods tend to be advanced and specific, and are usually more complicated than the schemes already discussed.

I. Bi-spectral Schemes

All recent operational meteorological satellites have provided image data in both visible and window infrared wavelengths. As signals from separate sensors, these data are independent. In Chapter 7 it will be seen that physically, too, there is a measure of independence. Infrared sensors provide information on temperature and thus (albeit indirectly) on heights of the tops of clouds. Visible sensors provide information on the thicknesses of clouds, their geometry and composition. By themselves these data sometimes fail to provide accurate information on where rain is falling, and where not. Together, perhaps they may succeed.

1. An example of bi-spectral thresholding

The problem is illustrated by a set of pictures from the GARP Atlantic Tropical Experiment. At the top of Fig. 6.1 are echoes measured by the 5 cm radar of the Canadian ship *Quadra,* over a circle of radius 210 km centered on the ship. The resolution of the image is 4 km, and it has been (digitally) remapped to the scale and projection of the satellite images. Next is the view provided by infrared sensors on the SMS-1 satellite, then the same view in visible. Resolution in infrared is about 8 km (though granularity in the infrared image is 4 km, owing to oversampling along lines and repetition of each line), in visible about 1 km.

Many scales and stages of convection are present in these pictures. Since a detailed description is given in Mower *et al.* (1979), here we need only note that there is a pair of lines extending west and south-west from the circular feature along the 21st meridian, and cloud coverage (over the radar area) is much greater than precipitation coverage.

The effects of threshold enhancement of the infrared image to simulate a radar echo display are apparent in Fig. 6.2, where the radar image has been superimposed on the infrared image. Agreement is best along the south-west line, where convection is fresh and vigorous, and not strongly sheared. The western line in the infrared picture is displaced northward, owing to a marked northward tilt of convection (Mower *et al.*), and the mature eastern circular mass is much larger.

Small echoes are not at all reflected in the infrared. Clearly no thresholding of the infrared image will produce more than a general correspondence between a satellite precipitation map and the radar precipitation map.

The match of visible and radar images is better for each of the main features. However, discrepancies appear where low and middle clouds are relatively bright (for example, toward the top centre and south-east of the south-west line).

At the bottom of Fig. 6.2 infrared and visible images have been thresholded separately, then combined. Precipitation is represented by the intersection (overlap region) of the separate thresholds. The improvement over visible data alone is modest. Large differences in areas of thick cirrus remain, but the remedy, raising enhancement thresholds, compromises agreement for the more shallow convective clouds. Thus two or more threshold pairs might be needed, each appropriate to a different stage of convective development. Additional examples (in colour) of image compositing may be seen in the 1979 article of Reynolds and Smith.

2. Early work

Systematic studies of this question go back to Lethbridge (1967), as

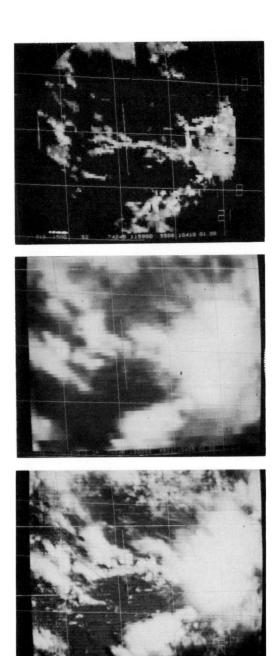

FIG. 6.1 Coextensive *Quadra* radar ppi image (top) and SMS-1 satellite images. One degree grid of latitude and longitude. Crosshair in each image is at 09°00′N, 22°40′W. Radar scan time, 1159 GMT; satellite scan start time, 1200 GMT; 2 September 1974. The radar beam was elevated 0·3° above the horizon. Note that the circular p.p.i. does not extend to the corners of the satellite image.

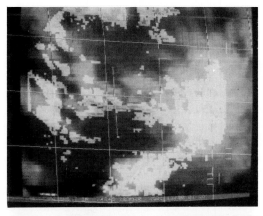

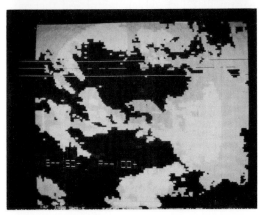

FIG. 6.2 Superimposed satellite and radar images from Fig. 6.1. Radar over
infrared (top); radar over visible (centre); radar over enhanced bispectral satellite
image (bottom). The bispectral satellite image was made by applying binary
enhancements to the visible and infrared images (at thresholds of 150 and 100 digital
counts, respectively), then adding the enhanced images. Black in the satellite–radar
composite image (bottom) means no rain satellite, no rain radar; white means rain
satellite, rain radar; dark grey means no rain satellite, rain radar; light grey means
rain satellite, no rain radar.

D

reported in Chapter 3. A precipitation index based on infrared radiance and albedo, and owing much to Lethbridge, was developed by Dittberner and Vonder Haar (1973) for studies of monsoon rainfall, but its usefulness was severely compromised by coarseness in available satellite and raingauge data. Differencing coextensive visible and infrared image data is a variant of cospectral analysis, which has been used by Reynolds *et al.* (1978) to distinguish various clouds and background surfaces, including thick precipitating clouds. In comparisons of coextensive visible and infrared data for Pacific cloud clusters, Gruber (1973a) found that a few cold centres were associated with relatively *dark* rather than bright cloud. (We shall come back to this work in Chapter 7.) Blackmer (1975) added radar in comparisons of satellite visible and infrared radiances for several thunderstorm outbreaks over the central and southern United States. He found a fairly consistent relationship between visible and infrared radiance (especially when convection was intense and growing) and, generally, an increase in average radiance with increasing rainfall rate (Table 6.1). However, levels of average brightness varied from day to day. Blackmer concluded that there was information in visible and infrared data together, but because of factors like stage of development there were no universal thresholds of visible and infrared radiance for isolation of precipitation.

Blackmer was handicapped by problems of data location, time differences, instrument calibration, and variable visible viewing and illumination geometry. These constraints were largely overcome in the recent study of Lovejoy and Austin (1979a), which addressed the particular question, "How well can rain areas be delineated in coextensive visible and infrared satellite images?"

TABLE 6.1   Values of brightness, radiance, and rainrates for selected small areas of cloud cover on 24 April and 7 May 1973.[a]

	Brightness[b]	Radiance[b]	Rainrate
24 April			
No echo	39·1	52·4	0·0 mm/h
Weak echo	46·9	56·7	0·76
Moderate echo	48·2	60·9	5·84
7 May			
No echo	28·5	27·1	0·0
Weak echo	34·0	29·7	0·76
Moderate echo	43·8	38·6	2·29

[a]From Blackmer (1975).
[b]Digital counts, range of 0 to 63.

3. Statistical population discrimination

According to Lovejoy and Austin's analysis, the problem reduces to finding some optimum boundary between m classes of objects distributed in n-dimensional space. If the objects are distributed statistically, this can be done by minimizing a *loss function* $f(x_1 \ldots x_m)$, where the $x_1 \ldots x_m$ are n-dimensional vectors representing the distribution of objects in the m classes. Specification of the loss function is based on Bayesian decision theory.

In the present case there are two classes (R = rain, N = no rain) in 2-dimensional space (infrared and visible), and four variables:

N_N—no rain assigned, no rain actual,

N_R—no rain assigned, rain actual,

R_N—rain assigned, no rain actual, and

R_R—rain assigned, rain actual.

The general loss function is

$$f = (l_R R_N + l_N N_R)/(N + R),$$

where $N = N_N + N_R$ and $R = R_N + R_R$, and l is a weighting factor, or *penalty* for incorrect assignment. If l_R and l_N are taken as unity,

$$f = (R_N + N_R)/(N + R)$$

is simply the fraction of errors. But the no rain class is typically very much larger than the rain class. l therefore may be weighted by the sizes of the sample s, as $l_R = (N + R)/R$ and $l_N = (N + R)/N$, so that the penalties for each class are inversely proportional to the size of the class. The loss function then becomes

$$f_1 = R_N/R + N_R/N.$$

If distributions of coextensive data are well behaved, for each picture pair there is a function g of visible and infrared radiance which corresponds to the minimum of f (or f_1). This function defines an *optimum boundary* between satellite rain and no rain classes. Once determined, g can be used to create a satellite rainfall map.

Individual maps were scored by means of a *correlation coefficient* and a *confidence limit*. Taking rain as unity, and no rain as zero, Lovejoy and Austin define a correlation coefficient ρ as

$$\rho = (R_R N_N - R_N N_R)/NR.$$

The confidence limit is based on the premise that for some purposes proximity is sufficient. Radar and satellite maps were divided into 40×40 km boxes (100 per map). Error for each box is defined as the difference between

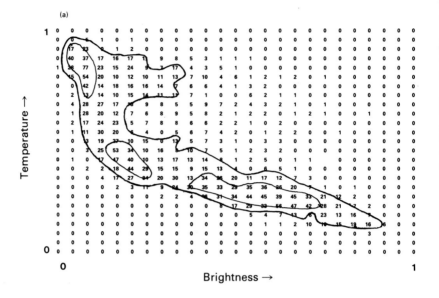

(a)

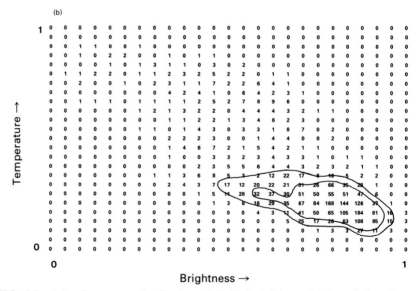

(b)

FIG. 6.3   Joint frequency distribution of SMS-1 visible and infrared data for a 400 × 400 km box centred at 09°00′N, 22°40′W in the eastern tropical Atlantic Ocean, 1300 GMT 5 September 1974. Data have been normalized to a scale 0–1. (a) No rain case (N class). (b) Rain case (R class). After Lovejoy and Austin, 1979a.

percent coverage for radar and satellite. That percentage corresponding to the smallest (in absolute terms) 75% of errors is the 75% confidence limit.

These techniques of population discrimination were applied to SMS/GOES imagery in two locations: the eastern Atlantic Ocean (site of GATE), and Montreal, Quebec and its environs. Satellite visible and infrared data were scaled to 25 levels. Visible data occupied the full scale on each picture, in effect accomplishing a simple normalization for illumination and viewing geometry. For each location, information on the actual distribution of rain was provided by a calibrated 5 or 10 cm radar. Twelve cases were analysed, all but one (from Montreal) predominantly convective. The three Montreal convective cases comprised sequences of four to six pictures.

Convective cases are illustrated by two joint frequency distributions for a day in GATE, one rain and one no rain (Fig. 6.3). The rain distribution is shaped like a comet, with the head cold and bright, and the tail trailing off towards higher temperatures and lower brightnesses. By contrast, the no rain sample is bi- or tri-modal. One centre is bright and cold, but displaced toward higher temperature and lower brightness. Generally similar relationships held for the other convective cases, though the distinctions usually were not as pronounced.

To overcome instability in the loss function f, an additional constraint—that the satellite rain area approximately equal the radar rain area—was added. The boundary function g for the image pair of Fig. 6.3 is shown in Fig. 6.4, and the satellite rainfall map with radar rainfall superimposed is

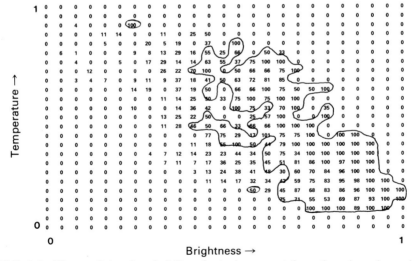

FIG. 6.4   The conditional probability of rain, in percent, from the rain and no rain arrays of Fig. 6.3. For any element this probability is all R as a percentage of all R plus all N. The 50% optimum boundary is sketched. From Lovejoy and Austin, 1979a.

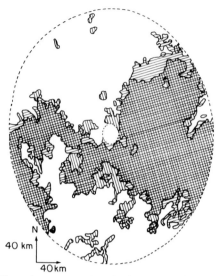

FIG. 6.5   The satellite rain map produced when the optimum boundary of Fig. 6.4 is applied to the same data. Vertical lines are radar rain areas, horizontal lines are satellite rain areas. The radar range is 180 km. Note that the percentage error (in this case 15·2) is the percentage of the striped areas to areas cross-hatched or blank (the latter represent satellite and radar agreement on no-rain). From Lovejoy and Austin, 1979a.

shown in Fig. 6.5. Rain boundaries differ in details, but whether from Montreal or GATE are substantially in agreement. The case illustrated gave superior results by most measures (Table 6.2), and the GATE cases overall tended to give smaller differences with radar. However, even the worst case demonstrates significant skill in delineating instantaneous rainfall patterns from satellite visible and infrared imagery.

TABLE 6.2   Data characteristics and error statistics for 5 September 1974 and ranges for the GATE and Montreal samples.[a]

	GATE Day 248 (5 September 1974)	Range for GATE	Range for Montreal
Cloud coverage (%)	85	11–85	73–95
Rain coverage (%)	45	2–45	8–25
Simple loss function f	0·15	0·02–0·18	0·06–0·20
Area weighted (inverse) loss function f_1	0·3	0·3–0·8	0·3–0·6
75% confidence limit (%)	15	1–15	5–15
Correlation coefficient ρ	0·7	0·2–0·7	0·4–0·7

[a]After Lovejoy and Austin (1979a).

The temporal variability of the optimum boundary function g was assessed by determining g for pooled GATE and Montreal data sets. Differences between radar rainfall maps and satellite rainfall maps made with the pooled g were measured in terms of the area error factor (the average of the larger value in the ratios of satellite and radar rain area) and the area root mean square error. The error factor was 1·3 and 1·4 for Montreal and GATE, respectively; root mean square error was 0·22 and 0·25. The differences probably are not significant; however, the fraction of correct rain (R_R/R) was significantly higher (0·65 vs 0·55) for GATE. This is attributed to the more distinct, predominantly convective, GATE rainfall regime.

In agreement with what was observed at the beginning of this chapter, when rainfall was mapped by a single optimum "spectral" threshold, the fraction of correct GATE rain was higher for visible than for infrared data (0·64 vs 0·53). Indeed, the visible spectral threshold did almost as well as the optimum threshold. This is consistent with the report of Reynolds et al. (1978) that thresholding on visible data alone was effective in defining precipitation in High Plains thunderstorms. On the other hand, for Montreal, with a larger variety of precipitation systems, infrared sometimes was superior to visible.

One stratiform case was treated. Rain and no-rain populations were practically indistinguishable, indicating that within areas of middle and low layer clouds there may be little information on the occurrence of rain in visible brightness and infrared temperature. The rain map for this case is shown in Fig. 6.6.

Lovejoy and Austin divided rain subsets of satellite visible and infrared data by low or high rain intensity. The heavy rain classes tended to be brighter and colder, especially for GATE; but only for the case of 5 September could the null hypothesis of identical populations be rejected. This implies, according to Lovejoy and Austin, that in a single image or pair most of the information is in distribution of rainfall rather than intensity; therefore, if average intensity is known (from climatology) the determination of distribution may be sufficient to provide a complete estimate of rainfall.

This idea was tested in a second paper. Lovejoy and Austin (1979b) measured echo area on each of 328 hourly *Quadra* radar ppi's. Echo area, assumed to be equal to rain area, then was correlated with volumetric rainrate (rain amount) and with echo average rainrate, for hourly ppi's and also for areas accumulated through periods up to 15 h long. The same correlations were calculated for 1233 summertime 1976 and 1977 cappi (constant altitude plan position indicator) scans from the 10 cm McGill radar near Montreal, Quebec. For both samples rain area was highly correlated with rain amount ($\rho \cong 0·9$), but (except in the GATE sample at 2 h and longer) area was weakly correlated with rain intensity ($\rho \cong 0·1$; Table 6.3).

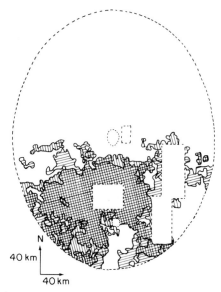

FIG. 6.6   Satellite rain map near Montreal for 13 September 1977, a day of heavy stratus cloud. The optimum boundary (not shown) was quite different from the convective case of Fig. 6.4. The percentage coverage of rain is 23·7%. Areas blocked out contain mountains and other clutter. Radar range is 180 km. From Lovejoy and Austin, 1979a.

TABLE 6.3   Correlations of echo area with rain amount (ρ_{VA}) and with rainrate (ρ_{RA}) for periods of various lengths.[a]

	Number of consecutive hours of data accumulated					
	1	2	4	7	10	15
GATE						
ρ_{VA}	0·91	0·91	0·91	0·91	0·94	0·93
ρ_{RA}	0·15	0·64	0·63	0·72	0·62	0·72
Number of sequences of accumulated data	328	146	59	28	15	8
Montreal						
ρ_{VA}	0·88	0·88	0·88	0·89	0·87	0·82
ρ_{RA}	0·06	0·06	0·05	0·10	0·17	−0·06
Number of sequences of accumulated data	1233	585	266	128	70	27

[a]After Lovejoy and Austin (1979b).

Thus Lovejoy and Austin concluded that determining rain *area* may be sufficient to achieve reasonable accuracy in an estimate of rain amount.

To assess the importance of rate, Lovejoy and Austin computed radar-area rain amount. This was defined as the product of echo area and the sample mean rainrate, averaged separately for GATE and Montreal echo areas. Estimated rain amounts were compared with individually measured amounts through root mean square difference, bias (mean ratio) and error factor statistics. As expected, each statistic declined with increasing period (Table 6.4). Root mean square (rms) difference at 1 h was only 44% for

TABLE 6.4   Root mean square difference (rms), bias and error factor between radar-echo rain amount and (radar) measured rain amount, for periods of various lengths.[a,b]

	Number of consecutive hours of data accumulated					
	1	2	4	7	10	15
GATE						
rms difference	0·41	0·37	0·33	0·33	0·22	0·20
bias	1·53	1·41	1·35	1·34	1·17	1·16
error factor	1·74	1·56	1·49	1·47	1·29	1·29
Montreal						
rms difference	0·44	0·42	0·39	0·34	0·32	0·27
bias	1·18	1·16	1·15	1·11	1·10	1·08
error factor	1·45	1·43	1·40	1·34	1·32	1·29

[a]Number of sequences of accumulated data is given in Table 6.3.
[b]After Lovejoy and Austin (1979b).

Montreal; at 15 h, 27%. GATE rms difference was slightly but consistently smaller. These errors, originating in rate, were found to be of the same magnitude as errors reported for the area brightness (cloud life-history) satellite techniques of Scofield and Oliver, Griffith and Woodley, Stout, Martin and Sikdar (Chapter 5) and Wylie (Chapter 6).

Radar-area rain amounts also were calculated for, and compared with, amounts estimated by the optimum boundary technique from satellite visible-infrared images—the 25 GATE and Montreal convective image pairs described previously. The same optimum boundary was used throughout; however, total satellite rain area was constrained to equal total echo area. Satellite rain area from each image pair was multiplied by the GATE or Montreal rainrate averaged for this limited sample. Root mean square differences between satellite and radar-area amounts were ~20% for

GATE, ~40% for Montreal. These were due solely to errors in the estimation of area by the optimum boundary technique. Thus they imply that errors in rain amount due to area determination were slightly smaller than errors due to use of a single sample-average rainrate.

For Montreal the rms error in determination of rain area by the optimum boundary technique (trained with coincident radar images) was 22%. The error due to use of a single rate averaged over a long period (the 1976 and 1977 sample) was 44%. Lovejoy and Austin estimated the error in radar determination of rain amount to be 5%. Then, if rms errors are normally distributed and independent, the total error of the optimum boundary technique is the root of the sum of their squares, or 49% for one visible-infrared image pair.

The 49% error in rain amount achieved by Lovejoy and Austin for one visible-infrared image pair with radar guiding the satellite determination of area should be compared with the 32% error achieved by Stout *et al.* (1978) for three infrared images in sequence and 15% overlap between satellite predictor data and radar calibration data. The sizes of the areas were the same and differences in location only slightly favor Stout. With no visible images, Stout *et al.* cannot be estimating area as well as Lovejoy and Austin (whose total rms error climbs from 49 to 84% with infrared images alone). Since performance is comparable, Stout *et al.* must be compensating through better estimates of rainrate. This can only come through cloud growth and selection, the factors not present in the scheme of Lovejoy and Austin.

4. Assessment

It is fair to conclude from these studies that, through spectral thresholding, satellite images can provide maps of rainfall which in convective regimes replicate at least mesoscale and synoptic features. Further, except perhaps in middle latitude baroclinic disturbances involving extensive stratiform cloud, visible images are superior to infrared images. The mapping of instantaneous rainfall is usually improved when visible and infrared images are combined in bi-spectral thresholding but the best results are achieved through statistical determination of an optimum boundary. If signal levels are constant, and visible images can be corrected for variable illumination, thresholds for deep convective regimes are fairly stable. However, samples treated so far are relatively small and quite limited in geographical extent. Lovejoy and Austin (1979) find that, by themselves, single pictures or picture pairs provide little information on rain intensity, at least for stratiform clouds and precipitation; however, the experiences of Scofield and Oliver, Griffith and Woodley, and Stout, Martin and Sikdar with deep convective clouds (Chapter 5) support a more sanguine view.

II. Convective Model Methods

The quest for a more elegant formulation of the relationship between rainfall and the characteristics of clouds, and the hope for improvements in accuracy by specification of additional cloud parameters has led some to methods incorporating models of convective processes. Two are discussed here: those of Gruber (1973b) and Wylie (1979). A third, more tentative possibility is outlined at the end of this chapter.

1. Parameterization of deep convection

The method of Gruber (1973) provides area estimates of widespread convective rainfall. It is based on Kuo's 1965 parameterization of convection over a grid box, which was developed as part of a hurricane model. The grid box is large relative to the cumulus clouds within it, and there is a net accession (positive flux) of water vapour due to low level convergence and evaporation. The atmosphere is conditionally unstable. All condensation occurs in deep cumulus clouds, which are undilute and identical. All condensed water is released as precipitation. Since the incoming water vapour must either be stored or condensed and precipitated, the rate of formation of cumulus cloud may, according to Kuo, be expressed as the ratio of inflow I to the sum of storage plus condensation Q, thus

$$c/\Delta t = I/Q.$$

Q is, from another point of view, the amount of moisture required to fill the grid with cumulus clouds. c is cumulus cloud cover, and Δt is a time scale related to the lifetime of a cumulus cloud. Since condensation through the depth of a cumulus cloud is given by

$$Q_1 = \frac{c_p}{gc} \int_{p_b}^{p_t} (T_s - T)\mathrm{d}p$$

the grid average rain rate is

$$R = \frac{1}{\Delta t}\frac{c_p}{g} \int_{p_b}^{p_t} (T_s - T)\mathrm{d}p = cQ_1/\Delta t. \tag{6.1}$$

p_b and p_t are pressures at the base and tops of the cumulus cloud, respectively; T_s is the saturation adiabat through cloud base, and T is the ambient temperature. The other terms have their conventional meaning. Equation 6.1 states that the average convective rainfall rate can be estimated from three parameters.

Gruber evaluated the time parameter Δt with simultaneous measurements of c, Q_1, and R. The observations came from the Florida cloud seeding

TABLE 6.5   Observations of parameters which determine convective lifetime Δt and calculations of Δt.[a]

Case no.	Average rainrate, R (cm 10 min^{-1})	Average cloud fraction, c	Latent heating due to precipitation, Q (g cm^{-2})	Ratio R/c (cm 10 min^{-1})	Calculated Δt (min)
1	0·006	0·013	1·26	0·46	27
2	0·024	0·046	2·00	0·52	38
3	0·007	0·018	1·50	0·39	38
4	0·023	0·055	1·26	0·42	30
5	0·030	0·071	1·50	0·42	36
6	0·014	0·037	1·20	0·38	32
7	0·030	0·042	1·26	0·71	18

[a]After Gruber (1973b).

experiments of June and July 1970. Fractional cloud cover c and rainfall R were measured with a gauge-calibrated 10 cm radar, and cumulus condensation Q_1 was determined from special soundings. Seven cases, each involving 30 or more measurements of cloud cover and rainfall, gave a modest range of Δt of 18 to 38 min (Table 6.5). The average was ~30 min. Predicted ratios of rainfall to cumulus cover agreed well with observed ratios, and the inferred Δt was consistent with direct observations of thunderstorm lifetime. Gruber therefore assumed a constant value of 30 min for Δt.

In the Florida sample there was much greater variability in c than in Q_1. Therefore though Q_1 ordinarily is determined from a proximate sounding, Gruber suggests in the tropics a climatological sounding may be adequate.

Gruber estimates fraction of cumulus from satellite infrared data and a sounding proximate to the estimate area. The level of zero buoyancy, determined from the sounding, determines a threshold infrared cloud temperature. Fraction of cumulus is cloud area colder than the zero buoyancy threshold temperature, divided by the estimate area.

Gruber tested his scheme on a 3 July 1970 squall line over Illinois and Indiana, for which soundings, radar scans, gauge rainfall, and a Nimbus-4 overpass were available. Fractional cumulus cover was 0·13 by satellite observations, in excellent agreement with radar echo coverage of 0·14, despite local discrepancies between cold clouds and echoes. Estimated average rainrate was 3·8 mm h^{-1}. Gauge-measured average rainfall rate from 33 stations was 2·5 mm h^{-1}, remarkably close to the satellite estimate; however, owing to the comparatively large spacing of stations (an average of 40 km) there is much uncertainty in this figure.

It is difficult on the basis of a single test to judge the performance of this technique. Considering formulation, the two assumptions of clouds in a single size and all condensate precipitating may be significant limitations. There also is ambiguity in the definitions of cumulus cloud cover c and cumulus lifetime Δt. Estimates of rainfall may prove to be sensitive to the temperature of zero buoyancy, as this temperature in the test case fell near an inflection in the histogram of satellite infrared temperature. Nevertheless, the test result is promising. Observations from satellites and soundings are combined in a scheme that manages at the same time to incorporate a significant measure of atmospheric physics.

2. A one dimensional cloud model

Much of the effort expended in developing methods for estimating rainfall from satellite data has concerned convection in tropical air masses. One might ask if these schemes may be applied elsewhere: in particular, can schemes like those of Griffith and Woodley (Section II, Chapter 5) and Stout, Martin, and Sikdar (Section I, Chapter 5), developed for use in the marine tropics, be applied in continental middle latitudes? Wylie (1979) has addressed this question through comparisons of cloud areas and rain rates from two distant sites, the 1979 GARP Atlantic Tropical Experiment (GATE) and Montreal and its environs.

Montreal cloud area was measured on GOES-1 infrared images. Rainfall was measured from constant altitude plane position indicator (CAPPI) displays of digital reflectivity from the gauge-calibrated 10 cm radar operated near Montreal by McGill University. CAPPI displays were remapped to the GOES scale and projection. Six cases were treated: three frontal-convective, from June; and three frontal-stratiform, from September; all 1977.

The GATE cloud areas and rainrates were the "calibration" data set of Stout *et al.* There particular clouds were measured: namely, all clouds entirely within the field of view of the radar which were believed on the basis of their satellite characteristics to contain significant rain. Because individual clouds rarely could be followed for long, the Montreal cloud areas and rainrate were measured over the field of view of the radar. The measurements, therefore, are not quite the same; however, the difference diminishes and probably becomes negligible at moderately cold threshold temperatures.

Rainrate averaged over cloud area at various thresholds is shown in Fig. 6.7. Two points stand out: GATE rainrates are higher than Montreal

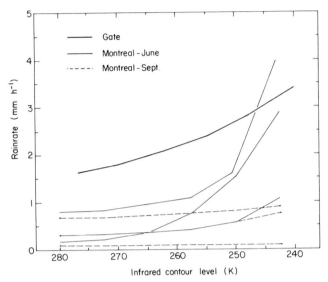

FIG. 6.7   Rainrates averaged over cloud areas measured on infrared satellite images, for various cloud threshold temperatures. Rain was measured by gauge calibrated radars. From Wylie, 1979.

rates, even at relatively cold threshold temperatures, and Montreal rates vary considerably, with convective cases tending to have higher rates. Therefore a scheme appropriate to GATE will overestimate rain at Montreal, and a constant factor of adjustment will leave a large variance.

Examining soundings from the GATE and Montreal areas, Wylie found that differences in average rainrate generally paralleled differences in average precipitable water (Table 6.6). The exceptions were cases of light rainrate, when the lower troposphere was exceptionally stable. To assess the effects of stability on precipitation, the Simpson and Wiggert (1969) one dimensional cloud model was run for GATE and Montreal cases with soundings representative of each case as the initial condition. Model precipitation for a 2 km bubble closely paralleled observed precipitation (Table 6.6). The ratio of model precipitation—for the region in question with a local sounding as initialization, and for the tropical region appropriate to the original scheme—becomes an adjustment factor for local stability. Wylie calculated stability adjustment factors for each of the six Montreal cases, and with the area term of Stout's GATE relation estimated precipitation for the area of the Montreal radar. Except in one case the stability factor substantially improved the satellite rain estimates. Overall bias was nil. Generally the satellite estimate was within a factor of two radar measurement. Wylie concluded that by incorporating a stability factor based on

TABLE 6.6 Estimated and observed precipitation and stability adjustment factors at Montreal.[a]

Case	Observed rainrate[b] (mm h⁻¹)	Precipitable water (mm)	(Montreal/GATE) Precipitable water	(Montreal/GATE) Model precipitation	Precipitation (10⁷ m³) Adjusted satellite	Precipitation (10⁷ m³) Radar	Bias	Error factor
GATE	2·2	54	—	—	—	—	—	—
Montreal								
29 June	1·0	38	0·7	0·5	24·6	26·5	0·9	1·1
2 June	0·9	22	0·4	0·3	8·9	8·8	1·0	1·0
26 Sept.	0·9	27	0·5	0·3	30·8	40·3	0·8	1·3
1 June	0·4	27	0·5	0·3	10·4	7·4	1·4	1·4
20 Sept.	0·4	27	0·5	0·0	3·2	14·6	0·2	4·6
16 Sept.	0·1	22	0·4	0·0	7·0	7·2	1·0	1·0
				Average	14·1	17·5	0·9	1·7

[a]After Wylie (1979).
[b]Average over cloud area defined by $T \leq 257 \cdot 5\ K$.

precipitation calculated from a one-dimensional cloud model, precipitation estimates based on relationships derived for relatively uniform tropical regions may be extended to more variable middle latitude continental rainfall regions.

In the late 1960s it occurred to D. N. Sikdar, looking at movies made from picture sequences of the ATS satellites, that expansion of an anvil atop a deep convective tower must be directly related to the upward transport of mass within the tower, hence also condensation of water vapour and precipitation from the tower. Sikdar developed a simple three-layer cumulonimbus model, with which he and colleagues successfully estimated vertical mass and heat transport in tropical and middle latitude thunderstorms (Sikdar *et al.*, 1970: Sikdar and Suomi, 1971). But estimation of rainfall by the technique remained elusive, partly because of the difficulty of acquiring adequate ground measurements of rainfall.

Sikdar's model contributed to the design of the methods described in Stout *et al.* (1979; see Section I, Chapter 5). Recently it was resurrected and improved by Martin and Sikdar (1979) and Lo *et al.* (1980) for the primary purpose of calculating upward transport of mass in tropical cumulonimbus clouds. A secondary product is rainfall.

Sikdar's model contained inflow, outflow and updraft layers. The innovations added by Lo are a variable depth of outflow and parameterized updraft velocity profile. Lo represents the expanding anvil as a stack of slabs, each of a uniform optical thickness. Starting at the top, upward mass transport is calculated for each slab, assuming conservation of mass and zero transport at the top of the highest slab. Depth of the anvil and convective core area are deduced from brightness gradients across the cloud. These are converted to thickness profiles by means of a relationship, established from plane parallel theory, between brightness and optical thickness. In microphysical terms, it is assumed for this purpose that the cloud consists of a cirrus cap on top of a cumulus stem.

Upward mass transport is largest at the base of the outflow layer, and there is confined to the area of the convective core. This establishes the maximum cloud vertical velocity. The profile of w above the maximum is determined by transport and area for each slab: it diminishes rapidly to zero at the top of the cloud. The profile of w from its maximum to the bottom of the cloud is modeled after Austin and Houze (1973). Condensate is based on the profile of w, which includes entrainment, together with ambient density and water vapour mixing ratio, deduced from soundings, as described by Austin and Houze. Precipitation is equal to the vertical integral of condensate, less evaporation.

Lo applied this model to seven cumulonimbus clouds grouped in the eastern Atlantic Ocean on 18 September 1974, and monitored as part of GATE by satellites, radars, aircraft and ship soundings. If evaporation is assumed to be one-half of condensation (an efficiency factor of 50%), cloud average precipitation rates calculated by the model for the four clouds also observed by radar vary from 0·30 to 2·45 mm h^{-1}, with an average of 1·38 mm h^{-1}. Equivalent rates calculated by the method of Stout *et al.* (1979) varied from 0·95 to 1·92 mm h^{-1}, with average of 1·36 mm h^{-1}. Radar rainfall rates ranged from 0·21 to 1·99 mm h^{-1}; the average was 0·98 mm h^{-1}. The relatively good agreement between anvil expansion and radar rainrates suggests that quite high accuracies are possible in satellite measurements of rain, through careful measurements of cloud area in conjunction with a conceptually simple model expressing essential optical, microphysical, dynamic and thermodynamic properties of the cloud.

With man-interactive data-processing systems coming into widespread use, and calibration of satellite signals becoming the norm rather than the exception, it seems likely that techniques of the kind described in this chapter will receive much attention over the next few years. Their strength is in the rigour by which information on rainfall is extracted from satellite imagery; their weakness, in the complexity of the algorithms needed to do so. Here it is appropriate to proceed to address the question of how rain is related to the visible and infrared characteristics of clouds viewed from space.

CHAPTER 7

Rainfall from Visible and Infrared Images:
A Physical Explanation

In summary form the physical basis of estimating convective rainfall from images in visible and infrared wavelengths may be stated as follows (Woodley and Sancho, 1971; Martin and Suomi, 1972; Scofield and Oliver, 1977a; Griffith *et al.*, 1978; Stout *et al.*, 1979):

(a) High brightness implies large cloud thickness, which implies greater probability of rain; and low temperature implies high cloud tops, which implies large thickness and greater probability of rain. Therefore,

(b) precipitating clouds can be distinguished from all others on the basis of brightness characteristics in visible images or temperature characteristics in infrared images, and the brightness (or temperature) of a precipitating cloud is a measure of precipitation intensity.

In no case do the developers of the methods described previously offer a satisfying explanation of how rainfall is related to the satellite cloud characteristics unique to their method. Yet quite apart from investigations of rainfall, important work has been done in parallel with rainfall studies which illuminates at least the central areas of this question. This chapter directly addresses the question of how rainfall is related to the radiance characteristics of clouds viewed by visible and infrared satellite sensors. It draws together statistical studies of the characteristics of clouds and some of the most relevant and original model studies of radiation transfer in clouds. Infrared and visible images are treated in separate sections, infrared first.

I. Infrared Images

Satellite infrared images are composed of measurements of radiant energy originating in the atmosphere or from land and water surfaces below. The intensity of this energy, integrated over all wavelengths, is, by the Stefan-Boltzmann law, proportional to the fourth power of temperature. "Temperature" refers to the medium in view. If the medium emits spectral radiant energy according to some temperature less than its thermal temperature, a second factor is introduced. This is the emissivity, ε, a measure of emission efficiency. Some surfaces in nature approach blackbody emission ($\varepsilon = 1$), but the rule is emission by "grey bodies", where $0 < \varepsilon < 1$.

At times a medium may transmit part of the radiation incident upon it. This is equivalent to the case of a radiometer simultaneously viewing two or more blackbody surfaces. The intensity it measures is a sum, which is always intermediate to the contributing blackbody intensities.

"Infrared" here refers to wavelengths between 8 and 12·5 μm. In this "atmospheric window" absorption (and re-emission) is strong for clouds and land and water surfaces and slight for the gaseous constituents of the atmosphere. For surfaces which are opaque and do not transmit radiation, measured intensity is closely approximated by the fourth power of temperature. For surfaces which are not opaque—such as some clouds—measured intensity is approximated by an "effective" emissivity times the fourth power of temperature. Effective emissivities—hereafter called cloud emissivities—vary widely. Cloud emissivity cannot be measured from observations at a single wavelength interval. Often it is simply assumed to be unity. Then the temperature calculated from observed intensity of radiation is called the "brightness temperature". Only if $\varepsilon = 1$ does the brightness temperature equal the blackbody temperature. These concepts are discussed further in Chapter 8.

The value of infrared measurements to rainfall estimation lies in the nearly universal condition of lapse of temperature with height through the troposphere. If temperature is known as a function of height, either by actual sounding or from climatology, the height of a cloud may be inferred from satellite infrared observations. Ordinarily, "grey" clouds are of no consequence in this context, because $\varepsilon < 1$ implies clouds which are thin (and therefore without precipitation). The more serious problem is distinguishing between cold clouds which are radiometrically thick but are confined to the upper or middle troposphere, and cold clouds which extend into the lower troposphere. The difference, often, is between cirro- or altostratus and the nimbostratus of a middle-latitude cyclone, or anvil cirrus and its parent thunderstorm.

A benchmark among studies of infrared cloud top temperature and convective precipitation is that of Arkin (1979). Arkin correlated 6 h fractional cloud cover above selected thresholds of temperature with 6 h rainfall. The area examined was a circle of radius 204 km centered at 08°30′N, 23°30′W, in the eastern Atlantic Ocean. This area covered the inner hexagon of the ships of GATE (the Global Atmospheric Research Program Atlantic Tropical Experiment), conducted in the summer of 1974. Cloud cover was based on infrared measurements of SMS-1, and rainfall on measurements of calibrated 5 cm radars. Three 3-week periods were studied, and (except for a single unexplained outlier of cloud cover), each was similar. Correlations increased with increasing threshold temperature, to 0·8 to 0·9 at a threshold temperature of ~235 K, corresponding to a cloud top height of 10 km, and then declined. At lower thresholds lag correlations were largest for rainfall leading cloud area, and remained high for lags to 2 to 3 h. Excluding the outlier, 75% of the variance in rainfall was explained by a linear function of cloud above 10 km.

Further analysis by Richards and Arkin (1979, 1981) has shown that the correlation of rainfall and cloud above 10 km is less sensitive to averaging period (from 1 to 24 h) than it is to averaging area (from 1/4 to 6 1/4 deg²). Thus, "for averaging areas of less than about 5000 km² the details of the convection may significantly reduce bulk correlations between precipitation and fractional cloud coverage based on [infrared] data" (Richards and Arkin, 1979), a conclusion consistent with correlations of Stout *et al.* (1979). Most of the variance in rainfall and cloud area occurred in synoptic periods (4–6 days) (Fig. 7.1). Coherence was largest for longer periods, large areas, and intermediate thresholds. Differences between periods did become significant for small averaging areas. Although Arkin and Richards advise against wholesale extension of their results to other locations, it seems fair to conclude that perhaps two-thirds of the variance in 6-hourly synoptic tropical oceanic rainfall can be explained by one simple measurement. To demonstrate real skill, a rainfall estimation scheme must offer better *accuracy* on these scales of time and area, or better *resolution* with no compromise in accuracy.

This can be achieved through procedures designed to discriminate between thick cirrus, and thick cirrus over deep stratus and cumulus clouds. One approach makes use of patterns in high resolution data: very cold (near tropopause temperature) cores, indicating a domed cloud top, are associated with cumulonimbus updrafts in anvil cirrus (Scofield and Oliver, 1977a; Adler and Fenn, 1979). The second approach makes use of changes appar-

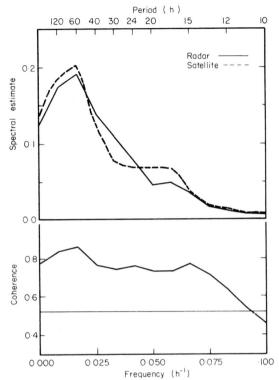

FIG. 7.1   Radar rainrate and satellite infrared cloud area colder than 235 K: spectra (top) and coherence (bottom). The averaging area is a box 2·5° on a side centred at 08°30′N, 23°30′W. Period is Phase III of GATE (30 August–19 September 1974). Data are hourly. Spectra are normalized by the total variance. The 99% level of significance for coherence is 0·53. From Richards and Arkin, 1979.

ent in a sequence of pictures. Rapid expansion of a cold (cirrus) cloud, amounting to at least a doubling of area in an hour, is characteristic of cumulonimbi (Sikdar *et al.*, 1970; Weickman *et al.*, 1977; Griffith *et al.*, 1978; Stout *et al.*, 1979; Adler and Fenn, 1979).

II. Visible Images

We saw a paradox in the observational studies of Chapter 3: despite a general association of precipitation with bright clouds, many bright clouds have no precipitation. This question might have lain dormant but for the emergence in the early 1970s of rainfall estimation schemes which depended on the strength of this association. Instead debate has intensified, and now ranges over theoretical as well as observational ground.

It is convenient to treat this question in two parts:

(a) What is the relationship of precipitation to cloud thickness? and
(b) How is thickness related to brightness?

1. Cloud thickness and precipitation

As Wexler (1954) prepared the speech which brought meteorology and satellites together, two former colleagues from the University of Chicago were flying through trade-wind cumulus clouds over the Caribbean Sea near Puerto Rico. Byers and Hall (1955) measured cloud height (and indirectly,

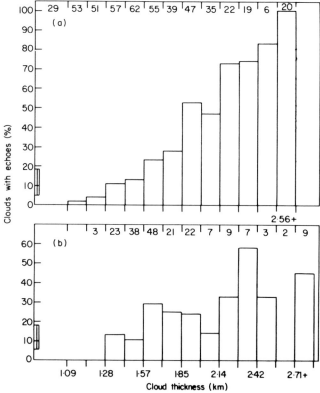

FIG. 7.2   Percentage of trade cumulus clouds with echoes as a function of thickness. Cloud base was assumed to be at 760 m (2500 ft). The range of cloud top temperature in °C is shown on the inside of the ordinate scale. Numbers of clouds in each interval are given across the top. Data were taken on missions flown during winter and spring, 1953–54. After Byers and Hall, 1955. (a) Marine clouds, 12 missions. (b) Island clouds, 3 missions.

thickness) and the presence or absence of echoes. The results for clouds over water (Fig. 7.2) show a steady increase in the probability of precipitation, from zero at a thickness of 1100 m (3500 ft) to 100% at a thickness of 2700 m (9000 ft) and greater. These are consistent with the observations of Squires (1958a), that, near Australia and Hawaii, marine cumulus 1800 m (6000 ft) or more in depth usually precipitate within ½ h.

Probabilities were lower and more variable for cumulus over Puerto Rico (Byers and Hall, 1955; Fig. 7.2b), New Mexico (Braham *et al.*, 1951), and south-east Australia (Squires, 1958a). Nevertheless, Battan (1953) concluded that thickness was more important than temperature in the formation of precipitation echoes in cumulus over Ohio.

With a vertically pointing 1·25 cm radar Plank *et al.* (1955) measured reflectivity in a 1-year sample of clouds near Boston, Massachusetts. Thickness in stratiform clouds was well correlated with occurrence of precipitation. Indeed, precipitation was present in all clouds of thickness greater than 3040 m (10 000 ft) or top temperature between -10 and $-20°C$. Coastal clouds were measured at Adelaide, South Australia, by Spillane and Yamaguchi (1962). In addition to cloud top temperature, the distinction between marine clouds (mainly onshore flow) and continental clouds (mainly offshore flow) was found to be important. Continental clouds produced substantial rain only when top temperatures were colder than $-15°C$ and thicknesses were greater than 4600 m (15 000 ft). Comparable marine clouds produced heavier individual rainfalls. Clouds greater than 4600 m in thickness comprised less than one-third of raining clouds, but produced almost 50% of rainfall (Fig. 7.3). Thickness was especially important for marine clouds warmer than $-10°C$. No cases of rain were reported from clouds less than 600 m (2000 ft) thick.

Therefore, it may be concluded that (a) as cloud thickness increases precipitation is more probable and its intensity is likely to be greater, and (b) the relationship is strongest for clouds warmer than $\sim -15°C$.

Explanations for these associations hinge on droplet growth rates and growth processes. In clouds warmer than $-15°C$, precipitation results from coalescence. Larger cloud thicknesses allow more time for growth before droplets fall out of a cloud (Plank *et al.*, 1955; Spillane and Yamaguchi, 1962). But it must be noted that other factors are involved, among them droplet concentration, evaporation and cloud temperature. Differences in precipitation probability between marine and continental clouds were attributed by Spillane and Yamaguchi to differences in concentrations of condensation nuclei and of droplets (also see Squires, 1958a,b), and to differences in evaporation below cloud base. Very high droplet concentrations (>400 cm^3) inhibit coalescence and formation of rain (Twomey, 1959). Evaporation of raindrops is proportional to saturation vapor deficit and fall

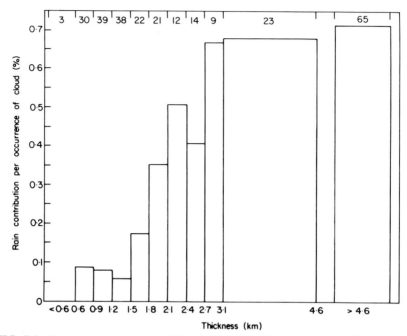

FIG. 7.3   Percentage of total rainfall as a function of cloud thickness. Numbers of rain days are given at the top of each column. From data in Spillane and Yamaguchi, 1962.

period (Kinser and Gunn, 1951; Spillane and Yamaguchi, 1962). Both droplet concentration and subcloud moisture and thickness are more variable over land than over water. Temperature governs process. At temperatures $\leq -10°C$ growth of hydrometeors is dominated by sublimation and collisions involving ice particles; that is, by Wegener-Bergeron rather than coalescence processes. Less important factors influencing precipitation are condensation nuclei, ice nuclei, updraft speed, and electrification (Simpson, 1975). Although important, thickness is only one of several factors determining production of rain in clouds.

2. Brightness and thickness

In the question of how thickness is related to brightness lies much of the controversy surrounding satellite estimates of rainfall. Erickson and Hubert struck the first spark in 1961 when they cautioned,

"Brightness alone cannot always imply cloud thickness. In more than one instance in this study [of cloud forms in TIROS pictures], large thunderstorms having vertical depths of the order of 30 000 feet appeared no brighter, or even less bright, than a rather dense overcast of much smaller vertical extent."

Before long, as we have seen, Conover (1962) countered the pessimistic view of Erickson and Hubert, and by 1970 the association of bright clouds with precipitation was the basis of proposals to estimate rainfall (Woodley and Sancho, 1970).

The brightness of solitary low-tropospheric layer clouds observed by the NOAA-4 satellite was studied by Kaveney et al. (1977). Thickness was inferred from pilot reports. A power curve of exponent 1·3 fit to 28 pairs of points explained 65% of the variance. Somewhat surprisingly, the correlation improved when maritime clouds were discarded: then 90% of the variance was explained and the exponent increased to a value of 1·6. In this study the largest thickness measured was 2500 m (8600 ft) and highest brightness was reached near 2100 m (7000 ft). The largest thicknesses measured by Neiburger (1949) in a study of the properties of Pacific stratus were ~600 m (2000 ft); otherwise the results were similar.

Gruber (1973a) reasoned that if bright clouds were thick, they should also be cold. He tested this hypothesis on Pacific cloud clusters scanned simultaneously by visible and infrared sensors of the polar orbiting NOAA-1 satellite. The relationship between brightness and temperature was weak, and it deteriorated as the threshold of brightness was raised. To explain why, Gruber called attention to observational and theoretical studies of clouds showing albedo approaching a maximum for thickness of little more than one kilometre. Clouds of modest thickness could be as bright as cumulonimbi.

Soon after, Reynolds and Vonder Haar (1973) and Griffith and Woodley (1973) published studies of the relationship of cumulus cloud height with brightness measured by satellite. Reynolds and Vonder Haar (1973) measured 26 Caribbean clouds from one day, and calculated a correlation coefficient of 0·9. Griffith and Woodley (1973) measured precipitating Florida clouds from one summer season. Their sample numbered 215, and the correlation they calculated was 0·7. In both studies heights were measured by radar, cloud base was assumed constant, and a significant linear correlation was claimed between cumulus cloud thickness and satellite cloud brightness.

Clearly there are circumstances—notably stratus clouds of small to modest depth, over water—when brightness is well related to thickness. However, the association weakens as depth increases. In respect to deep convective clouds, the evidence is contradictory. Confusion mounts if studies directly relating brightness to precipitation are considered: Blackmer (1975) found a modest correspondence of brightness with precipitation over the central United States (and much variability from case to case); Cheng and Rodenhuis (1977) reported a poor correlation (~0·3) for a single case near Miami, Florida. To resolve these conflicts, we must address a more fundamental question: what ensemble of factors governs cloud brightness?

III. The Factors Governing Cloud Brightness

Here we have, fundamentally, a problem in optics, which again requires an excursion into the theory of electromagnetic radiation.

We wish to describe the aggregate effects of light scattering by particles of a size comparable to or slightly larger than the wavelength (λ) of incident light. The wavelength interval of incident light is constant. The particles may vary in size distribution, shape, composition, orientation, and in their distribution in space. For water clouds the important variables are size distribution (the number of particles as a function of their size) and cloud geometry (the arrangement of particles in relation to the incident beam of solar radiation). It is convenient to treat this problem in two parts: single scattering and multiple scattering.

In single scattering we are concerned with the interplay of a small volume of cloud droplets with an incident beam of light. The theory used is that of Mie for isotropic, homogeneous spheres (Hansen and Travis, 1974).

Incident radiation is extinguished in passing through the scattering volume, either by absorption and conversion to another kind (or wavelength) of energy, or by redirection (scattering) out of the incident beam. The diminution of the incident beam I along a path ds through the volume is:

$$dI/I = k_{ex}\, ds$$

The volume extinction coefficient, k_{ex}, is composed of two parts: an absorption coefficient k_{ab} and a scattering coefficient k_{sc}. k_{sc} is proportional to the effective scattering cross section of the particles (σ_{sc}) and to their number in the volume (n). (In Mie theory the effective scattering cross section σ_{sc} depends on two parameters: the complex index of refraction n_c and a size parameter, $x = 2\pi r/\lambda$. We are justified in assuming λ is a constant because the range of drop size radius r in natural clouds is much greater than the range of wavelength λ encountered in visible imaging meteorological satellites.) The volume extinction coefficient increases with increasing mean particle size and increasing number density. The probability $\tilde{\omega}$ that scattering will occur (instead of absorption) is equal to the ratio of scattering and extinction coefficients. For water clouds, absorption is small (Van de Hulst, 1957). Many scattering experiments assume $\tilde{\omega} = 1$. Extinction then is given by

$$dI/I = -k_{sc}\, ds$$

The direction of scattering, Ω, is expressed through the scattering phase function $P(\Omega)$, which is normalized to an integral value of one. The scattering phase function is symmetric about the incident axis, but has a very sharp peak in the forward direction. An asymmetry factor, g, expresses the shape of P. g ranges from $+1$ to -1. Positive g means forward scatter dominates back scatter; negative g, the reverse. If $g = 0$, scattering is isotropic. The strength of the forward peak, and the number of secondary peaks at intermediate angles, increases with increasing mean diameter of droplets (for a constant wavelength of incident light) and decreasing variance of droplet diameter. Scattering phase functions for water clouds are given in Hansen (1971), Hansen and Travis (1974), and Deirmendjian (1969). These are based mainly on analytic rather than observed distributions, in part because of the difficulty of measuring drop size distributions in nature.

Single scattering theory describes the fate of incident light in a very small part of a natural cloud. The aggregate effect on the incident beam of all parts of a cloud is treated through the theory of multiple scattering. Here the dimensions of the cloud and its orientation with respect to a given incident beam become important. Extinction coefficient, single scatter albedo, and scattering phase function determine how much energy is subtracted from an incident ray passing through a small depth of cloud, the probability that this energy will be scattered or absorbed, and the probability that if scattered it will be sent in a particular direction. Cloud geometry (dimensions and orientation) determines the probability of additional scattering events. Theory therefore provides a history of scattering and absorption events within the cloud. Where this history intersects the top and sides is found the answer we seek: intensity and direction of light leaving the cloud.

Often the optical dimensions of a cloud are of more interest than its geometrical dimensions. These may be expressed in terms of optical thickness τ, where

$$\tau = \int_s k_{ex} \, ds$$

s is the (geometric) depth of the cloud. τ is a measure of the total fractional amount of light extinguished as a beam passes through s. Thus

$$dI/I = -\sec\theta_0 \, d\tau,$$

where θ_0 is the solar zenith angle. Integrating over s, we have

$$I/I_0 = e^{-\tau \sec\theta_0},$$

a form of Beer's equation. Rearranging terms yields

$$\tau = -\cos\theta_0 \ln I/I_0.$$

Optical thickness for a typical cloud of 1 km depth is of the order of 20, so it is apparent that extinction of light in atmospheric clouds is rapid (see Twomey *et al.*, 1969).

The composition of clouds is often described in terms of liquid water content (M). M is expressed as a mass concentration with units of g m^{-3}.

Whereas τ is proportional to number density and the *square* of droplet radius, M is proportional to number density and the *cube* of droplet radius. If the shape of the droplet size distribution is held fixed (r = constant), optical thickness and liquid water content are linearly related.

Multiple scattering problems may be solved in several ways, depending upon the answers sought and resources at hand. Excellent summaries are given by Irvine and Lenoble (1974) and Hansen and Travis (1974). The present concern is what these solutions have to say about effects of the principle variables: cloud drop size distribution and geometry.

The simplest case is the *plane-parallel cloud*—a horizontally homogenous cloud layer of infinite lateral extent. Radiation (irradiance) incident on the top may be absorbed or may emerge from the bottom or from the top. The ratio of upward emergent to incident irradiance is called *reflectance*. (Directional reflectance, often used in descriptions of finite clouds, implies incident (beam) irradiance of a particular zenith direction. Albedo usually means reflectance averaged over all incident wavelengths.) The ratio of downward emergent irradiance to incident irradiance is called *transmittance*.

The plane parallel (or semi-infinite) cloud has become a reference for judging radiative transfer in more complicated clouds. We illustrate its properties through results of Twomey *et al.* (1967).

Twomey *et al.* calculated reflectance as a function of geometric thickness for a cloud drop distribution of 6 μm average radius and 1 μm standard deviation. Liquid water content was 0·2 and 0·4 g m^{-3}, and two solar zenith angles were represented (Fig. 7.4). Reflectance increased rapidly in hyperbolic fashion for both values of liquid water content, implying little additional increase above a thickness of 0·7 km. The effect of absorption (nonconservative scattering) was to increase the rate of hyperbolic convergence. Changing the solar zenith angle had little effect (though at large zenith angles reflectance would approach unity), and the effects on reflectance of changing drop size distribution radius and variance were small for $\tau > 100$.

Texture of a semi-infinite (plane parallel) cloud layer has the effect of decreasing albedo, according to studies by Weinman and Swarztrauber (1968) and Wendling (1977). The decrease is a consequence of the exponential relationship of intensity to optical thickness: thinning a layer by $\Delta\tau$ affects transmission more than thickening the layer by the same $\Delta\tau$. Asymmetry in the scattering phase function increases the disparity. Wendling's calculations show that for a constant average thickness, albedo differences increase with deepening striations and height of the sun (but they would decrease if average optical thickness were increased).

The effects on *radiance*, the rate of radiant energy flow per unit area per unit solid angle, are similar, except that radiance (brightness, as measured by

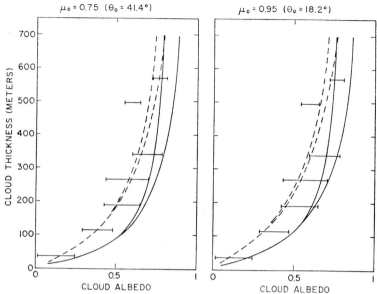

FIG. 7.4   Directional reflectance versus thickness for semi-infinite clouds. Solid curves correspond to a liquid water content of 0 40 g m⁻³, and the dashed curves to 0·20 g m⁻³. Of each pair of curves, the right-hand member represents conservative scattering and left-hand member a fractional absorption per scattering of $1·2 \times 10^3$. Horizontal bars are Neiburger's (1949) experimental results (with standard deviations). Left: solar zenith angle of 41·4°; right: solar zenith angle of 18·2°. From Twomey *et al.*, 1967.

a satellite radiometer) from the striated cloud is larger than plane parallel radiance in the antisolar direction with inclined illumination. Wendling tested two drop size distributions, one peaking at $r \sim 5$ μm, and one peaking at $r \sim 2$ μm, both appropriate to stratus. Differences in albedo and radiance at small zenith angles were comparable to differences resulting from changing the geometry of the cloud top; however, these also decreased with increasing optical thickness. Therefore, at optical thicknesses typical of precipitating clouds, texture and drop size distribution may be of secondary importance in explaining cloud brightness.

To keep the geometry simple, most finite clouds have been modelled as cubes or cuboids. The first accounts of the properties of cubical clouds were from McKee and Cox (1974, 1976). A Monte Carlo model simulated conservative scattering in a cloud of Deirmendjian type C.1 (cumulus) drop size distribution with 0·063 g m⁻³ liquid water content. In their first paper McKee and Cox presented calculations of bi-directional reflectance for three solar zenith angles, and optical thicknesses up to 73·5 ($\Delta X \sim 1·5$ km). The second paper treats radiance averaged over a face of the cube. This

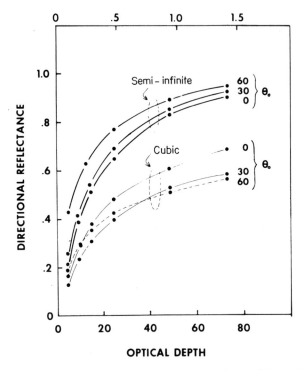

FIG. 7.5   Directional reflectance of semi-infinite clouds and finite cubic clouds for three solar zenith angles. The upper abscissa is the geometrical thickness (km) for a $0 \cdot 2$ g m^{-3} model cloud. From McKee and Cox, 1974.

simulates the intensity a satellite radiometer might sense if its field of view matched the projected area of the cloud face. McKee and Cox found that much energy was lost through the sides of a cubic cloud: as much as half the incident energy for overhead sun even at $\tau = 73$. Thus bi-directional reflectance was smaller for cubic than plane parallel clouds at all incident angles and optical thicknesses (though cubic-bi-directional reflectance approaches plane parallel bi-directional reflectance for overhead illumination as optical thickness becomes large). For overhead sun the difference at $\tau = 50$ was $0 \cdot 22$ (Fig. 7.5), with the cubic cloud reflecting 74% as much light as the semi-infinite cloud. Zenith radiance of the cubic cloud was one-half that of the semi-infinite cloud. McKee and Cox concluded that relationships between brightness and rain might reflect the slower approach of cubic clouds toward a limiting maximum radiance: in clouds like those measured by Griffith and Woodley (1973) brightness may be most sensitive to width.

The effect of varying drop size distribution was tested several years later by McKee and Klehr (1978) with a similar model. Directional reflectance and radiance were compared for Deirmendjian C.1 and C.3 distributions (cumulus and mother-of-pearl, respectively). For clouds of optical thickness 60, geometry was a far more important factor than drop size distribution.

The transition from cubic response to a plane parallel response has been explored by Davies (1978), McKee and Klehr (1978) and Reynolds *et al.* (1978) in a series of experiments on rectangular parallel-piped (cuboid) clouds. Davies extended calculations of reflectance to optical thicknesses of 200 ($\Delta Z \simeq 4$ km, for a typical cumulus drop size distribution with $M \sim 0{\cdot}2$ g m^{-3}). Directional reflectances were calculated with a Monte Carlo mode, and with a new, computationally faster "finite analytical" model based on a (delta-Eddington) approximation to the scattering phase function. Finite analytical model results departed from Monte Carlo results only where the optical path length was short (Fig. 7.6a).

Where cuboidal clouds were taller than they were wide (the case of "pillar" clouds), the plane parallel assumption failed (Fig. 7.6a). Where they were very much wider than tall (stratiform clouds), the plane parallel assumption was tenable. The ratio of width-to-height (X/Z) had to be at least ten before results approached the plane parallel: for thick clouds and inclined illumination, considerably more than ten (Fig. 7.6b). For $X/Z \lesssim 1$ with zenith sun, changes in directional reflectance were achieved only through increases in width. For $X/Z \lesssim 1$ with 60° illumination, increasing thickness *decreased* directional reflectance.

The effect of changing sun angle (Fig. 7.7) was least for small optical thicknesses. As τ increased, side illumination caused increasing departures of finite from plane parallel results. Differences were large at all zenith angles for $\tau = 200$. Absorption ($\tilde{\omega} < 1$) decreased the differences between finite and plane parallel results, however, Davies noted that modest measured cloud albedoes, even in very thick clouds, could be explained by results of the finite model, without invoking absorption.

Reynolds *et al.* (1978) compared Monte Carlo results with geostationary satellite (SMS-2) observations of a field of natural cumulus clouds 1·5 to 2 km thick, and confirmed the importance of width. Monte Carlo calculations of brightness for 2 km thick clouds of variable width generally agreed with satellite brightness plotted as a function of measured cloud width. More field observations are needed to resolve lingering uncertainties in the scattering properties and the heights of the natural clouds.

Still more realistic cloud geometries have been modelled. Busygin, Yevstratov and Feigel'son (1973) calculated the albedoes of cumulus clouds configured as cylinders, paraboloids and truncated spheres (Fig. 7.8). Albedo was largest for the truncated sphere, but remained significantly below plane

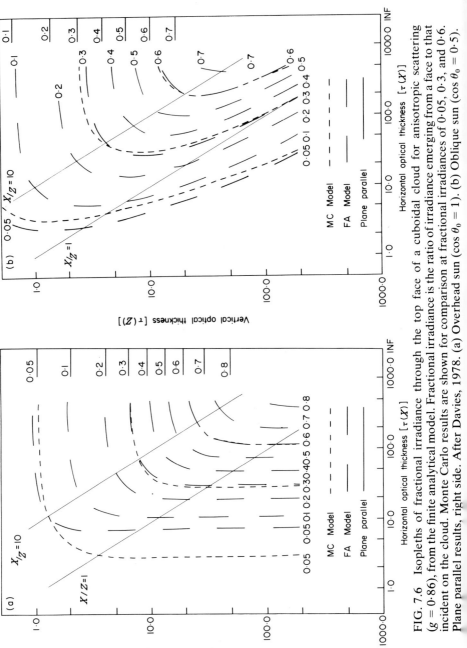

FIG. 7.6 Isopleths of fractional irradiance through the top face of a cuboidal cloud for anisotropic scattering ($g = 0·86$), from the finite analytical model. Fractional irradiance is the ratio of irradiance emerging from a face to that incident on the cloud. Monte Carlo results are shown for comparison at fractional irradiances of 0·05, 0·3, and 0·6. Plane parallel results, right side. After Davies, 1978. (a) Overhead sun ($\cos \theta_0 = 1$). (b) Oblique sun ($\cos \theta_0 = 0·5$).

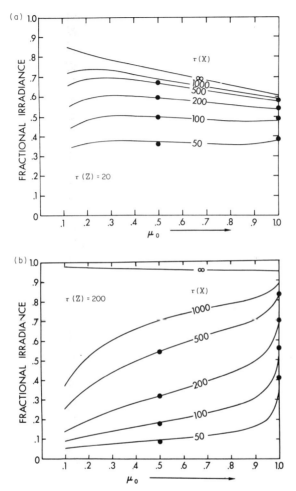

FIG. 7.7   Fractional irradiance through the top face as a function of illumination angle for a range of cloud widths. Scattering is anisotropic ($g = 0\cdot86$) and absorption is zero. Dots show Monte Carlo results. From Davies, 1978. (a) Optical thickness of 20. (b) Optical thickness of 200.

parallel albedo even to $\tau \sim 100$. The upright cylinder was also treated by Barkstrom and Arduini (1977), with diffusion theory. Reflectance of diffuse radiation on the top (equivalent to insolation over a day) increased markedly with increasing radius for radii less than twice cloud thickness. Increasing average drop size radius from 10 to $21\cdot5$ μm, while decreasing number density from 100 to 10 cm^{-3} (keeping M constant at $0\cdot4$ g m^{-3}), decreased reflectance at large radii by nearly 20%.

E

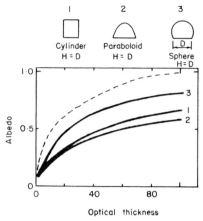

FIG. 7.8   Cloud albedo versus optical thickness as calculated for three cloud shapes by the Monte Carlo technique. The dashed line is albedo of a plane parallel cloud. Solar zenith angle is 30°. After Busygin, Yevstratov, and Feigel'son, 1973.

A most interesting variant of the cuboidal cloud is the *cuboidal stack,* a turret on a slab, which simulates the tower rising from a lower cloud deck. McKee and Klehr (1978) treated this case by the Monte Carlo technique. Width to height dimensions of the turret were varied from 1/2 to 6 with a maximum Z optical thickness of 80. Slab dimensions were kept constant in the ratio of 2:2:1, with a constant Z optical thickness of 60. With overhead sun, the reduction of zenith radiance from the plane parallel cloud to 2:2:1 slab was 32% (Fig. 7.9). Adding a short turret (1:1:1/2, with $\tau(Z) = 20$) to the slab *reduced* radiance by 12%. With the tall turret (1:1:2, $\tau(Z) = 80$) radiance was reduced 35% from the slab case. The turret therefore was dark relative to the slab, and became darker as its height increased. Radiance differences decreased with increasing viewing zenith angle, and reversed at large zenith angles. The opposite case, of a bright turret against a relatively dark slab cloud, is possible for a squat turret nearly as broad as the slab below. McKee and Klehr offer these results as a possible partial explanation of Gruber's paradox (see p. 109).

The causes of this surprising behaviour were explored through radiation budgets for a variety of configurations. As the turret rose, fractional reflected light from the top of the turret and slab decreased, and fractional reflected light from the sides increased, in almost exact proportion. Increasing side area increased the likelihood of lateral loss of light. As the turret spread, fractional reflected light from the top decreased slightly, then increased slightly. Similar, somewhat weaker patterns held for directional reflectance. McKee and Klehr concluded that these effects are most perti-

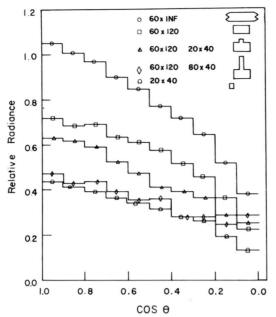

FIG. 7.9   Relative radiance from semi-infinite, slab, and turret clouds as a function of viewing angle. Indicated cloud geometries (top right) are in units of optical thickness. The sun is overhead. From McKee and Klehr, 1978.

nent to cumuliform clouds. Cumulus towers may at times be dark, and interstices may at times be bright.

Mosher (1979) has treated complex cloud geometries of large optical thicknesses and also inhomogeneous clouds by combining Monte Carlo and adding techniques. Mosher's clouds are composed of cubes, for which transfer functions describing the apportioning of radiative flux incident on a side were obtained from Monte Carlo computations. The fluxes of radiation across the sides of the model cuboidal cloud are determined by summing radiation fluxes coming from the sides of each cubical "building block". All fluxes leave in a direction normal to the side, thus directional resolution is very limited, and the building block model is least reliable where optical thicknesses are small. Departures from Monte Carlo results for cubic clouds drop below 10% for $\tau \gtrsim 25$.

Upward fluxes across the top face of a $10 \times 10 \times 10$ building block cloud of optical thickness 500 are shown in Fig. 7.10. Plane parallel and cuboidal fluxes tended to be closest at the centre of the cube. The difference diminished slowly with increasing solar zenith angle from a maximum of 9% at $\theta_0 = 0°$. Increasing the solar zenith angle uniformly reduced upward flux, except along the sunlit edge, where initially flux actually increased.

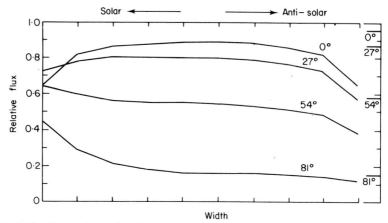

FIG. 7.10   Upward relative flux across the top of a cuboidal building block cloud, for a range of solar zenith angles. The cloud is composed of ten cubes in each dimension. It is illuminated on the top and left sides. Total optical thickness is 500. Relative flux for a plane parallel cloud of the same Z thickness is shown at the right. After Mosher, 1979.

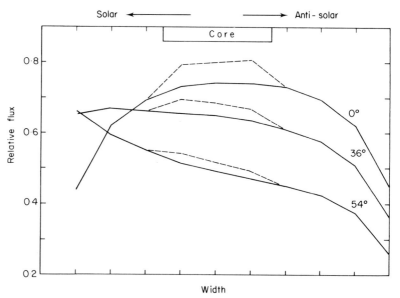

FIG. 7.11   Upward relative flux across the tops of homogeneous and inhomogeneous cubic clouds. The homogeneous cloud (solid line) has an optical thickness of 100. The inhomogeneous cloud (dashed line) has a core of doubled optical thickness in the top centre half. Three sun angles are represented. From Mosher, 1979.

A "core" of high liquid water content was added to a $10 \times 10 \times 10$ cubical cloud of optical thickness 100 by doubling the optical thickness of the top centre $3 \times 3 \times 5$ cubes. Flux increased over the core, by 8 to 9% for the cloud with overhead sun (Fig. 7.11). The flux anomaly subsided with increasing solar zenith angle. It also subsided, quite rapidly, as the core was submerged, dropping to 0·5% with 30 units' optical thickness of normal cloud above.

Calculations of flux from slabs with turrets agreed with results from McKee and Klehr: towers are dark. Mosher calculated flux across a cube with a short tower and without (Fig. 7.12). The tower caused a marked reduction of overall flux, which was largest near the edges of the tower. The dark tower tended to have a darker ring along its edge. Increasing the solar zenith angle destroyed this symmetry. The illuminated edge became much brighter, and the shadowed edge and base cube became much darker. Doubling tower liquid water content, under zenith sun, increased flux at the centre of the tower to the background (no tower) level, but edge flux remained low. Thus even towers with increased liquid water content would tend to show dark rings when illuminated from above.

Mosher also modelled a cloud in nature. This was a cumulus/cumulus congestus cloud line over the eastern tropical north Atlantic Ocean, which was viewed by aircraft and satellite cameras during the GARP Atlantic Tropical Experiment. The cloud was 26 km long, 1 to 1·6 km thick, 5·5 to 6 km wide at the base, and 1·7 km wide at the top. The aircraft flew across the long axis. One airplane at 1·13 km, within the cloud, measured up and

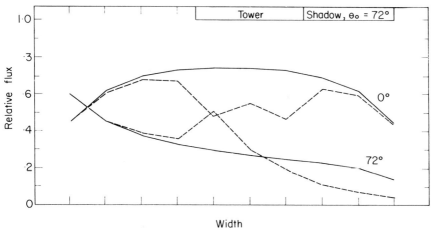

FIG. 7.12   Upward relative fluxes across the top of a cubic cloud of optical dimensions 100^3 (solid lines) and across the cubic cloud with a $30 \times 30 \times 20$ tower on top (dashed lines). The two cases are sun overhead and sun at 72°. After Mosher, 1979.

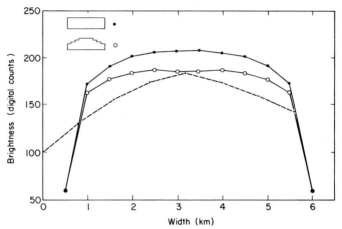

FIG. 7.13   Brightness across a cumulus line as observed by a satellite (dashed line), and as modelled by a cuboidal cloud (solid circles) and a double staircase cloud (open circles). Extreme optical dimensions of both model clouds are 200 × 200 × 60. After Mosher, 1979.

down visible flux, upwelling long wave flux, and liquid water content. A second airplane at 5·5 km, well above the cloud, provided photographs. Cloud top brightness was measured at ~1 km resolution by visible sensors on the geostationary satellite SMS-1. The model cloud had a liquid water content of $0·1$ g m^{-3}, a volume scattering coefficient of 40 km^{-1}, and zenith sun. Sea surface albedo was included in the model run. Isotropy of emerging fluxes was assumed for comparison of model results with satellite observations. With the cloud modelled as a simple slab, building block cloud brightness substantially exceeded satellite observed brightness (Fig. 7.13). However, configuring the cloud as a double staircase brought model and observed brightnesses into much closer agreement. This configuration also improved agreement between measured and calculated interior fluxes.

Most of nature's clouds lie between the extremes of the semi-infinite cloud and the isolated cloud: they are finite, and they have neighbours. The question of how cloud interactions affect radiative transfer has been addressed by Barkstrom and Arduini (1977) and by Aida (1977). Barkstrom and Arduini calculated transfer between an inner cylindrical cloud and an outer annular cloud, both 1 km thick. The fraction of side flux entering to side flux leaving diminished rapidly with increasing separation. At a distance of 2 km, the "lining albedo" was <20%. Aida calculated transfer between a central cloud and eight equally spaced neighbouring clouds. Each cloud was a cube, with properties appropriate to cumulus. Radiative transfer was calculated by the Monte Carlo technique. Cloud amount was kept constant at 0·5, but cloud size distribution was varied from uniform to exponential, the latter based on Plank's (1969) observations of cumulus clouds over Florida. Intercloud reflection increased the reflectance of the central cloud by a few percent. Reflectance tended toward plane parallel with increasing solar zenith angle and increasing uniformity of cloud distribution.

All the studies cited above dealt with liquid water clouds. Few considered the effects of a finite background albedo. Only one (Mosher) treated inhomogeneities of cloud liquid water content. The geometries, it may be argued, are really not much like cloud shapes in nature. But simplifications notwithstanding, through these studies a great deal has been learned about light scattering in clouds. In the context of rainfall estimation, the main factors governing brightness are

(a) thickness,
(b) the area-to-volume ratio, and
(c) orientation in the solar beam.

These factors, collectively called geometry, determine optical path within the cloud. Drop size distribution and liquid water content are secondary in importance. Background albedo may be significant over land areas. Interactions apparently are important only when clouds are close together. Apparently phase is not an important variable (though the importance of ice to precipitations processes gives this question special significance).

Three studies (Twomey *et al.*, 1967; Reynolds *et al.*, 1978; and Mosher, 1979) compared calculations with observations. More are needed, for an adequate evaluation of factors important to satellite cloud brightness will come only through experiments involving both field observations of precipitating-cloud radiant, geometrical, and microphysical properties and model calculations of radiative transfer.

CHAPTER 8

Properties of Microwave Radiation in the Atmosphere

Previous chapters have described measurements made in the visible and infrared window regions of the atmosphere. In the microwave region, there is a third window through the atmosphere, in which even clouds are partly or wholly transparent. In this as well as in the shorter wavelength windows transparency is a function of wavelength. The significance of this condition to meteorology is that observations of microwave energy at selected wavelengths provide information on the structure of the atmosphere: profiles of temperature and water vapour, or the liquid water content of clouds can be retrieved (Staelin, 1969; Paris, 1971). The problem, as Fraser (1975) points out, is to find interactions between constituents of the atmosphere and the field of radiation which are observable, sensitive to variations in the parameter of interest, and unambiguous.

For most meteorological purposes, the microwave region is defined as the interval from about 3 to 300 GHz (1 GHz = 10^9 cycles s^{-1}; in wavelength, this interval extends from 10 to 0·1 cm). It merges with infrared radiation at higher frequencies, and with radio waves at lower frequencies. The wavelengths used in most meteorological radars also lie within the microwave interval; these will be a concern in Chapter 14. For the present our interest is not active but *passive* sensing—the measurement of natural radiation.

Microwave radiation may be absorbed, reflected or scattered. What happens to it depends on elements of the Earth and atmosphere, whose influence depends in turn on temperature, dielectric state and microphysical properties such as roughness or size and shape. Active constituents at microwave frequencies are the water, ice and land surface of the Earth, the gases O_2 and H_2O of the atmosphere, and particles of various kinds, most notably cloud droplets and hydrometeors. Methods of inferring rainrates from satellite observations of microwave radiation are described in Chapter 9. This chapter is concerned with general properties of microwave radiation in relation to rain.

I. Brightness Temperature and Polarization

Microwave radiation is characterized by intensity and polarization. The intensity of thermal emission I is given in general by the emissivity ε and the Planck (blackbody) function $B(T)$. Thus

$$I_\lambda = \varepsilon_\lambda B_\lambda(T) = [\varepsilon_\lambda(2\pi c^2 h/\lambda^5)]/(e^{hc/k\lambda T} - 1), \qquad (8.1)$$

where h is Planck's constant, k is Boltzmann's constant, c is the speed of light, and T is thermal temperature. Because microwavelengths are long (in the sense that $hc/\lambda \ll kT$), in microwave radiometry thermal emission by the Planck function can be approximated by the Rayleigh-Jeans formula,

$$B_\lambda = 2kT/\lambda^2.$$

Then intensity is a linear function of temperature, and it is convenient to define a *brightness temperature* $T_{b\lambda}$ such that

$$T_{b\lambda} = \varepsilon_\lambda T \qquad (8.2)$$

Thus brightness temperature is that blackbody temperature which corresponds to radiant energy of intensity I_λ.

Emissivity is a complicated function of the dielectric constant of a medium. For liquids and solids it depends also on the roughness, relative to incident wavelengths, of the surface and thus also on the angle of incident radiation. Emissivity is known quite well for gases and for calm water. It is not so well known for the more complicated cases of rough water and land.

At a point in space the electric field vector of a simple electromagnetic wave will trace an ellipse. Polarization describes the eccentricity and orientation of this ellipse and the direction the vector rotates. The *plane of polarization* is the plane of the major axis, projected in the direction of wave propagation. *Circular polarization* and *linear polarization* are the two components of the ellipse: together they determine its shape, orientation and direction. Usually polarization refers to the group of simple waves comprising a beam of radiation. Whereas a simple wave is always polarized, a beam is polarized only to the extent that the electric field vectors trace lines of the same orientation or circles of the same direction. In microwave radiometry polarization ordinarily means the linear aspect. Often it is measured by the difference in brightness temperature between the horizontal and the vertical components of intensity, i.e. $T_{bH} - T_{bV}$, with vertical being defined as the local normal to the surface (Savage, 1976).

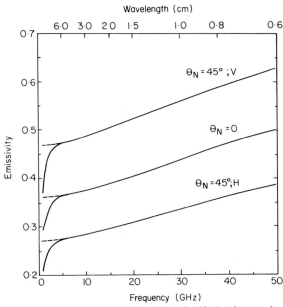

FIG. 8.1   Emissivity of water at 20°C for nadir and 45° viewing angles and vertical and horizontal polarizations. The emissivity of fresh water is shown by the dashed lines, that of a 3.5% solution of salt water by the solid lines. After Wilheit, 1972.

The polarization of reflected, transmitted and emitted microwave radiation depends on molecular or crystalline properties and the surface roughness of a medium. Like emissivity, it is a function of angle of view. Substances (like water) of large dielectric constant produce highly polarized light at oblique viewing angles (Shifrin and Ionina, 1968; Paris, 1971; Wilheit, 1972; see Fig. 8.1). At 37 GHz the calculated brightness temperature difference of horizontally and vertically polarized radiation emerging through a standard atmosphere exceeds 30 K (Weinman and Guetter, 1977; Rodgers *et al.*, 1979). Radiation emitted by atmospheric gases is *not* polarized; neither is there significant polarization in radiation emitted and scattered by cloud droplets and raindrops (Savage, 1976; Weinman and Guetter, 1977; Tsang *et al.*, 1977). Differences in polarization are a primary basis for distinguishing areas of dry land, wet land, and rain over land in schemes to be discussed in Chapter 9.

II. Sources and Sinks of Microwave Energy

Contributions to the intensity $I(\lambda)$ sensed by an orbiting satellite are treated here in terms of the terrestrial surface, the gaseous atmosphere and cloud particles.

1. Surface emission and reflection

The emissivities (and, hence, reflectivities) of surface materials are quite variable (Staelin, 1969; Shifrin et al., 1968). Water is a case of special interest. Its emissivity—which may be calculated from the Fresnel relationships (Paris, 1971; Hollinger, 1973)—is small: ~0·4 at low microwave frequencies (Fig. 8.1). At some wavelengths, including the 19 GHz window, the decrease of ε with increasing thermal temperature compensates increasing blackbody emission, so that brightness temperature is nearly constant over a wide range of surface water temperatures (Staelin, 1969; Wilheit, 1972). The effect of salinity on ε is significant only at very low microwave frequencies, however roughness, including foam, increases ε at all wavelengths (Wilheit, 1972; Nordberg et al., 1969, 1971).

The emissivities of soils and vegetation are much higher than those of water, but not as well documented. Dry land emissivities are generally taken to be between 0·85 and 0·95 at wavelengths of 1 to 1·5 cm (Wilheit, 1972; Savage and Weinman, 1975; Rodgers et al., 1979). Shifrin et al. (1968) report emissivities for vegetated land surfaces in this range. The most significant variable in the emissivity of land surfaces is water, which lowers ε approximately in proportion to its fractional surface content (Moore, 1975). Thus the emissivity of rain-wetted vegetation may be such as to lower T_b by ~15°C or more (Rodgers et al., 1979), and puddled soil may have the emissivity of a lake or sea. According to Rodgers et al. (1979) even dew may be significant.

2. Absorption by the gaseous atmosphere

Emission and absorption by the gaseous atmosphere is dominated by water vapour and diatomic ozygen. The absorption characteristics of these gases have been summarized by Staelin (1969), Paris (1971), Derr (1972), Waters (1976), Fraser (1975) and others. Absorption by H_2O and O_2 is due to energy transition states of molecular dipole moment. These produce rotational spectral lines at frequencies proportional to the energy difference between the initial and final states of a transition. The rotational lines of H_2O and O_2 are "pressure broadened" in the atmosphere owing to the presence of other gases; there is also a slight dependence on temperature. Absorption

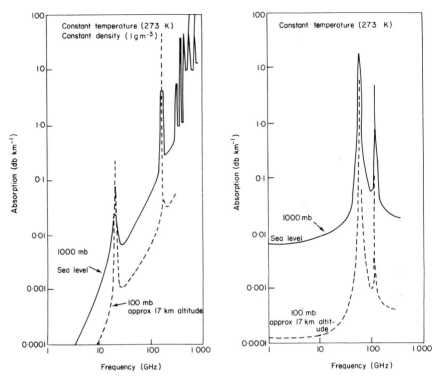

FIG. 8.2   (left) Absorption spectrum of water vapour from 1 to 300 GHz for two pressures at constant temperature. From Fraser, 1975.

FIG. 8.3   (right) Absorption spectrum of atmospheric oxygen from 1 to 300 GHz for two pressures, at constant temperature. From Fraser, 1975.

spectra for H_2O and O_2 in an atmosphere of nitrogen are shown for two pressures in Figs 8.2 and 8.3. Water vapour (Fig. 8.2) has major absorption lines at 22·237 and 183·3 GHz. These are superimposed on a steeply sloping profile, covering five decades of absorption across the interval from 1 to 300 GHz: absorption by water vapour is negligible at low frequencies, very strong at high frequencies. Oxygen has two major peaks, one near 60 GHz resulting from many close lines, and one at 118·75 GHz resulting from a single line. In contrast to water vapour, absorption outside the main spectral lines is relatively flat, and gaseous absorption at low microwave frequencies is due mainly to oxygen. To a first approximation, total absorption depends upon mass integrated along the radiometer line of sight (Lovejoy and Austin, 1980). For a well mixed gas such as oxygen this is mainly a function of path length. In the case of water vapour, however, local variations often are large and may significantly modulate line of sight absorption.

3. Absorption and scattering by cloud particles

In Chapter 7 we examined the question of how light is scattered by cloud droplets. The theory for microwave interaction with cloud particles is the same: wavelengths and certain boundary conditions are not. In the case of microwave interactions with cloud particles, local emission and absorption may be significant sources and sinks of microwave energy, the range of wavelengths is large compared with the range of particle sizes, and the complex index of refraction is a variable also.

When an electromagnetic wave interacts with a spherical dielectric, electromagnetic energy may be redirected (scattered) or transformed to mechanical energy (absorbed). The relative importance of scattering and absorption is a function of the complex index of refraction of the sphere $n_c(T,\lambda)$, and the size parameter $x(r,\lambda)$. *We wish to describe this scattering and absorption across the microwave spectrum, for representative ranges of cloud drop size distributions and atmospheric temperatures, and ice and liquid phases.*

In the ensuing discussion, microwave interactions with a single drop are considered first, then interactions with an ensemble of drops. Authoritative treatments of this subject may be found in Gunn and East (1954), Shifrin and Chernyak (1968), Paris (1971), Schwiesow (1972), Hansen and Travis (1974), Savage (1976), and Fraser (1975), the papers from which the present discussion is largely drawn.

(a) *Scattering and absorption for a single drop*

Following the theory of Mie (1908) for scattering and absorption of electromagnetic radiation by spheres, we define *effective cross sections* for extinction and scattering, σ_{ex} and σ_{sc}. Both are functions of the complex index of refraction n_c and wavelength λ. The product of σ_{ex} and I_λ is the power intercepted and extinguished by a single drop.

Dividing the effective cross sections by the geometric cross section gives the (dimensionless) *Mie efficiency factors* for single drops,

$$Q_{ex}(n_c,\lambda) = \sigma_{ex}/\pi r^2,$$

and

$$Q_{sc}(n_c,\lambda) = \sigma_{sc}/\pi r^2,$$

which find analogies later in extinction and scattering coefficients for ensembles of drops.

The electromagnetic field (Maxwell) equations are solved (by expansions in dipoles and multipoles) for radiation interacting with a dielectric sphere of radius r and index of refraction n_c. Then

$$Q_{ex} = \frac{2}{x^2}\sum_{l=1}^{\infty}(2l + 1)\text{Re}[a_l(n_c,x) + b_l(n_c,x)], \qquad (8.3)$$

and

$$Q_{sc} = \frac{2}{x^2}\sum_{l=1}^{\infty}(2l + 1)\,[|a_l(n_c,x)|^2 + |b_l(n_c,x)|^2]. \qquad (8.4)$$

The coefficients a_l and b_l depend only on the two boundary conditions which are set in taking values for n_c and x. Q_{ab}, the Mie efficiency factor for absorption, is equal to the difference, $Q_{ex} - Q_{sc}$.

For size parameter $x \ll 1$, that is, wavelength long compared with drop diameter, Eqs 8.3 and 8.4 reduce to

$$Q_{ex} = 4x\,\text{Im}|-K| + (8/3)x^4 \cdot |K|^2 + \ldots \qquad (8.5)$$

and

$$Q_{sc} = (8/3)x^4 \cdot |K|^2, \qquad (8.6)$$

and the Mie efficiency factor for absorption therefore is given by

$$Q_{ab} = 4x\,\text{Im}|-K| \qquad (8.7)$$

K is equal to $(n_c^2 - 1)/(n_c^2 + 2)$. The long wavelength limit defines the *Rayleigh regime*, for which scattering and absorption are described by Eqs 8.6 and 8.7.

n_c, the complex index of refraction of a medium, is equal to the square root of the dielectric constant ε; thus it too is a function of temperature and wavelength. n_{Re} the real part of n_c, is the refractive index of the medium (and therefore governs scattering); n_{Im}, the imaginary part of n_c, is the absorption coefficient (Gunn and East, 1954).

For water in the microwave region the complex index of refraction is adequately given by the formula of Debye (1929; also see Fraser, 1975), which is plotted (for three temperatures) as n_{Re} in Fig. 8.4a and as n_{Im} in Fig. 8.4b. The real part of the index of refraction is very large relative to the imaginary part at microwave frequencies less than ∼10 GHz; at the high frequency end of the microwave spectrum the differences are much smaller. (The effects on scattering and absorption can be seen in Figs 8.5a, b and c, below.) Temperature effects are relatively small.

For ice the complex index of refraction is independent of wavelength in the microwave region, and only the imaginary part varies with temperature (Gunn and East, 1954). The refractive index of ice approaches that of water at high frequencies (Fig. 8.4a), but the absorption coefficient for ice is smaller than that of water by three orders of magnitude.

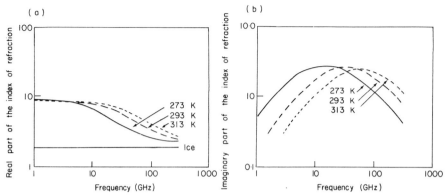

FIG. 8.4   Index of refraction of pure water for frequencies from 1 to 300 GHz, as a function of temperature. From Fraser, 1975. (a) Real part, with ice from Gunn and East, 1954. (b) Imaginary part.

Fraser (1975) has calculated Mie efficiency factors for extinction and scattering and the coefficient for Rayleigh extinction for a range of drop sizes; these are shown in Figs 8.5a, b, and c for 3·0, 30 and 300 GHz, respectively. We note first the relative simplicity of scattering and extinction curves over the Rayleigh regime. Neglecting the wavelength dependence of K, scattering and absorption here are related as $(r/\lambda)^3$. Since (by definition) $\lambda \gg r$, only at the short wavelength/large droplet limit of the Rayleigh regime is scattering comparable to absorption: elsewhere absorption dominates completely. We observe further in Fig. 8.5a that at low frequencies (large wavelengths) only heavy rain (large r) extinguishes microwave radiation to a significant degree, and at low frequencies the Rayleigh regime embraces the whole range of rain and cloud drop sizes. On the other hand, at high frequencies (Fig. 8.5c) all rain-clouds extinguish microwave radiation, and the Rayleigh regime is limited to cloud drop sizes.

(b) *Scattering and absorption in a small volume of drops*

Since scattering and absorption can be calculated for a single drop, it is possible by integration to determine corresponding coefficients for ensembles of drops (providing there is no coherence from drop to drop in the phase of scattered light, a condition which is met for microwave radiation interacting with rain clouds). If in a volume dV the drops are of variable size, the distribution of drop sizes is $N(r)$, and the number of drops of radius r in the interval dr is $dN(r)/dr$. Then the drop concentration n is given by

$$n = 1/dV \int_r \frac{dN(r)}{dr}\, dr,$$

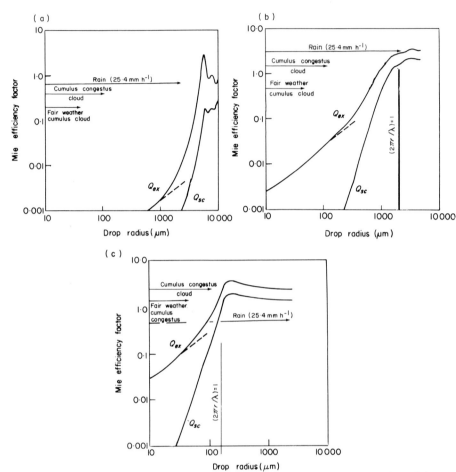

FIG. 8.5   Mie efficiency factors for scattering and extinction as a function of drop radius. Dashed lines show Rayleigh extinction. Temperature is 273 K. From Fraser, 1975. (a) 3·0 GHz. (b) 30 GHZ. (c) 300 GHz.

and the volume coefficient for extinction is

$$k_{ex}(\lambda, N) = \frac{1}{dV} \int_r \sigma_{ex}(r, \lambda) \frac{dN(r)}{dr} \, dr \qquad (8.8)$$

Similar expressions hold for k_{sc} and k_{ab}, and, again, $k_{ex} = k_{sc} + k_{ab}$.

Over most of the Rayleigh regime $\sigma_{ex} \cong \sigma_{ab}$. But from Eq. 8.8, the absorption cross section of a single drop is proportional to r^3. Therefore $k_{ex} \doteq k_{ab} = M \cdot f(\lambda, T)$. Thus in the Rayleigh regime the effect of cloud particles on microwave radiation is a function only of the liquid water content and temperature of the cloud, and the wavelength of the radiation.

Outside the Rayleigh regime, which is the more general case, we must express $dN(r)/dr$ analytically and by numerical methods solve the full Mie equations for extinction and scattering coefficients and the scattering phase function $P(\Omega)$. Savage (1976) has observed that scattering phase functions for hydrometeors over the range of microwave radiation are not strongly peaked either in forward or backscatter directions, and therefore may be approximated by relatively few terms of an expansion. Further, it is convenient to express solutions for extinction coefficients and phase functions in terms of liquid water content, which requires an explicit relationship of M and drop size distribution. Thus it is common for solutions of the Mie equations for rain-clouds and microwave frequencies to be based on the 1948 Marshall-Palmer drop size distribution.

In microwave frequencies the most complete published tabulation of extinction, scattering and absorption coefficients and scattering phase func-

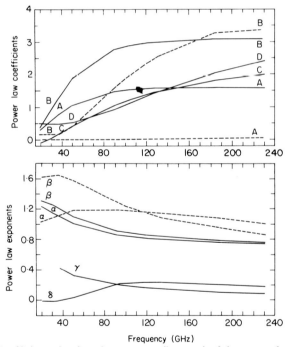

FIG. 8.6   Coefficients (top) and exponents (bottom) of the power law relations for absorption (A, α), extinction (B, β), and the first two coefficients of the Legendre function for the scattering phase function $(C, \gamma; D, \delta)$, as a function of frequency. Coefficients of absorption and extinction have dimensions (km^{-1}), others are dimensionless. Solid lines are for water, dashed lines for ice (absorption and extinction only). Water and ice particles are modelled by a Marshall-Palmer distribution, with $T = 0°C$. After Savage, 1978.

tions is that of Savage (1978). Savage approximated the scattering phase functions by Legendre polynomials, and expressed the Legendre coefficients (for the phase functions) and extinction, scattering, and absorption coefficients as power law relations in liquid water content of form $(Z \cdot M^\zeta)$, where Z is the power law coefficient and ζ is the power law exponent. Mie coefficients were calculated for seven wavelengths across most of the microwave spectrum. Liquid water content in the calculations ranged from 0 to $1 \cdot 63$ g m^{-3}. The corresponding range of Marshall-Palmer rainfall rate was 0 to 32 mm h^{-1}.

Power law coefficients and exponents for volume coefficients of extinction and absorption and the first two Legendre coefficients of the scattering phase function for water cloud of Marshall-Palmer drop size distribution at $T = 0°C$ are plotted in Fig. 8.6. Power law coefficients and exponents for extinction and absorption approach as asymptote near 100 GHz above which extinction is primarily due to scattering. The second Legendre coefficient, corresponding to power law coefficient C and exponent γ, is a measure of g, the asymmetry of the scattering phase function. Forward scatter dominates backscatter. In the Rayleigh regime near 30 GHz they are comparable, and at very low frequencies backscatter slightly exceeds forward scatter.

For a Marshall-Palmer distribution of ice at 0°C (Fig. 8.6), absorption is small at all frequencies. Scattering increases with frequency, and can become significant for size parameters approaching one.

III. A Complete Description—The Radiative Transfer Equation for Microwaves

Various sources and sinks of microwave radiation in the Earth–atmosphere system are related and united in the equation of radiative transfer. The general problem of radiative transfer was solved by Chandrasekhar (1960). It is treated for microwaves in the cloudy atmosphere by Volchok and Chernyak (1968), Paris (1971), Snider and Westwater (1972), Fraser (1975), Savage (1976) and Wilheit et al. (1977).

For a medium in thermodynamic equilibrium, the change of microwave intensity I_λ over a distance ds in the direction (θ, ϕ) is given by

$$\frac{dI_\lambda}{ds} = -(k_{ab} + k_{sc}) I_\lambda + k_{ab}B_\lambda$$
$$+ k_{sc}/4\pi \int_0^{2\pi} \int_0^\pi P_\lambda'(\theta_s, \phi_s, \theta, \phi)I_\lambda(\theta_s, \phi_s) \sin\theta_s d\theta_s d\phi_s \qquad (8.9)$$

$P_\lambda'(\theta_s, \phi_s, \theta, \phi)$ is an "inverse" scattering phase function which describes the relative contribution of each polar angle θ_s and each azimuthal angle ϕ_s to the energy scattered in the direction (θ, ϕ):

$$\int_0^{2\pi} \int_0^{\pi} (P_\lambda'/4\pi) \sin\theta d\theta d\phi = 1.$$

Equation 8.9 says that the gradient of I_λ along s is determined by the balance of energy *lost* by absorption and by scattering out of the direction of I, and energy *gained* by thermal emission and by scattering into the direction of I. Here k_{ab} includes absorption by oxygen and water vapour.

Using the definitions for opacity,

$$\tau(1, 2) = -\int_1^2 (k_{ab} + k_{sc}) ds,$$

and single scatter albedo, $\tilde{\omega} = k_{sc}/(k_{ab} + k_{sc})$, Eq. 8.9 may be integrated. Then

$$I_\lambda(\theta, \phi, Q) = I_\lambda(\theta, \phi, \tau) e^{-\tau} + \int_0^\tau J_\lambda(\theta, \phi, \tau') e^{-\tau'} d\tau', \qquad (8.10)$$

where

$$J_\lambda(\theta, \phi, \tau') = (1 - \tilde{\omega}) B_\lambda + \tilde{\omega}/4\pi \int_0^{2\pi} \int_0^{\pi} P_\lambda'(\theta_s, \phi_s, \theta, \phi) I_\lambda(\theta_s, \phi_s) \sin\theta_s d\theta_s d\phi_s.$$

Paraphrasing Snider and Westwater (1972), Eq. 8.10 states that at a point Q outside the medium the intensity of radiant energy in the direction (θ, ϕ) is the sum of emissions at all points upstream of Q reduced by the factor $e^{-\tau'}$ to account for extinction by the intervening medium.

If $k_{sc} = 0$, there is an analytic solution to Eq. 8.10 (Savage, 1976). According to Staelin (1969) scattering in clouds is negligibly small for $\lambda > \sim 30r$. Fraser (1975) has calculated single scatter albedo as a function of wavelength for drop size distributions representative of very thin fair weather cumulus and very dense cumulonimbus clouds (Fig. 8.7). In the cumulus cloud scattering is negligible (i.e., $\tilde{\omega} \leq 0.1$) at all wavelengths. In the cumulonimbus cloud, which is raining at a rate of 150 mm h^{-1}, it is small only for $\lambda > \sim 3$ cm. Citing Stepanenko (1968a) and Wilheit et al. (1977), and using a criterion of $\tilde{\omega} < 0.2$, Lovejoy and Austin (1980) conclude that scattering can be neglected for all clouds if $\lambda > 0.5$ cm, and for $R < 10$ mm h^{-1} if $\lambda > 1.0$ cm. Thus the analytic solution applies to transfer of microwave radiation not at all where rain is heavy and only at longer wavelengths where rain is light.

For the case of negligible scattering, and assuming temperature in the atmosphere (T) to be constant, Lovejoy and Austin (1980) show that the brightness temperature measured by a satellite radiometer viewing the Earth at a zenith angle θ can be expressed as

$$T_b = \varepsilon_s T_s e^{-\tau_0} + T(1 - e^{-\tau_0}) + (1 - \varepsilon_s) T(1 - e^{-\tau_0}) e^{-\tau_0}. \qquad (8.11)$$

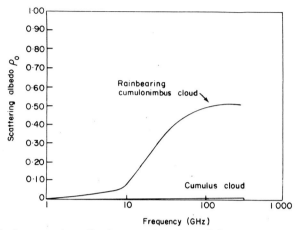

FIG. 8.7 Single-scattering albedo as a function of frequency computed for a cumulus cloud (liquid water content of 0.01 g m^{-3} and mode drop radius of 10 μm) and for a cumulonimbus cloud (80 g m^{-3} and 400 μm). From Fraser, 1975.

Equation 8.11 follows from Eqs 8.1, 8.2, and 8.10 if $\tilde{\omega} = 0$ and $\varepsilon = \varepsilon_s$. A term involving $(1 - \varepsilon_s)$ is added to account for reflection from the surface of the Earth. Subscript s represents surface values and τ_0 is the microwave optical thickness integrated along the line of sight, i.e. total absorption. This form of the radiative transfer equation, as Lovejoy and Austin point out, demonstrates that radiometric temperature can be very sensitive to surface emissivity, and is determined to first order by total absorption. Indeed, the error in T_b made by assuming constant T is, by Lovejoy and Austin's estimate, no more than $\pm 10\%$.

From a sensitivity analysis, Lovejoy and Austin (1980) conclude that the contribution of oxygen to total absorption can be calculated to an accuracy of $\sim \pm 1\%$, about the size of the error in the radiometric measurement of temperature. The contributions of water vapour is less certain, but even by climatology might be determined to an accuracy of $\pm 12\%$. With contributions of oxygen and water vapour accounted for, the remaining absorption may be expressed as the sum of rain and cloud terms, that is

$$\tau_0 = k_{ab}(R)h + K_{ab}\tilde{M} \qquad (8.12)$$

$k_{ab}(R)$ is the absorption coefficient for rain and h is the rain layer thickness along the line of sight; K_{ab} is the total cloud water absorption coefficient and $\tilde{M}$ is the total cloud liquid water content along the line of sight. τ_0, k_{ab}, and K_{ab} all depend upon wavelength. To a good approximation the absorption coefficient for rain may be expressed in power law form, as $k_{ab} = AR^\alpha$. It is apparent, then, that total absorption is a function of three variables: rain-rate, rain layer thickness, and total cloud liquid water content.

To solve for τ_0 three measurements are needed, which in principle can be made at three different wavelengths. But Lovejoy and Austin show that the accuracy of the solution depends critically on the size of the difference in the exponent α as wavelength changes. Depending on dropsize distribution the range of α across the microwave window (from 0·8 to 30 cm) is only 0·2 to 0·6, and the likely error in determination of α is 0·2. Thus Lovejoy and Austin conclude that even with multiple microwave channels it is "unlikely that the rainfall rate can be estimated without recourse to modelling the relative cloud-rain contributions".

If $k_{sc} > 0$, through the scattering term intensity at a point along s depends upon intensity in every direction around the point, and there is no analytic solution. This is the general case for microwave transfer through rainclouds. A numerical solution is required, as, for example, by the methods of Chandrasekhar (1960; also see Savage, 1976, and Tsang et al., 1977), or by Monte Carlo techniques (Paris, 1971; Weinman and Davies, 1978).

1. Plane parallel clouds

The conditions under which solutions to the present problem have been achieved are the following (Savage, 1976):

(a) the atmosphere is horizontally homogeneous and in thermal equilibrium;
(b) the Earth-atmosphere system is viewed from space; and
(c) contributions to upwelling radiation $I_\lambda(\theta) = f(T_{b\lambda}, \theta)$ may come from the surface of the Earth, the gaseous atmosphere, cloud droplets, or raindrops.

The calculations by Rosenkranz (1978) of the zenith opacity of the atmosphere at a range of microwave wavelengths illustrate the cumulative contributions of oxygen, water vapour and cloud to extinction (Fig. 8.8). Where extinction is due only to absorption by oxygen, the curve resembles those of Fig. 8.3. In the middle and upper curves, for oxygen plus water vapour and for oxygen, vapour and cloud, the relative contributions of vapour and cloud are evident, as well as the frequency intervals best suited for their measurement. From the upper curve we may also infer the frequency intervals best suited for measurement of rain: less than 20 GHz and close to 30 GHz.

Savage (1976) offers a solution to the radiative transfer equation for the simple case of an isothermal raincloud of optical thickness 10 at the top of the atmosphere, with surface emissivity of unity (approximating "black" soil) and surface temperature of 300 K. Calculations were made in the

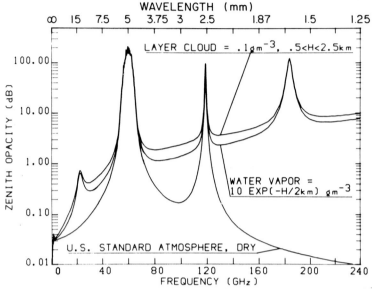

FIG. 8.8   Zenith opacity of the atmosphere as a function of frequency. Opacity due to oxygen (lower curve), opacity with 20 kg m^{-2} water vapour added to the oxygen (middle curve), opacity with 0·2 kg m^{-2} stratus cloud added to the oxygen and water vapour (upper curve). After Rosenkranz, 1978.

37·0 GHz microwave window. The results, shown in Fig. 8.9, are as follows:

(a) limb darkening is large at nadir angles $> \sim 30°$, however, it is well behaved and therefore predictable;

(b) the effect of changing cloud temperature is an equivalent change of brightness temperature;

(c) the effect of changing the asymmetry factor g from the extremes of zero asymmetry (a Rayleigh phase function) to 0·66 asymmetry (strong forward scatter) is to increase T_b; and

(d) increasing single scatter albedo $\bar{\omega}$ has the same effect as decreasing g—both changes act to decrease T_b.

The largest effect is that of $\bar{\omega}$, which is governed (through liquid water content) by precipitation rate R and phase, and by gaseous absorption. The asymmetry factor g also is important; g is governed by R and phase. Cloud temperature is a function of altitude. Precipitation rate, thickness, and phase also determine cloud optical thickness. Savage concludes, "The single scatter albedo appears to be the most important single radiation variable; the hydrometeor phase seems the most important atmospheric variable."

The radiative transfer equation was also solved by Savage for a more complex and realistic case of rainclouds in depths of 1·52, 4·57, and 7·62 km

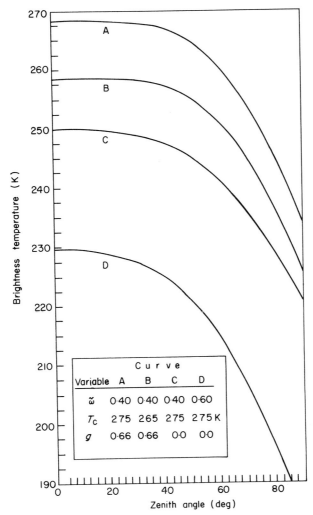

FIG. 8.9 Upwelling brightness temperatures for four combinations of radiative variables. Surface temperature = 300 K; surface emissivity = 1·0; cloud optical depth = 10·0. $\tilde{\omega}$ is a single scatter albedo, T_c is isothermal cloud temperature, and g is asymmetry factor. From Savage, 1976.

(5000, 15 000, and 25 000 ft), Marshall-Palmer or Best drop size distributions in water or ice, a standard atmosphere lapse of temperature, atmosphere to 19·80 km (65 000 ft), and surface emissivities appropriate to land (0·9) or sea water (calculated from Fresnel formulas). Calculations of T_b were made for seven wavelengths and three zenith angles, and plotted for $\theta = 50°$ as contours of T_b in respect to rain-cloud thickness and liquid water

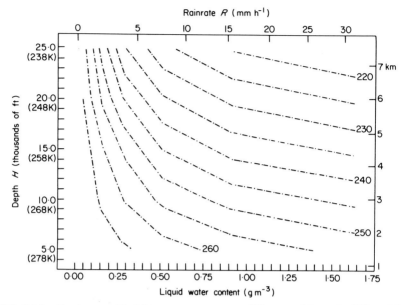

FIG. 8.10 Contours of brightness temperature (K) at zenith angle 50° and frequency 37·0 GHz. The hydrometeors are described by the Marshall-Palmer distribution; the underlying surface temperature is 288·2 K, with surface emissivity 0·90. Independent variables are liquid water content, rainrate, and rain-cloud depth. The ambient atmosphere is the 1962 Standard Atmosphere. Temperature at rain-cloud top is printed below cloud top height on the left-hand scale. The liquid water content is uniform from the Earth surface to the top of the rain-cloud. From Savage, 1976.

content/rainrate. Results at 37·0 GHz for a Marshall-Palmer distribution of water drops with a surface temperature of 288·2 K and emissivity of 0·9, shown in Fig. 8.10, are as follows:

(a) at small liquid water contents (rainrates), some radiation comes from the surface;

(b) where surface radiation is not important, cloud temperature governs brightness temperature; and

(c) the sensitivity of T_b to changes of M decreases as M increases.

The second of these conclusions is consistent with the result of Lovejoy and Austin (1980), that the thickness of the rain layer has a large effect on brightness temperature (see Eq. 8.12). Lovejoy and Austin assessed this effect through Gorelik and Kalashnikov's (1971) effective rain layer height H, which is the integral of the absorption coefficient through the depth of the rain layer, normalized by k_{ab} at the surface. For $R < 7$ mm h^{-1}, and wavelengths of 0·8 and 1·6 cm, Gorelik and Kalashnikov found H to vary by

0·9 km about its mean of 1·6 km. Lovejoy and Austin calculated H at the same wavelengths for echoes observed by the Quadra radar in the eastern Atlantic Ocean. Both echo height and H were highly variable. As expected, echo height generally increased with rainrate; however, owing to large height gradients in R for heavy rain, H actually *decreased* with increasing rainrate, and in fact was always smaller than either echo height or freezing level. According to Lovejoy and Austin's figures, the uncertainty in rainrate resulting from the variability in H is of order ±50%. Because scattering enhances contrast in rainclouds over a surface of high emissivity, we might expect the uncertainty in H to produce larger errors in R for clouds over land, and indeed Savage's results (see Fig. 8.10) seem to confirm this.

The cloud component of liquid water was not modelled in Savage's calculations because, according to Savage (1976), some cloud size particles are implicit in the Marshall-Palmer distribution. Recently Lovejoy and Austin (1980) assessed the relative contributions of cloud droplets and raindrops to total cloud layer absorption. Although the observations are scant, in the Russian literature at least there is evidence that total cloud water content is as large or larger than rain water content. Assuming equal contents, Lovejoy and Austin calculated the ratio of cloud and rain absorptions at three window wavelengths and three rainrates. At $R - 10$ mm h^{-1}, and consistent with observations reported by Gorelik *et al.* (1971), cloud absorption was 30 to 40% as large as rain absorption. Savage (1976) found that the effect of adding a non-scattering cloud layer above the rain layer was to counteract scattering and increase brightness temperature by an amount proportional to the thickness of the cloud layer. A similar effect would be expected if cloud were added to the rain layer, but because this would change the single scatter albedo, the size of the change cannot be easily predicted.

Differences arising from Marshall-Palmer and Best drop size distributions could be assessed only for the over-water case. There, because the Best distribution is weighted toward larger drops which scatter more effectively, Best brightness temperatures tend to be colder than those based on a Marshall-Palmer distribution. The difference is as large as 10°C for $R = 16$ mm h^{-1} at 19·35 GHz.

Subject to these uncertainties, Savage's plots tell us that *if we know the depth of a raincloud over a surface of (known) high emissivity, the brightness temperature may be uniquely related to rainrate.* However, over a surface of (known) low emissivity, brightness temperature is a double valued function of rainrate. (If, in Fig. 8.10, emissivity were reduced to 0·5, say, the contours of T_b would break downward as $R \rightarrow 0$.) Furthermore, in both cases, the association of brightness temperature with rainrate becomes more and more ambiguous at large rainrates, and the effects of cloud water content and departures from the Marshall-Palmer drop size distribution remain to be established.

2. Finite clouds

In aircraft observations of a line of raining cumulus clouds at 19·35 and 89·5 GHz, Savage (1976) found a tendency for rainrate inferred from 19·35 GHz brightness temperature to vary *inversely* with rain height inferred from 89·5 GHz brightness temperature. Savage noted also that whereas his model calculations of T_b at 37·0 GHz agree well with those from Wilheit's model, at medium rain rates both models predicted much warmer temperatures than were observed (see Fig. 9.9b). The discrepancies, Savage concluded, probably were due to departures from plane parallel conditions. More recently Weinman and Davies (1978) observed that Savage's (1978) extinction coefficient at 37·0 GHz (Fig. 8.6) indicates, for $R < 8$ mm h^{-1}, some loss of radiation through the sides of clouds a few kilometres in diameter. In the same paper Weinman and Davies evaluated this loss for observations of the Nimbus-6 Electrically Scanning Microwave Radiometer by calculating microwave radiation at 37 GHz emerging from hydrometeor clouds of finite dimensions.

Like the finite cloud of Davies (1976, p. 115), the model cloud is cuboidal. The usual assumptions attending calculations of microwave radiative transfer in hydrometeor clouds are made: Marshall-Palmer distribution of drop size, Rayleigh-Jeans approximation for Planck radiation, linear profile of temperature. The source term in the equation of radiative transfer (see Section III, Chapter 7) includes absorption by O_2 and H_2O and scattering and absorption by raindrops. Both the intensity $I(x, y, z, \theta, \phi)$ and the phase function for scattering into the beam (P) are approximated as expressions in Legendre functions [of the kind $I_0 (x, y, z) + I_z (x, y, z) \cos\theta + I_x (x, y, z) \sin\theta \cos\phi + I_y (x, y, z) \sin\theta \sin\phi + \ldots$]. For P these reduce to an expression in the asymmetry factor g. Further manipulation yields a second order differential equation in I_0. Boundary conditions for irradiance on each side of the cloud are applied. The resulting equations then are solved by finite Fourier transformations to obtain an expression for $I_0 (x, y, z)$. Further development is confined to radiation upwelling from the three regions of the cloud defined by the front face and top and their projections on the rear face and bottom. Assuming the Earth to be Lambertian, and the cloud to be homogeneous, an integral equation for $I_l(x, y, z, \theta, \phi)$ on the surface of the cloud is derived. I_l has an analytic solution. To facilitate comparisons with ESMR-6, I_l is averaged over the cloud's front and top surface.

This analytic finite cloud model was checked by comparing its intensities with those from a Monte Carlo model. The Monte Carlo model is similar to that described by Davies (1976), except that photons originate within the cloud. Calculations include several phase functions and two single scatter albedoes. The ground neither reflects nor emits. Weinman and Davies

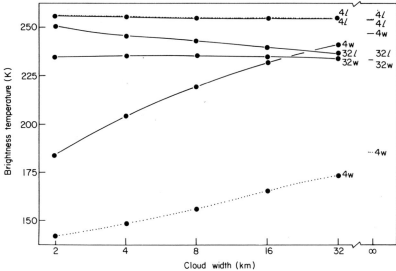

FIG. 8.11 Mean brightness temperatures of 4·57 km hydrometeor clouds over land (l) and over water (w). Liquid phase, solid lines; ice phase, dotted lines. $\varepsilon = 0.9$ (land), 0·462 (water); $T_s = 288$ K, $\partial T/\partial Z = 6.56$ K/km. From tables of Weinman and Davies, 1978.

present their results in tables of brightness temperature (at a nadir angle γ of 50°) as a function of cloud optical thickness in vertical and horizontal directions. Except for a few of the wider clouds, analytic and Monte Carlo brightness temperatures agree to within the error limit of the Monte Carlo calculations.

Changes in rainrate for a cloud of fixed depth also were studied with the analytical model. The cloud is ice or water. Here it exists over land or water, of temperature 288 K and emissivity 0·9 for land, 0·462 for water. Temperature in the cloud decreases at a rate of 6·56 K km^{-1}. Scattering and absorption parameters for water and ice hydrometeors are taken from Savage (1978). Results of this experiment are summarized in plots taken from Weinman and Davies' tables (Fig. 8.11). Adding a background source of microwave radiation diminishes the difference between finite and plane parallel radiances. Cooling in finite clouds is significant only for light precipitation over water. At high rainrates over land the relationship reverses: the finite cloud is warmer than the plane parallel cloud.

We will turn next to applications of this theory to observations of terrestrial microwave radiation from satellites.

CHAPTER 9

Passive Microwave Methods

Technology as well as physics imposes certain constraints on microwave observations from satellites. This discussion of microwave observations of rain begins with comments on the main engineering problems. These are followed by a brief history of microwave instrumentation on satellites, then discussions of methods based on two instruments which stand out for their contributions to rainfall meteorology.

I. The Observing Problem

According to Staelin (1969), the defining parameters in microwave observations are sensitivity, absolute accuracy (calibration), angular resolution (direction response) and spectral response. Sensitivity—commonly expressed as noise equivalent temperature difference (NEΔT)—is directly proportional to the sum of antenna (scene brightness) temperature and receiver noise temperature, and inversely proportional to the square root of the bandwidth times the integration (dwell) time. Thus even with a noise-free receiver, there are immutable limits on the nature of microwave observations: improved sensitivity (smaller NEΔT) is achieved at the cost of larger bandwidth or longer dwell time, the former compromising spectral resolution, the latter compromising observational coverage. For most antennae, beamwidth (at 1/2 power, in radians) is equal to 1·3 (λ/antenna diameter). Suppressing radiation from outside the main beam (sidelobes and stray radiation) is achieved by increasing antenna diameter. Thus, maintaining a small field of view—that is, high ground resolution—at large wavelengths

144

requires very large diameter antennae. Scan is required for coverage to the sides of the satellite track. Antenna and receiver calibration is required for inferences of rainrate.

In present satellite microwave observing systems, according to a recent analysis by Staelin *et al.* (1978), critical components are the receiver and antenna. In receivers the main problem is increasing sensitivity, especially in millimetre and submillimetre ranges; in antennae it is achieving scan while maintaining resolution.

II. History

COSPAR, the Committee on Space Research, assessed possibilities for measuring rain from satellites, and in 1967 endorsed observations of passive microwave radiation as potentially the most viable approach. A year later the USSR launched Cosmos 243, the first meteorological satellite to carry a microwave instrument (Stoldt and Havanac, 1973). The radiometer on Cosmos 243 (and that on Cosmos 384, launched in 1970) had four channels (Table 9.1), and measured, among other things, the liquid water content of clouds. Although the operating lifetime of each instrument was less than 2 weeks, the observations they transmitted were sufficient to construct latitudinal profiles of liquid water content in the atmosphere (Gurvich and Demin, 1970; Basharino *et al.*, 1971; Shtern, 1972; Stoldt and Havanac, 1973).

Nimbus-5 in 1972 carried the first American orbiting microwave instruments: a Nimbus-E Microwave Spectrometer (NEMS) and an Electrically Scanning Microwave Radiometer (ESMR-5). NEMS, primarily a sounder, was of little use in measuring rainfall because of its 180 km resolution and limited coverage. ESMR-5 is an imaging radiometer. It was followed in 1975 by a similar radiometer on Nimbus-6. Nimbus-6 also carried a scanning version of the Nimbus-5 microwave spectrometer, the Scanning Microwave Spectrometer Experiment, or SCAMS. Though orbit-to-orbit coverage was nearly complete, the ground resolution of SCAMS was no better than 145 km, still a significant constraint on usefulness for rain measurement. The Scanning Multichannel Microwave Radiometer on Nimbus-7 and Seasat, both launched in 1978, employs a new antenna design which allows conical scanning. Resolution was improved to 20 km at the smallest wavelength. Tiros-N, launched in 1978, and NOAA-6 and Block-5D, launched in 1979, contain coarse-resolution microwave sounding instruments. In the decade of the seventies the Soviet Union also orbited a number of microwave radiometers, on satellites of the Meteor and Meteor-2 series.

The usefulness of these recent instruments for rain measurements remains

TABLE 9.1   Characteristics of microwave sensors on meteorological satellites.[a]

Satellite	Year of launch	Instrument	Channels (wavelength, cm)	Polari-zation	Pointing direction[b] (deg)	Scan	Beam resolution (km)
Cosmos-243	1968	4-channel radiometer	8·5, 3·4, 1·35, 0·8	no	0	no	~
Cosmos-384	1970	(see Cosmos-243)					
Nimbus-5	1972	Nimbus-E microwave spectrometer (NEMS)[c]	1·11, 0·955, 0·558, 0·546, 0·510	no	0	no	1
		Electrically scanning microwave radiometer (ESMR-5)	1·55	no	0	yes, ±50°	2
Nimbus-6	1975	Scanning microwave spectrometer experiment (SCAMS)[d]	1·35, 0·95, 0·57, 0·49, 0·46	no	0	yes, ±43·2°	1
		ESMR-6[e]	0·81	yes	40	yes, ±35° (conical)	20
Nimbus-7	1978	Scanning multichannel microwave radiometer (SMMR)[f]	0·8, 1·4, 1·7, 2·8, 4·6	yes, all	42	yes, ±25° (conical)	2
Seasat	1978	(see Nimbus-7)					
Tiros-N	1978	Microwave sounding unit of Tiros-N operational vertical sounder (MSU of TOVS)[g]	0·596, 0·558, 0·546, 0·517	no	0	yes	1
NOAA-6	1979	(see Tiros-N)					
Meteor (operational series: −1, −2, −3, . . .)	1974 to 1978	VHF polarimeter	0·8	yes			
Meteor (experimental series; −18, −25, −28)[h]		3-channel radiometer	0·8	M-18 only		yes	24
			1·35	no		no	
			8·5	no		no	
Meteor-2 operational series		penetrating radiation radiometer					
Block-5D	1979	Special system microwave temperature sounder (SSM/T)	0·504 to 0·594 (total of 8)	yes	0	yes	~

[a]Sources are Stoldt and Havanac (1973), World Meteorological Organization (1975), Gloerson and Hardis ( ) Staelin et al. (1978), and Smith et al. (1979).
[b]Along the subpoint, measured from nadir.
[c]Primarily a sounder.
[d]Improvement of NEMS.
[e]Improvement of ESMR-5.
[f]Evolved from NEMS and SCAMS, primarily for ocean surface measurements.
[g]11 fields of view per swath.
[h]900 km orbit.

to be determined. At the time of writing the outstanding contributions of satellite microwave instruments to measurements of rainfall are those of ESMR-5 and ESMR-6, to which we turn now.

III. ESMR-5

1. Characteristics

The instrument design and expected scientific returns of ESMR-5 are described by Wilheit (1972). ESMR-5 was tuned to a frequency of 19·35 GHz (1·55 cm), which is just outside the 22·235 GHz water vapour line and well outside any oxygen absorption lines (see Figs 8.2, 8.3 and 8.9). Gaseous absorption could be expected to be small, and extinction of radiation in water clouds could be expected to obey the Rayleigh formulas: that is, extinction would result mainly from absorption (see Fig. 8.5b). Clouds with rain, however, would scatter a significant fraction of upwelling radiation, mainly in the rearward hemisphere. Every 4 seconds the antenna of ESMR-5 scanned across the subpoint track of the satellite, from 50° to the left of nadir to 50° to the right, in 78 steps. Each scan step yielded a "footprint" (the area of the Earth seen by the antenna during each observation), which varied in size from 25×25 km at nadir to 40×160 km at the left and right limits of the scan. Footprints overlapped along scans, along the subpoint track and laterally from one orbit to the next; thus the whole Earth was observed once every 12 h.

2. Rain mapping—the 1972 Wilheit model

It was expected of ESMR-5 (Wilheit, 1972) that over land the contrast between rain-clouds and background would be low and dominated by variations in surface emissivity. Over the oceans, because emissivity is small at 19 GHz (0·4) and an inverse function of thermodynamic temperature, it was expected that contrast would be much higher. Wilheit (1972) suggested a linear relation, $T_b = 125$ K $+ 6\cdot8q$ (g cm^{-2}) $+ 300\bar{M}$ (g cm^{-2}), between brightness temperature at 19·35 GHz and vertical column measurements of water vapour (q) and liquid water content of clouds not raining ($\bar{M}$), and concluded the vapour and cloud water terms are comparable. However, calculations and aircraft tests of the ESMR instrument had also established a dependence of brightness temperature at 19 GHz on sea surface roughness and foam (Paris, 1969; Nordberg et al., 1971). Thus, as Nimbus-5 was being readied for launch, Wilheit (1972) concluded, "the 19·35 GHz brightness temperature cannot be unambiguously interpreted in terms of any one atmospheric variable".

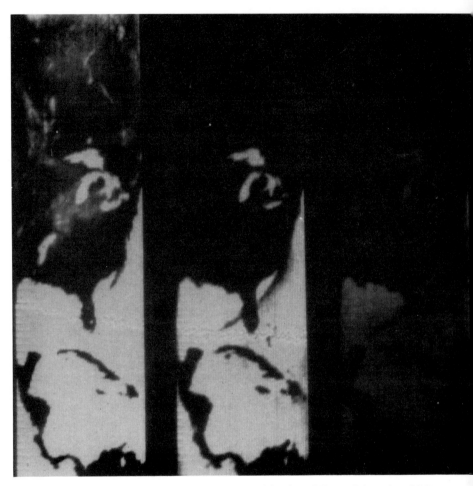

FIG. 9.1 (a-d)   Cloud and rain systems over North and Central America, 22 January 1973. From Allison, Rodgers, Wilheit and Wexler, 1974. (a) ESMR-5 photofacsimile image, orbit 569. Grey scale range of T_b values (left) 254–290 K, (centre) 194–256 K, (right) 138–210 K.

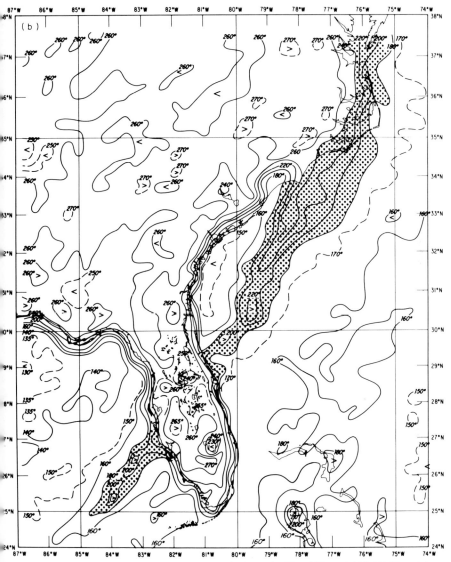

FIG. 9.1 (b)   ESMR-5 grid print map analysis. T_b in K, 1639 to 1643 GMT, orbit 569. Dotted area shows $T_b > 180$ K over water.

F

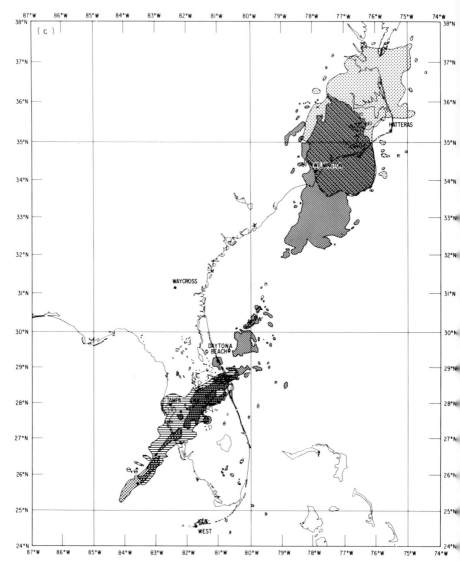

FIG. 9.1 (c)   Hand-rectified National Weather Service WSR-57 weather radar images from Key West, Tampa, and Daytona Beach, Florida; Waycross, Georgia; and Wilmington and Hatteras, North Carolina; from 1637 to 1643 GMT. Darker grey hatching indicates overlapping radar echoes.

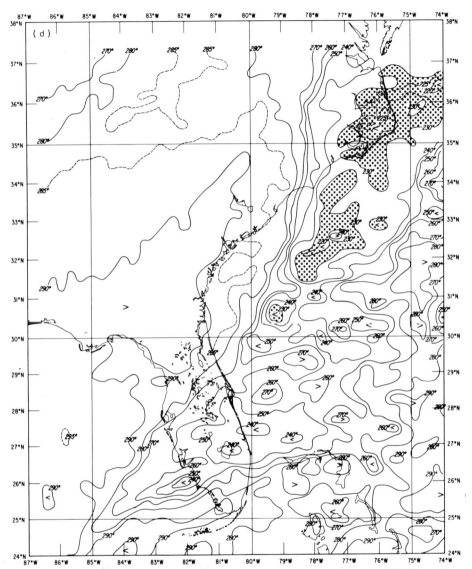

FIG. 9.1 (d)   Nimbus-5 Temperature Humidity Infrared Radiometer (THIR) grid print map analysis. T_b in K, 1639 to 1643 GMT, orbit 569, 11 μm. Dotted area shows $T_b < 230$ k.

Simple comparisons of ESMR-5 imagery with imagery from visible and infrared wavelength radiometers and maps of radar rainfall and surface synoptic conditions established, over water, an unmistakable association of areas of relatively warm brightness temperature with areas of rain (Theon, 1973; Wilheit et al., 1973, 1976). Synoptic cloud features also were associated with area of high T_b, but covered a much larger area. (Falling snow seemed to be transparent at this frequency (Wilheit et al., 1976).) An example, from Allison et al. (1974a), is shown in Fig. 9.1. In this swath from the Great Lakes south to the Caribbean Sea, land surfaces stand out starkly against the radiometrically cold oceans and large lakes (Fig. 9.1a). Areas of colder brightness temperatures not associated with permanent bodies of water were present over the eastern United States and central Canada. The most distinct of these (in the lower Mississippi Valley) was found to be a consequence of reduced surface emissivity due to wet soils. Over the oceans one feature is present: a narrow arc of warmer brightness temperature from Cape Hatteras on the East Coast into the eastern Gulf of Mexico. This feature was 60 to 80 K warmer than ocean temperatures around (Fig. 9.1b); it coincided with a narrow band of radar echoes (Fig. 9.1c), and in fact represented a rain band with a cold front. The ESMR-5 warm band was associated with colder cloud tops (Fig. 9.1d), but was much less extensive. It was interrupted over Florida, even though radars there measured high rainfall rates (the break in the radar line between Florida and Hatteras coincided with a gap in spacial coverage).

Allison et al. (1974a) correlated ESMR-5 T_b with rainrates measured by radar over the ocean near Miami, Florida, and compared their result with a model calculation. Rainfall increased with increasing brightness temperature, but much less so than predicted by theory: the difference reached 50 K at 15 mm h^{-1}. This difference, they concluded, is due partly to partial filling of the ESMR-5 field-of-view (to which we will return later) and partly to inadequacies in the treatment of scattering in the model.

3. Tropical storms—the Gaut and Reifenstein Model

Tropical storms of the Pacific Ocean were examined by Allison et al. (1974b). Their study introduced a radiation transfer model (developed by N. E. Gaut and E. C. Reifenstein) relating satellite brightness temperature at 19·35 GHz to rainrate. For the larger raindrops scatter in this model is treated as an equivalent absorption, thus "there is no scattering source function and no order of multiple scattering". Because of extinction by raindrops below the freezing level the model curve of brightness temperature as a function of rainrate was markedly asymptotic in R, approaching a limit of 280 K at $R \sim 15$ mm h^{-1} (see the "no scattering" curve in Fig. 9.3).

Allison *et al.* therefore distinguished no more than four categories of rain-rate: nil, 0–2, 2–7, and >7 mm h^{-1}.

They examined rain in five storms, of which one, Ava, is shown here. Ava was an intense east Pacific hurricane from 1973. ESMR-5 pictures in three enhancements are compared with NOAA-2 visible and infrared pictures in Figs 9.2a,b. Rainfall, calculated from the Gaut and Reifenstein function, is mapped in Fig. 9.2c. In Ava heavy rain was concentrated in a ring close to the eye. An open cyclonic spiral band of lighter rainfall emerged from the north-east side of the ring and trailed off toward the south and south-west. Despite relatively warm infrared cloud tops away from the storm, moderate rain was indicated in this band. A marked boundary encircling the storm some distance to the north (Fig. 9.2a, left) was interpreted as the northern limit of low level moisture trapped in the storm's circulation.

ESMR-5 captured an evolutionary view of storm rain structure in Nora, a 1973 storm of the west Pacific, as it grew from a weak low to a near-record typhoon. During the deepening period rain bands became increasingly concentric. At maturity the rain pattern of Nora was more compact and more tightly coiled than that of Ava. At all stages banding was much more prominent in microwave than in corresponding infrared images.

From the five cases studied, Allison *et al.* (1974b) concluded that rain areas of tropical storms can be delineated from ESMR-5 observations and crude rainrates can be inferred; in addition, from ESMR-5 it is apparently also possible to map areas of abundant low level water vapour. However, with a single measurement there is no way to distinguish water vapour from light rain, and the absence of a source function for scattered microwave radiation implies rain estimates which are too small, especially at larger rates.

4. The 1977 model of Wilheit, Chang, Rao, Rodgers and Theon

Curves of 19·35 GHz brightness temperatures and rainrate were published by Wilheit *et al.* in 1977. Their rain-cloud was modelled as a stack of many extensive, optically thin layers bounded on top by a variable (1 to 5 km) freezing level.

Raindrops were represented by a Marshall-Palmer distribution, modified to include cloud droplets in the 0·5 km thick layer below the freezing level. Rainrates were calculated from the Marshall-Palmer distribution and Wald-teufel's (1973) expression for drop fall-velocity at the surface. To make calculated rainrates independent of altitude in radiative transfer calculations, the drop-size distribution was scaled downward by the factor $(\rho/\rho_0)^{0.4}$, where ρ is density of air and the subscript indicates sea level. The rain-cloud model resembles that of Kessler and Atlas (1959), except that Kessler and

Atlas suggest increased water content above the freezing level. But this water, it is observed, is "primarily in the form of ice, which has no consequential effect on the brightness temperature" (Wilheit *et al.*, 1977).

Temperature within and above and below the cloud has a constant lapse of $6.5°C\,km^{-1}$. Humidity decreases below the freezing level from 100% to 80% at the surface; above it follows the 1962 US Standard Atmosphere. The ocean was assumed to be calm. Reflectance was approximated by Fresnel formulas, with a Lambertian distribution of reflected radiation.

Curves of brightness temperature as a function of rainrate were calculated iteratively, first for absorption, then for absorption plus scattering. Both upwelling (anabatic) and downwelling (katabatic) cases were treated; the upwelling T_b curves for nadir viewing are shown for five freezing levels in Fig. 9.3. Each curve has a characteristic "S" shape, which is a consequence of the transparency of clouds to 19 GHz microwave radiation at very small rates ($< \sim 1$ mm h^{-1}) and the opacity of clouds (resulting from a strong backscatter) at high rates (>20–50 mm h^{-1}). Neglecting scattering gives significantly warmer brightness temperature at higher rainrates. It is apparent from the no scattering curve in Fig. 9.3 that at 19 GHz absorption contributes very much the largest part to changes of brightness temperature with rain. Therefore the curve of brightness temperature should be insensitive to the distribution of drop sizes, except perhaps at high rainfall rates. This was confirmed by calculations of brightness temperature for the Sekhon and Srivastava (1971) distribution (Fig. 9.3), which at high rainrates has fewer large drops than the Marshall-Palmer distribution.

(a)

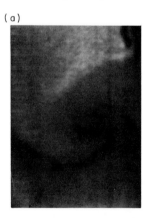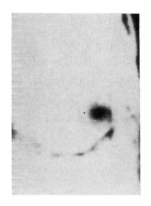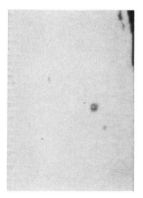

FIG. 9.2   Hurricane Ava; 7 June 1973, eastern Pacific. From Allison *et al.*, 1974b. (a) ESMR-5 photofacsimile pictures, data orbit 2936. Grey scale range of T_b values (left) 138 to 210 K, (centre) 194 to 266 K, (right) 254 to 290 K. (b) NOAA-2 Very High Resolution Radiometer (VHRR) visible and infrared images and Nimbus-5 THIR image. (c) ESMR-5 rainrates for Hurricane Ava, 1903 to 1913 GMT.

(b)

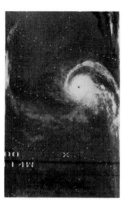

NOAA - 2VHRR IMAGES OF HURRICANE "AVA"
REV 2942, JUNE 7, 1973

NIMBUS 5 THIR (11μm)
D/O 2396 D 7 JUNE 1973

VIS   IR

(c)

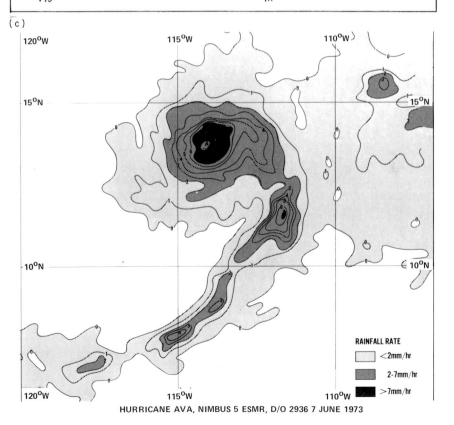

RAINFALL RATE

☐ <2mm/hr

▨ 2-7mm/hr

■ >7mm/hr

HURRICANE AVA, NIMBUS 5 ESMR, D/O 2936 7 JUNE 1973

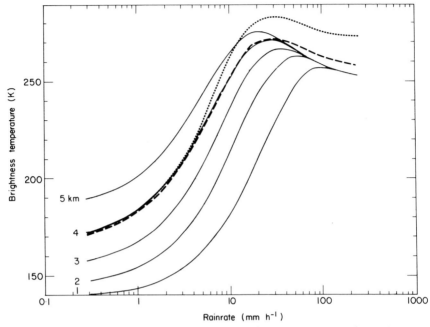

FIG. 9.3   Calculated brightness temperature at 19·35 GHz (1·55 cm) as a function of rainrate for freezing levels of 1–5 km. The dotted line shows T_b for a 4 km freezing level with scattering excluded from the model; the dashed line shows T_b for a Sekhorn-Srivastava particle size distribution. After Wilheit et al., 1977.

The Wilheit model was tested with satellite and ground radiometers. The satellite test matched four ESMR-5 passes against maps of over-ocean rainrate calculated from reflectivity measured by the Miami, Florida, National Weather Service WSR-57 (10 cm) radar. Radar rainrate was averaged in 2:1 weighting for the inner 1·5 dB contour circle and outer 1·5 to 3·0 dB contour ring of the ESMR-5 antenna pattern (Fig. 9.4). The ground test, conducted at Goddard Space Flight Center, Greenbelt, Maryland, matched gauge rainfall against brightness temperature measured by two radiometers (tuned to 19·35 and 37.0 GHz), which were pointed 45° from the zenith. For all tests the freezing level was close to 4 km.

Radar rainrates tended to be higher than rates predicted by the model (Fig. 9.5), but generally within a factor of two of model rates. Most of these departures were attributed to local saturation in the brightness temperature–rainrate relation at high rainrates: brightness temperature cannot increase at rates above ~20 mm h^{-1}. Gauge rainrates and corresponding brightness temperatures averaged by rainrate interval are plotted as equivalent nadir 19·35 GHz T_b values over water in Fig. 9.5, in original

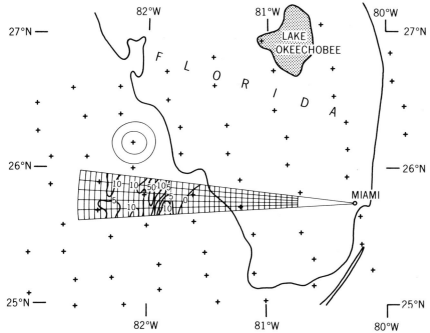

FIG. 9.4   A portion of the WSR-57 radar data for 25 June 1973. The cross hatching shows the range and azimuth resolution of the radar. Isopleths of 0, 5, 10 and 50 mm h^{-1} rainrate are indicated. The crosses indicate the locations of the beam centres of the corresponding ESMR-5 data. The 1·5 dB (inner oval) and 3 dB (outer oval) contours are shown for a typical ESMR beam position. From Wilheit et al., 1977.

form for 19·35 GHz in Fig. 9.6a and for 37·0 GHz in Fig. 9.6b. Observations and theory are in close agreement at low rainrates. At high rates model brightness temperatures are warmer than observed temperatures. The difference is smaller at 37 GHz, but appears at lower rates.

The most important shortcomings in this model, according to Wilheit and his colleagues, are unrepresentativeness and errors in the measurement of rain by gauge and (particularly) radar, discrepancies in the registration of radar rainfall and ESMR-5 brightness temperature, approximations in averaging brightness temperature over the ESMR field of view, non-linearities in the relation of brightness temperature and rain, and effects of wind speed and viewing angle. These problems notwithstanding, the authors concluded that from observations of ESMR-5, rain at rates up to 20 mm h^{-1} may be inferred to an accuracy of factor 2. Recently this estimate has been called into question. As has been reported in Chapter 8, Lovejoy and Austin (1980) find effective rain layer heights for clouds in the tropics to be much

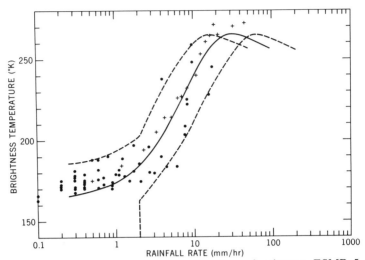

FIG. 9.5 Brightness temperature as a function of rainrate: ESMR-5 versus WSR-57 radar (dots) and inferred from ground-based measurements of brightness temperature and direct measurements of rainrate (crosses). The solid line is the calculated brightness temperature for a 4 km freezing level. The dashed lines represent departure of 1 mm h^{-1} or a factor of 2 in rainrate (whichever is greater) from the calculated curve. From Wilheit *et al.*, 1977.

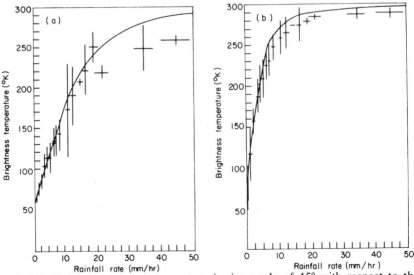

FIG. 9.6 Brightness temperature at a viewing angle of 45° with respect to the zenith as a function of directly measured rainrate. At each point, the height and width of the cross represent two standard deviations in the corresponding dimensions. The solid line is the theoretically calculated curve. From Wilheit *et al.*, 1977. (a) 19·35 GHz (1·55 cm). (b) 37·0 GHz (0·81 cm).

smaller than freezing level heights, and highly variable. Additionally, column-integrated cloud liquid water content ($\tilde{M}$) in Wilheit's model is an order of magnitude smaller than $\tilde{M}$ as measured and as modelled elsewhere. Lovejoy and Austin show that as a result of these conditions the Wilheit model (and others like it) may significantly underestimate rain at higher rates. Applications of the Wilheit model are discussed in Chapter 10.

5. Non-beam filling and other problems

With a footprint of 600 to 6000 km^2 and a variable gain across the footprint, it might be expected that the ESMR-5 measurements of rainrate would not only be very significantly smoothed, but also depart substantially from actual rainrate averaged over the footprint. Lovejoy and Austin (1980) calculated the fraction of footprint area covered by echo in plan scans of the *Quadra* radar at a station in the eastern Atlantic Ocean. (Following Stepanenko (1969b), they refer to this fraction as the duty factor.) For areas 44 ×20 km in size, the duty factor, when rain was present, averaged 0·29: indeed, the probability of smoothing, and potential for error, is substantial.

Smoothing is not the only problem resulting from a large field of view. Early in the analysis of ESMR-5 data, for example, in the papers of Allison *et al.* (1974a,b), it was realised that significant errors might result when rain is variable across the ESMR field of view owing to the diminishing contribution of rainrate to increases of field-of-view brightness temperature at intermediate and high rates. At very high rates, as observed by Wilheit *et al.* (1977), increases in rainrate contribute nothing (or, for relatively warm sea surface temperature, even contribute negatively) to field-of-view brightness temperature. The size of this error is proportional to the gradient of rain within the field of view, and its effect is always the same: a reduction of ESMR-5 rainrates below actual rates.

Let us first consider the smoothing aspect of the field-of-view problem. The beam pattern of the microwave antenna introduces an uncertainty in the satellite rainrate which is proportional to the inhomogeneity of rain across the radiometer footprint. Lovejoy and Austin (1980) assessed this uncertainty by simulating the brightness temperatures which would be measured by a satellite radiometer having an elliptical field of view 44 km in length by 20 km in width, a nadir pointing angle, and a wavelength of 1·55 cm. Using the relation of Wilheit *et al.* (1977) for a 4 km freezing level, rainrate in 4 × 4 km resolution *Quadra* radar scans was transformed to brightness temperature. These brightness temperatures were integrated over the antenna beam pattern (down to 6 dB) for 30 000 adjacent areas, each sized to the radiometer field of view, then were converted to a field-of-view average rainrate by means of the same relation. Comparing the simulated satellite rainrates with the actual (radar) rainrates, Lovejoy and Austin found (after removing a 31 % bias) rms errors of 180% for the field-of-view area, 26% for the 200 km radius radar scan area. Errors were largest for small duty factors, but exceeded 50% even for a 0·95 field-of-view duty factor.

The error due to non-linearities in the relation of rainfall to brightness temperature has been treated by several groups.

Smith and Kidder (1978) considered differences of rainrate on the scale of individual ESMR-5 fields of view by mapping ESMR-5 brightness temperatures measured during the Global Atmospheric Research Programme Atlantic Tropical Experiment (GATE) into the cartesian system of the GATE radars. The resolution of the remapped ESMR-5 images was 28 km; that of the radar images, 4 km. Brightness temperature was converted to rainrate by the algorithm of Wilheit *et al.* (1977), as modified by Kidder (1976). Using 0·13 mm h^{-1} as a threshold for rain, fractional rain coverage was calculated for fields of view from seven orbits of nadir scan angle <30°. For each of 542 fields of view with 0·2 or greater rain coverage, average radar rainrate was calculated, then subtracted from the ESMR-5 rainrate. The results, averaged by intervals of 0·1, are plotted in Fig. 9.7. ESMR-5 rainrates are systematically smaller than radar rates, and the difference monotonically increases as the field of view fills with rain. Smith and Kidder attribute the trend to the unfilled field of view effect. Both effects are significant, for the gauge-measured, GATE-average rainrate for the area represented in this analysis was 0·47 mm h^{-1} (Hudlow and Patterson, 1979). ,·

That the difference of ESMR-5 and GATE radar rainfall is largely a problem with ESMR is a conclusion of another study, made by Austin and Geotis (1978). The first part of their study involved instantaneous 1° × 1°

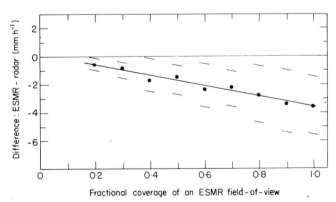

FIG. 9.7   Mean difference between ESMR-5 and radar rainrates for the GATE area between 30 August and 19 September 1974. From Smith and Kidder, 1978.

square resolution ESMR-5 rainrates prepared for the GATE archive (*GATE Data Catalogue*, 1975) and 1 h average radar rainrates for the GATE B-scale hexagon, which occupied a circle 3° in diameter. The ESMR rainrates were averaged over a 2° × 2° block centred on the hexagon. Limiting ESMR cases to nadir scan angles of 30° or less, 31 rainrate pairs were found. For the sample as a whole, the ESMR-5 rainrate was 0·47 mm h^{-1}, radar rainrate, 0·76 mm h^{-1}, but with large variance.

In the second part of their study Austin and Geotis compared 2- to 7-day average ESMR-5 rainrates from the 1976 Atlas of Rao *et al.* (see Chapter 10) of areal resolution 4° latitude by 5° longitude, with the average of rain measured by gauges on three to five GATE ships within each of two of the 4° × 5° grid boxes. For nine rain pairs in each box ESMR-5 average rainrate was 25 to 40% less than gauge rainrate, again with large variability.

Austin and Geotis assessed the non-beam filling contribution to errors in ESMR rainrate by simulating an ESMR-5 measurement of brightness temperature across fields of rainfall known (to 1 km resolution) from radar. The simulated ESMR measurement covered blocks 28 km on a side. There were 16 blocks for each radar observation time, and 13 times, all from the same day, thus a total of 208 simulated ESMR-5 observations. For each block they calculated an average radar rainrate and the fraction of area associated with rain at each of 13 intensity intervals, the means of which ranged, in approximate geometric progression, from 0 to 200 mm h^{-1}. A brightness temperature was determined for each interval rainrate, using the Wilheit model for a freezing level of 4 km (see Fig. 9.3). The average of the interval brightness temperatures, each weighted according to fraction of area covered, gave a simulated brightness temperature for the block, and this, from the Wilheit model, yielded a simulated ESMR-5 rainrate. These rates are plotted against actual (radar) rainrates in Fig. 9.8. Above a block-average rainrate of 1 mm h^{-1}, the simulated ESMR-5 rainrate is much less than the actual rate. (The overestimate at rainrates <1 mm h^{-1} is not considered significant because it is common to assign a threshold of brightness temperature below which ESMR-5 rainrates are set to zero.) As would be expected for convective rain, the size of the negative error increases with increasing average rainrate. For rates >3 mm h^{-1} it averages 100%. For the sample as a whole, the mean simulated ESMR rainrate is 64% of the actual rate; 43% if field-of-view rates <1 mm h^{-1} are set to zero. This compares well with the bias found in the much larger sample studied by Lovejoy and Austin: a difference of 31% between simulated and actual rainrates, with simulated rates tending to be smaller.

Austin and Geotis attribute this error to small cells of intense rain. As a measure of the effect of saturation, for each block they calculate the percentage of rain contributed at rates >20 mm h^{-1}. Although highly variable, the

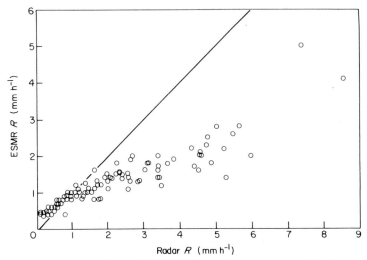

FIG. 9.8   Computed values of ESMR-indicated rainrates as a function of radar mean rainrates, in 28 km × 28 km areas. From Austin and Geotis, 1978.

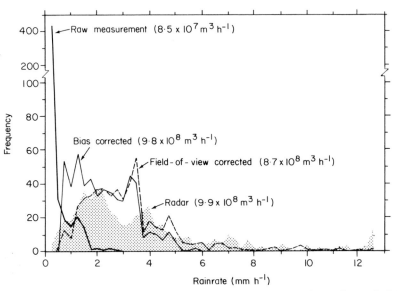

FIG. 9.9   Rainrate distributions based on 542 field of views from the period 30 August to 19 September of GATE. Lines represent ESMR-5 rates; stipling, radar rates. Numbers in parenthesis are the integrated volumetric rainrates. After Smith and Kidder, 1978.

fraction contributed by saturation rates increases with increasing block-average rate up to about $2 \cdot 5$ mm h^{-1}. These results are consistent with and help to clarify those of Smith and Kidder. They lead Austin and Geotis to conclude that the ESMR-5 estimates of GATE rainfall are about 40% low, and non-beam filling is a very significant source of error in rain estimates from ESMR-5.

Although there is a tendency when confronted with novel data to forget that other information exists, the use of microwave, visible and infrared imagery together was advocated by Allison *et al.* (1974b). Kidder and Vonder Haar (1976) and Smith and Kidder (1978) have addressed the non-beam filling problem from this point of view, namely, can ESMR field of view rain coverage be estimated from the higher resolution observations of geostationary satellites? For the simple dichotomous case of zero rainrate and constant non-zero rainrate, Smith and Kidder derive an expression for brightness temperature $T_{b,R}$ in the raining part of the ESMR field of view,

$$T_{b,R} = [T_b - (1 - f_R) \cdot T_{b,N}]/f_R. \tag{9.1}$$

f_R is the fraction of the field of view covered by rain, $T_{b,N}$ is brightness temperature in the absence of rain, and T_b is, as usual, the satellite observed brightness temperature. In applicatons of this equation, $T_{b,N}$ is inferred from a frequency distribution of T_b for an area large compared to the spacing and dimension of rain systems and f_R from individual or combined visible and infrared images; then field-of-view rainrate is estimated from

$$R(\text{fov}) = R(T_{b,R}) \cdot f_R. \tag{9.2}$$

The potential value of the approach was tested with the 542 transformed 28×28 km ESMR-5 fields of view of $f_R \geq 0 \cdot 2$ described previously (see p. 160). These and the corresponding radar rates are plotted as frequency distributions of average rainrate in Fig. 9.9. ESMR-5 rates are weighted heavily toward light rain. Two corrected ESMR-5 distributions are also shown in Fig. 9.9. The first, labelled "bias", attempts to account for the offset between ESMR-5 and radar rates. The ESMR rates for each interval of fractional rain coverage are increased by the difference between ESMR and radar rainrates for that interval. This greatly improves the match of ESMR and radar distributions. The second corrected distribution, labelled "fov", attempts to correct for non-beam filling as well as offset. ESMR field-of-view rainrates are calculated from Eqs 9.1 and 9.2, using *radar* to estimate fractional rain area, then are subjected to the bias correction. This results in a further improvement in the match of ESMR and radar distributions, though not as dramatic as for the bias correction.

Lovejoy and Austin (1980) also assessed the improvement that might be made by determining the duty factor for each field-of-view. Their results are much less optimistic: significant improvement only for low duty factors, with practically no difference in the range of duty factor (0·4 to 0·6) that contributes most of the rain. Surprisingly, the error in a rain estimate based solely on rain area times the sample mean rainrate is not very much larger than that of the simulated microwave rain estimate with known duty factor.

Contributing to the errors in ESMR-5 rainrates for GATE is the small number of samples over periods of major rain events—never more than two per day, and often not even one per day. Lovejoy and Austin (1980) used 18 days of hourly *Quadra* radar data to assess the effect of infrequent sampling. For this record and for both field-of-view and scan areas, extending instantaneous rates through 1 day yielded a larger error than assuming a constant rainrate. Sampling twice per day brought the two scan area error curves together, partly by eliminating a contribution from a strong diurnal cycle in *Quadra* rainfall; climatology still surpassed instantaneous views on the field of view scale. Thus Lovejoy and Austin conclude, "estimates are needed far more frequently than every 12 h in order for the satellite to give significantly better results than climatology".

Austin and Geotis noted the occurrence of greatly enhanced absorption and scattering in radar "bright bands" of melting snow under the anvils of mature and decaying convective systems, and therefore recommended adding a bright band to the microwave rainfall model. They also recommended that comparisons of ESMR-5 and local measurements of rainrate be extended, climatologies of tropical rainstorms be expanded, and that other microwave-lengths be considered. This leads us to a discussion of ESMR-6.

IV. ESMR-6

1. Characteristics

According to Wilheit (1975), the three important differences between ESMR-6 and ESMR-5 are wavelength, polarization and scanning geometry. The receiver of ESMR-6 is tuned to a higher frequency, 37·0 GHz (0·81 cm). Increasing frequency from 19·35 to 37·0 GHz approximately doubles sensitivity to water vapour and oxygen, but triples sensitivity to liquid water droplets (see Figs 8.2, 8.3 and 8.6). On ESMR-6 both polarization components are measured. It was expected that polarization difference would be useful in making estimates of water vapour and liquid water content over the oceans (Wilheit, 1975), but so far its main use has been in rain estimation over land. Lastly, ESMR-6 has a conical scan which maintains a constant incident angle (50° ± 1°) to the Earth's surface. Scan is from 35° to the right of the spacecraft track to 35° to the left, and sweeps are repeated every 5·33 s. The half-power (3 dB) field of view is ~35 × 25 km, with the long axis aligned with the spacecraft track. Cross track coverage is 1272 km, about half the ground distance between consecutive crossings of the equator. Only above 60° latitude is coverage continuous.

2. Evolution of models

A practical consequence of measuring microwave radiation at a higher frequency had been anticipated by Paris (1971). From Monte Carlo model calculations of brightness temperature Paris predicted that enhanced scattering at 37 GHz would result in significant contrast between rain and rain-free areas over land. This was confirmed theoretically and observationally by Savage and Weinman (1975), who used Savage's rain-cloud model (described in Chapter 8) to calculate 37·0 GHz and 19·35 GHz brightness temperature for a 4·57 km thick rain-cloud over land and sea (Fig. 9.10). At 19·35 GHz the increase of T_b over water is very much greater than the

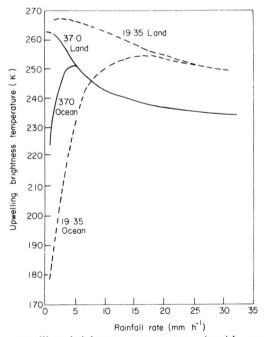

FIG. 9.10   The upwelling brightness temperature (zenith angle = 48·6°) of a 4·47 km rain-cloud over land (emissivity = 0·90) and ocean surfaces (emissivity 0·42 at 19·35 GHz, 0·47 at 37·0 GHz). The 19·35 GHz values are the dashed lines; 37·0 GHz values are solid lines. From Savage and Weinman, 1975.

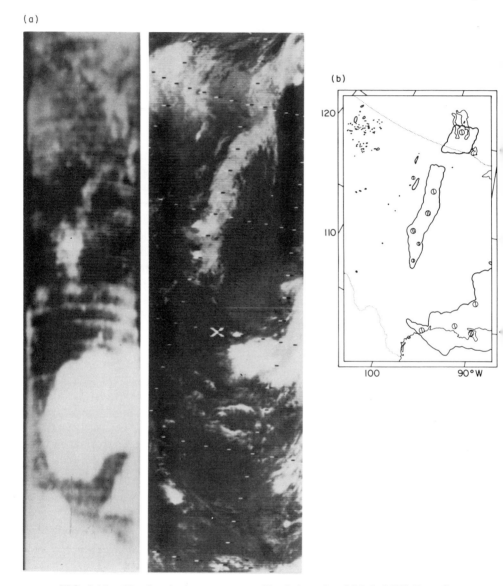

FIG. 9.11   Cloud and rain systems over North America, 31 July 1975. From Savage and Weinman, 1975. (a) Nimbus 6 views. Left, ESMR-6 radiation, interrogation orbit 663, approximately 1735 GMT. Central America and Yucatan Peninsula at lower centre; North American central plains top centre. Lighter shades represent colder radiative temperatures. Right, Nimbus 6 THIR 11·5 μm imagery cropped to coincide with ESMR-6 data on left. Computer gridded X near centre is 30°N, 96°W. Lighter shades represent lower radiative temperatures. (b) Part of National Weather Service Summary Chart for 1735 GMT covering the area shown in (a).

decrease over land (though, owing to non-beam filling, ESMR-5 observations of both changes were smaller than expected). At 37·0 GHz the increase over water reverses near 4 mm h^{-1}; however, the curve for land is single valued and its range is larger by half than the range of the curve for 19·35 GHz. A thicker rain-cloud would enhance the effect of scattering and lower T_b, decreasing the contrast over water and increasing the contrast over land. Contrast would be diminished in partly filled field of views; nevertheless Savage and Weinman anticipated that organized frontal and tropical rain systems of light to moderate rain would fill the ESMR-5 field of view and give a "true" brightness temperature.

A pass of ESMR-6 across the centre of North and Central America demonstrates the validity of these deductions. The cold feature over the central US in the ESMR image of Fig. 9.11a corresponds with a frontal squall line in a matching infrared image and with a band of showers and thunder showers in the radar summary map nearest the time of the ESMR pass (Fig. 9.11b). This pass also demonstrates some of the problems of using ESMR-6 to infer rainrates: very little contrast over the northern Gulf of Mexico, where infrared and radar indicate significant rain; and patterns over land, for example, in southern Canada, which have little apparent connection to rainfall.

We saw early in this section, in the case from ESMR-5 presented by Allison *et al.* (1974a), how water could lower brightness temperature by wetting the surface and decreasing emissivity. In terms of brightness temperature alone there is no distinction between reduced temperature resulting from surface flooding and reduced temperature resulting from heavy rain. However, as Savage (1976) has observed, the ambiguity may be resolved through polarization. Whereas microwave radiation emitted and reflected from a flooded land surface is known to be uniformly polarized, because hydrometeors have no preferred orientation the radiation emitted and scattered from rainclouds is expected to be randomly polarized.

The degree to which radiation at 27 GHz from rainclouds is polarized was calculated by Weinman and Guetter (1977). The calculation assumes a cloud (of 4·57 km depth) and boundary conditions like those of Savage's model (p. 138). Scattering is approximated by a Rayleigh phase matrix, i.e. $\lambda \gg r$, and the equations of transfer of polarized upwelling radiation are solved for a range of zenith/nadir angles. The results are summarized in Fig. 9.12. Polarization is significant only for calm water, and there only at rainrates less than ~ 4 mm h^{-1}. Although rough water and land show negligible polarization, the curve for brightness temperature over rough water behaves like the curve of T_b over calm water (see Fig. 9.10). Therefore, as illustrated in Fig. 9.13, the three cases of land, rough water and calm water are represented by combinations of scalar brightness temperature and polar-

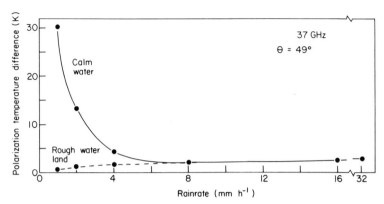

FIG. 9.12   Vertical polarization brightness temperature minus horizontal polarization brightness temperature for a 4·57 km rain-cloud over land, over rough water,
and over calm water. From tables of Weinman and Guetter, 1977.

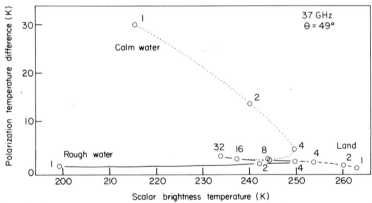

FIG. 9.13   Scalar brightness temperature versus polarization temperature differ-
ence $(T_V - T_H)$ for a 4·57 km rain-cloud over land, rough water and calm water.
Numbered circles along each curve indicate rainrates. From tables of Weinman and
Guetter, 1977.

ization temperature difference which at low rainrates are unique, but which
diminish with increasing rainrate and become indistinguishable at rates of
about 8 mm h^{-1}.

A simple application of discrimination by polarization is described in a
1979 paper by Rodgers *et al.* Their modest goal was to test the value of
polarization as a discriminator between raining clouds and wet ground.
Working with eight ESMR-6 passes over the south-east US, Rodgers and his
collaborators used standard radar and surface synoptic observations to
classify areas observed by ESMR-6 in moisture categories of dry (no rain for

at least 24 h), wet (immediately upstream of rain areas), and raining. ESMR areas also were classified in temperature categories of warm (surface thermal temperature $T_s > 15°C$) and cool ($5° \leq T_s \leq 15°C$). Vertically and horizontally polarized brightness temperature is plotted for each moisture category in Fig. 9.14, first for warm surfaces (a), then for cool surfaces (b). In both parts the bivariate means and standard deviations of moisture categories are indicated; in (a), Fisher linear discriminant lines separating the three populations are also drawn. Where surfaces are warm, the populations are distinct, as would be expected from Fig. 9.13. It is apparent, and confirmed by statistical tests, that their distributions are Gaussian. The distinctions diminish to insignificance for a cool surface (b). This is mainly a consequence of much lower brightness temperatures in dry areas.

If populations are distinct, at least for warm surfaces, one might wish to know how best to make the discrimination. The answer given by Rodgers *et al.* is through a supervised algorithm based on Bayes' classifier— "supervised" because the algorithm is trained on data. This is like the classification used by Lovejoy and Austin to map rain from visible and infrared images (Section I, Chapter 6), except the variables here are three in number. Three quadratic likelihood functions in T_V and T_H, one for each category, are derived from the training samples of wet, dry, and rain, irrespective of surface temperature. A temperature pair (T_V, T_H) is assigned to a category by the maximum of the likelihood functions. An expression also is given for the confidence of the assignment. The Bayesian algorithm classified the training sample data with an accuracy of 79%, very close to the 77% calculated from theory. It is most reliable for dry areas (92 and 82% actual and theoretical accuracies), least reliable for wet areas (58 and 72% accuracies). The algorithm also was tested on independent data from a ninth pass of ESMR-6 over the south-eastern US by generating maps, at 70 and 80% confidence levels, of ESMR-6 rain and wet ground. The map at 80% confidence is shown in Fig. 9.15a; the corresponding map of rain from radar and synoptic observations in Fig. 9.15b. Although there is general agreement in the main rain area, this area is short on the east at 80% confidence, and short on the south-west at 70% confidence (not shown). A second test when conditions were dry over the south-east US gave a dry classification. However, a third test, of a nighttime pass under clear conditions, gave a rain classification, This was attributed to dew on vegetation, which apparently lowered the surface emissivity. Rodgers and his collaborators concluded that warm synoptic-size rain systems could be mapped over land, if vegetation was not covered with dew. The distinction was weakest between areas of rain and wet land.

Weinman and Guetter (1977) have pursued this question through the model calculations of polarization temperatures discussed earlier. They find

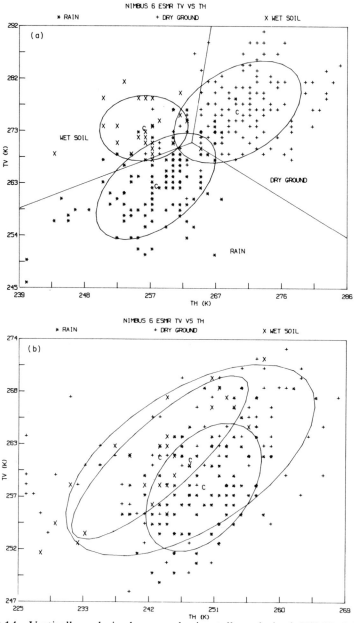

FIG. 9.14 Vertically polarized versus horizontally polarized ESMR-6 T_b for categories of rain, wet soil and dry ground. From Rodgers *et al.*, 1979. (a) Surface thermal temperatures greater than 15°C. (b) Surface thermal temperatures between 5° and 15°C.

that at 37 GHz calculated land (unpolarized) and calm water (polarized) temperatures are approximately related (Fig. 9.13) as

$$T_b^{\text{land}} = T_V^{\text{calm water}} - 1 \cdot 2(T_H^{\text{calm water}} - T_V^{\text{calm water}} + 2 \cdot 5°)$$

Applying a relationship of this form to a single ESMR-6 pass, that shown in Fig. 9.11a, and adjusting coefficients until contrast between land and water was eliminated, Weinman and Guetter determined an empirical transformation,

$$\tilde{T}_b = T_V - 1 \cdot 5(T_H - T_V + 2 \cdot 5°), \tag{9.3}$$

in which the polarization coefficient is enhanced because the open water viewed by ESMR-6 is not quite calm. Equation 9.3 restores the data to an equivalent land background viewing situation, and in doing so removes the double value ambiguity in the brightness temperature curve for rain clouds over water. Its effectiveness is demonstrated in two maps, one of the polarization difference in ESMR-6 temperatures (Fig. 9.16a), and one of transformed brightness temperatures $\tilde{T}_b$ (Fig. 9.16b). Major features in Fig. 9.16a correspond with the Gulf of Mexico and lakes such as Winnipeg, Winnipegosis, Manitoba, Cedar, and others. Part of the Gulf along the Louisana-Mississippi coast is obscured by rain. In Fig. 9.16b two major cold areas over the northern Gulf and the central Great Plains match major rain systems recorded by radars (see Fig. 9.11b). The lack of correspondence of some smaller cold areas with echoes was thought to reflect gaps in radar coverage.

The problem of variable surface emissivity was addressed by Hall *et al.* (1978) through congruous infrared observations as a supplement to ESMR-6. For this experiment the brightness temperature relation of Weinman and Guetter (1977, Eq. 9.3) was modified to

$$\tilde{T}_b = \begin{cases} T_V + 1 \cdot 5\,(T_V - T_H - 2 \cdot 5) & T_V > T_H + 2 \cdot 5 \\ T_V & T_H \le T_V \le T_H + 2 \cdot 5 \\ T_H & T_V < T_H \end{cases} \tag{9.4}$$

ESMR-6 radiances were reduced through Eq. 9.4 to an equivalent land background, then were subjected to a series of screenings. The first identified regions of probable rain ($R > 4$ mm h^{-1}) according to the condition, $\tilde{T}_b < \varepsilon T_S$, where ε is taken to be 0·9, and T_S is a linear function of latitude, specified from climatology. The second identified (for regions of probable rain) those $\tilde{T}_b$ associated with cloud top temperatures colder than freezing. This was accomplished with infrared data: the condition is $T_{ir} < 272$ K. The brightness temperatures so identified were then converted to one of five rain rate categories using the table of Weinman and Davies (1978; see p. 141) for a plane parallel rain cloud over land. The assignment

(a)

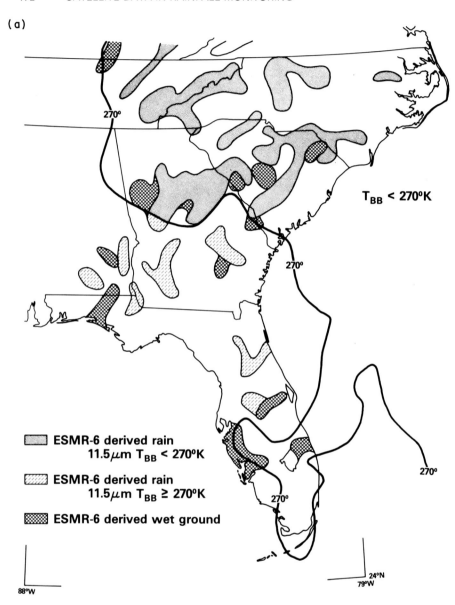

ESMR-6 derived rain
11.5μm T_{BB} < 270°K

ESMR-6 derived rain
11.5μm T_{BB} ≥ 270°K

ESMR-6 derived wet ground

FIG. 9.15 Rainfall patterns over the southeast United States, near 1630 GMT on 14 September 1976. From Rodgers *et al.*, 1979. (a) From ESMR-6 using the Bayesian classifier with a confidence level of 80%. (b) As delineated by the WSR-57 radar and hourly rainfall reporting stations.

(b)

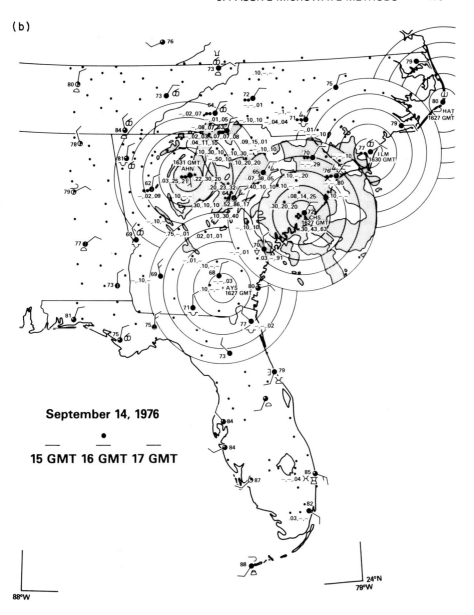

September 14, 1976

•

15 GMT 16 GMT 17 GMT

88°W

24°N
79°W

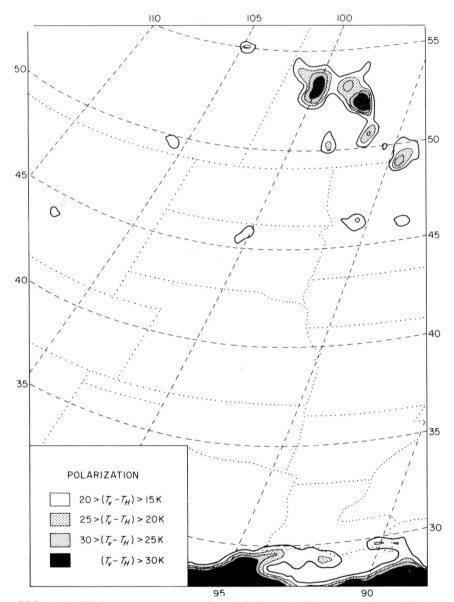

FIG. 9.16   Brightness temperatures from ESMR-6, Gulf Coast and Central Plains of North America, interrogation orbit 663, about 1735 GMT, 31 July 1975. From Weinman and Guetter, 1977. (a) Polarization temperature difference $(T_V - T_H)$. Various lakes in the northern United States and Canada are evident as is the Gulf of Mexico. (b) Brightness temperature transformed to an equivalent land background.

(b)

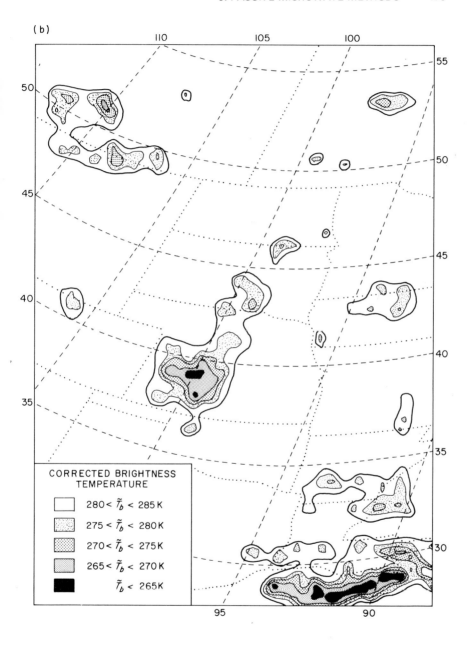

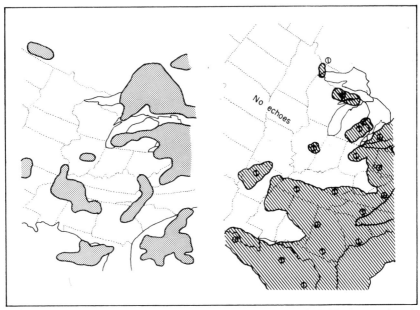

FIG. 9.17   Radar map simulated from Nimbus 6 ESMR and THIR observations for 1700 GMT 20 July 1975 (left); National Weather Service radar summary for 1735 GMT (right). After Hall *et al.*, 1978.

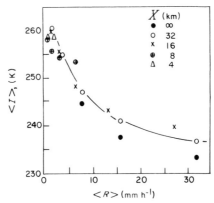

FIG. 9.18   Mean brightness temperature versus mean rainfall rates averaged over the 300 km² footprint of a microwave radiometer. The footprint area contains one 4·57 km-high rain-cloud above land. The various points correspond to clouds with the horizontal dimensions indicated. From Weinman and Davies, 1978.

of observations to the highest categories depends upon an infrared determination of the proximity of the cloud top to the tropopause.

An example of a rain map generated by applying this scheme to ESMR-6 and Nimbus infrared data is shown in Fig. 9.17. On this day convective rain was scattered and widespread east of the Mississippi River, thus the hand drawn radar summary chart (left) shows a greater dispersion of rain systems than the microwave-infrared map (right). In general microwave-infrared reports of rain are grouped with surface reports of rain. Little can be inferred about intensity.

We have seen how the temperature measured when rain-clouds fail to fill a radiometer's field of view b ecomes biasee towards lighter rainrates. To assess this error for the finite cloud Weinman and Davies (1978) placed a cuboidal cloud of 4·57 km depth in the centre of an ESMR-6 field of view, then calculated radiometric temperature and average actual rainrate for a range of cloud sizes and cloud rainrates. The results indicate that for medium and high rainrates, uncertainty in the size of a cloud leads to a factor of two uncertainty in average rainrate (Fig. 9.18).

VI. The Future

There remain major obstacles in the path to routine inference of rainfall from observations of passive microwave radiation. Yet, considering the solid theoretical and experimental progress made in this first decade of microwave observations, it seems likely that passive microwave detectors will be pivotal in future systems for monitoring rainfall from satellites. To give such observations the credibility required of operational systems substantive work is needed in three areas: theory and modelling, improved instrumentation, and synthesis of data.

Efforts in modelling should be directed toward a more realistic representation of the range of natural rain clouds. The aspects singled out in critical discussions of extant models (by, for example, Savage (1976), Wilheit et al. (1977), Austin and Geotis (1978), Weinman and Davies (1978), Lovejoy and Austin (1980)) are cloud water content and supercooled rain, which may be important in convective storms of strong updrafts, the melting layer or bright band, the height of the rain layer, and finite cloud dimensions. Austin and Geotis have also argued for a better climatology of rain systems. This should be aimed particularly at documenting drop size distributions and creating statistical models of the spectrum of sizes of rain clouds.

Improvements needed to instruments, particularly ESMR, become apparent from a reading of papers based on ESMR data. ESMR is plagued by a fair share of the bugs and defects attending operation of most experimental instruments—including noisy data, day–night biases, scan bias and poor location accuracy. Design improvements have been recommended by Savage and Weinman (1975), Savage (1976), Geotis and Austin (1978), and Hall *et al.* (1978). Larger antennae are needed to ease the critical problem of non-beam filling. Since requirements for complete and timely Earth coverage conflict with those for small beam resolution, low noise level, and constant incident viewing angle, for operational systems a family of satellites is needed. Addressing the problems of variable (land) surface emissivity and non-beam filling, Hall *et al.* (1978) advocate two additional microwave channels: one (of low resolution) at 6 GHz to measure surface brightness temperature, and one (of high resolution) near 90 GHz to measure the heights and areas of rain clouds. To solve the problem of timely sampling, they also suggest microwave observations from geostationary altitude. The 25 m antenna required for 5 km resolution at 37 GHz would be feasible if it could be assembled in space.

Quite recently Staelin *et al.* (1978) described a "baseline observatory system" for high resolution passive microwave observations. The baseline observatory would employ a multichannel microwave spectrometer on a low orbit satellite to meet general needs for microwave data, including measurements of rainfall. Atlas *et al.* (1978) anticipated measurements of rainfall through the Nimbus-7/Seasat Scanning Multichannel Microwave Radiometer, which would be improved through enlarging the antenna (to ~10 m diameter), extending the scan, and adding channels at 1·4, 55, and 94 GHz.

A few tentative steps have been taken toward combining passive microwave and shorter wavelength data. Of particular promise are the estimation of fractional field of view coverage from visible and infrared imagery and maintaining spatial and temporal continuity between infrequent microwave observations by means of geostationary visible and infrared imagery. Visible-infrared imagery might also be used to select an appropriate brightness temperature model.

PART III

Satellite Rainfall
Monitoring Applications

CHAPTER 10

Rainfall Inventories

I. Introduction to Part III

In Part I the significance of rainfall as an environmental element was
outlined and discussed, and rainfall monitoring problems were reviewed. In
Part II the development of satellite techniques for the improved monitoring
of rainfall was summarized. Here in Part III our attention naturally proceeds
to examine the kinds of results which have been obtained, in the context of
several broad fields of application. In this chapter our concern is with the
demonstrated utility of satellite rainfall data to make valuable contributions
to the funds of information from which patterns and associated processes of
rainfall may be deduced. This involves both the macro- and mesoscales.
Since broader scale studies usually involve longer unit periods of time the
emphasis in the first place will be dominantly climatological; in the second
place more emphasis will be given to rainfall patterns and processes of
meteorological, rather than climatological, significance.

II. Macroscale Inventories of Rainfall

At the largest scale most statements of rainfall based on conventional
sources of data have been either globally comprehensive but very generalized (e.g. estimated mean precipitation values for "land" and "sea" areas in
models of the hydrological cycle of the Earth–atmosphere system), or more

detailed and accurate in the spatial sense but fragmentary and incomplete in their coverage of the world (e.g. typical maps of "global precipitation" in atlases, where rainfall patterns over the continents are usually depicted fully—though inherently with varying degrees of detail—whilst rainfall over the oceans is commonly ignored). This is because it is clearly not possible to assess rainfall over open water in the same way as, or with equal facility to, that over land. Although raingauge reports may be available from some islands, there is often little direct comparability between rainfall assessments over land masses on the one hand and over even neighbouring waters on the other as a result of different evaluation procedures and/or the different physical processes which are involved.

Over open water it has been common practice to estimate rainfall from the comparative frequencies of cloud types and hydrometeors observed by ships and land-based stations (see e.g. Tucker, 1961), but such approaches have been fraught with difficulties, and resulting assessments of oceanic rainfall as mapped by different workers have sometimes shown large measures of disagreement (Clapp and Posey, 1970). Perhaps the main point of agreement amongst meteorologists is that, compared with the spatial distribution of rainfall over land, rainfall over the open ocean tends to be less complex on climatological time scales.

Something of this dichotomy between rainfall over land and sea has affected satellite rainfall monitoring also, not so much because of any postulated differences between the behaviour of rain clouds as because of problems of image calibration and satellite system performance. Many island and coastal raingauge stations are of doubtful value for rain-cloud calibration offshore because of the significance of local rainfall stimuli. Coastal radar is difficult to use as a calibration tool for similar reasons. The most useful oceanic rainfall measurements are those from coral atolls and other low islands, although these are mostly confined to the tropics, and it is generally accepted that even the humblest tropical islands act as "hot spots" for cumuliform convection. So far as satellite system performance is concerned, we have seen in Chapter 9 that ESMR-type passive microwave indications of rainfall have been evaluated more readily and widely over sea than over land—a fact which has not been unwelcome in view of the outstanding gaps in previous inventories of rainfall over the major oceans of the world.

When all these points are taken together, it is not surprising that the more extensive satellite-based studies of rainfall—those broad enough to interest the climatologist whose concern involves major areas of the globe through extended periods of time—have covered ocean areas rather than continents or substantial sections of continental land-masses. This fact is clearly evidenced by the results summarized below.

III. Macroscale Rainfall Mapping over the Oceans

The first attempt using satellites to elucidate climatological rainfall patterns over major ocean areas was reported by Barrett (1971), relating to the region of the "tropical Far East". This was bounded by 30°N and S, and 90° and 180°E. Although it included most of Australia, parts of the south-east Asian mainland, and many large islands in Indonesia and Melanesia, the area was four-fifths oceanic, embracing parts of the Indian and Pacific Oceans. However, this demonstration study, based on analyses of nephanalyses drawn from ESSA visible images (see Section II, Chapter 4), was restricted to two high season months, namely July 1966 and January 1967, and was of essentially·coarse (2½° grid intersection) resolution.

Despite its limited nature, the tropical Far East study was promising in that it revealed some features in accordance with established climatological knowledge (e.g. the dry zone penetrating westwards from the eastern equatorial Pacific: see Fig. 10.1a), as well as others of a more novel kind. In particular, the map for January 1967 (Fig. 10.1b) contains a number of roughly parallel, symmetrically spaced bands of heavy rainfall, extending northwards from the ITCZ between 100 and 150°E. Very similar phenomena were found in the same region, but for other months and years, in the analyses of Tiros MRIR data by Leigh (1973) evidencing cloud structures in the upper troposphere, and for autumn rainfall in the conventional rainfall study of Dorman and Bourke (1979). Thus there is support for the belief that such patterns are both real and recurrent features of the climatology of the eastern Indonesian region, though their time and space extents, and their physical stimuli, have yet to be established.

Much broader and more continuous satellite rainfall studies of the tropical Pacific have been undertaken since then by Kilonsky and Ramage, using the technique summarized in Section IV, Chapter 4. Results of this approach have been reported for the period from May 1971 through April 1973 (Ramage, 1975; Kilonsky and Ramage, 1976). It is expected that this method, which is simple and inexpensive, will be extended to all the tropical oceans for every month onwards from May 1971, with a substantial break only from March 1978 through January 1979 when problems with the satellites interrupted the preparation of mosaics of images by NESS (Ramage, NASA-NCAR-NOAA TV Colloquium, 15 June 1979).

Representative products from this work are included here as Figs 10.2 and 10.3. Maps of highly reflective clouds (Figs 10.2a–c) are interpretable in rainfall terms insofar as monthly frequencies of such clouds have been shown to be highly correlated with corresponding rainfall totals observed at coral island stations. December 1971 (Fig. 10.2a) appears to have been a much wetter month over the tropical Pacific than December 1972

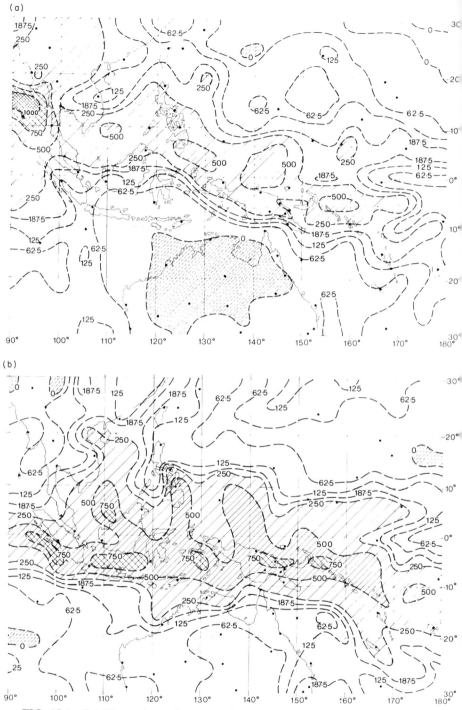

FIG. 10.1   Satellite-estimated precipitation, tropical Far East, (a) July 1966; (b) January 1967 (in mm). From Barrett, 1971.

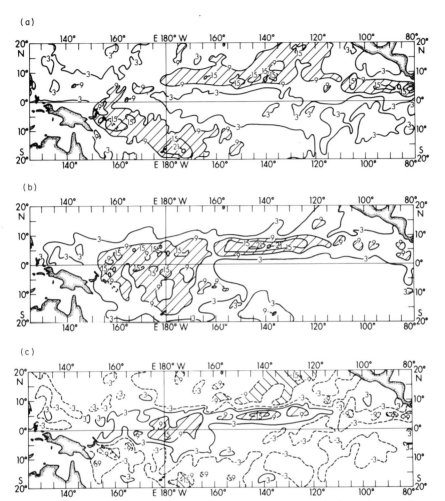

FIG. 10.2   The frequency of highly reflective clouds at 1° lat/long intersections across the tropical Pacific Ocean, determined from NESS Mercator mosaics of polar-orbiting satellite visible imagery for (a) December, 1971; and (b) December 1972; and (c) December 1972 minus December 1971. From Ramage, 1975.

(Fig. 10.2b). Ramage (1975) put the difference at 65%: the first was a non-El Niño year, the second a year of an intense El Niño phenomenon. It is evident that in December 1972 the east-west rainfall maximum in the Central Pacific was both closer to the equator and more intense than in December 1971, whilst the "doldrum" maximum in the western Pacific extended some 20° further east. These differences are shown in Fig. 10.2c, which further reveals *negative* rainfall changes from 1971 to 1972 west of

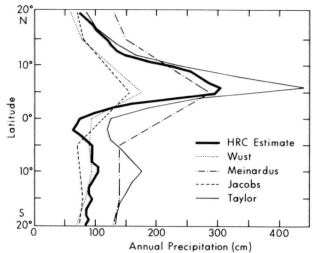

FIG. 10.3   Five estimates of the annual precipitation for the Pacific Ocean between 20°N and 20°S, including that from an analysis of satellite-observed highly reflective cloud (HRC) (see Figs 4.7 and 10.2). From Kilonsky and Ramage, 1976.

145°E. Ramage concludes that this pattern suggests "an equatorial vertical circulation difference between December 1971 and December 1972 consisting of two cells whose rising branches cover the central region, and whose sinking branches reduce rainfall in the east and west" in line with the earlier suggestion by Krueger and Gray (1969). Ramage (1975) also proposed that, through such analyses, "extensive weather satellite data have removed much of the observational uncertainty of earlier studies", for "the frequency of highly-reflective clouds can now be used with some confidence in deducing open ocean rainfall" (Kilonsky and Ramage, 1976). This assessment receives some support from Fig. 10.3, showing a cross-section of highly-reflective cloud estimates for the Pacific Ocean between 20°N and 20°S compared with rainfall estimates from four other (conventionally based) estimates. It has received further, independent, support more recently through the findings of comparative studies by Garcia (1979) (see Chapter 4).

The GARP Atlantic Tropical Experiment of 1974 stimulated two efforts to map rainfall across the equatorial and northern tropical Atlantic Ocean. Garcia (1979) used NESS Mercator mosaics to infer rainfall for periods of 19 days and longer. He found, as we saw in Chapter 4, that the results quite closely match those of Griffith and Woodley.

As has been described in Chapter 5, from one-hourly geostationary infrared data Griffith, Woodley, Griffin and Stromatt inferred 6 h rainfall over a $\frac{1}{3}° \times \frac{1}{3}°$ grid stretching from 5° to 50°W, 5°S to 22°N. From their

GATE satellite rain atlas we present here rain accumulated for phases and for the whole period of GATE.

The Phase 1 map, for 7 June–16 July (Fig. 10.4a), shows rain confined to a zonally oriented belt with its axis near 12°N over West Africa and near 5°N over the Atlantic. Rain is heaviest and the belt is widest near the coast. During Phase 2, from 28 July to 15 August (Fig. 10.4b), and Phase 3, 30 August–29 September (Fig. 10.4c), the rain belt moved progressively northward, especially over the ocean. Rains practically ceased along the equator but spread far into the trade wind belt of the north Atlantic, even developing a secondary axis of heavy rain near 13°N (Fig. 10.4c). For the summer period 27 June–20 September (Fig. 10.4d) rain was heaviest, exceeding 1600 mm, along the south-west coast of West Africa. Little or no rain fell along the south of the equator, except over South America west of 42°W. Rainfall over water was found to closely match patterns of sea surface temperature: 95% of all rain fell where sea surface temperature was 26°C or higher. Latitudinal profiles of rain for the summer period of GATE (Fig. 10.5) are Gaussian in form, except for a northward skew in the westernmost profiles. Gradients are very large, especially at 10°W.

Rain fell preferentially during nighttime hours over West Africa and the north-eastern corner of South America and during daytime hours in the eastern Atlantic. This, as we saw in Chapter 4, accounted for the largest discrepancy between the results of Garcia and those of Griffith and Woodley.

We turn lastly in this section to two applications of the model of Wilheit *et al.* (1975, 1977) for the translation of ESMR-5 microwave brightness temperatures into rainfall (see Section III.4, Chapter 9). The first is the effort by Kidder and Vonder Haar (1977) to infer seasonal frequency of rain over the world's tropical and subtropical oceans. Kidder and Vonder Haar selected 0.75 mmh^{-1} as a threshold for rain and distinguished light, medium, heavy, and very heavy classes of rainrate. The brightness temperatures corresponding to each class varied by latitude according to observed zonal mean freezing levels.

Applying the detection threshold to two months of ESMR-5 data (22 December 1972 through 26 February 1973), Kidder and Vonder Haar produced the total rain frequency map shown in Fig. 10.6a. The precipitation frequency map of McDonald (1938), based on ship observations, is shown in Fig. 10.6b for comparison. Major features in the tropics are well represented in the ESMR-5 map. Small differences, such as those in the Equatorial Pacific, may be due to real precipitation anomalies in the ESMR-5 map period. The large differences at higher latitudes (for example, in the north-west Pacific) are not so easily explained. Kidder and Vonder Haar suggest these may be due to increases in surface emissivity resulting from foam

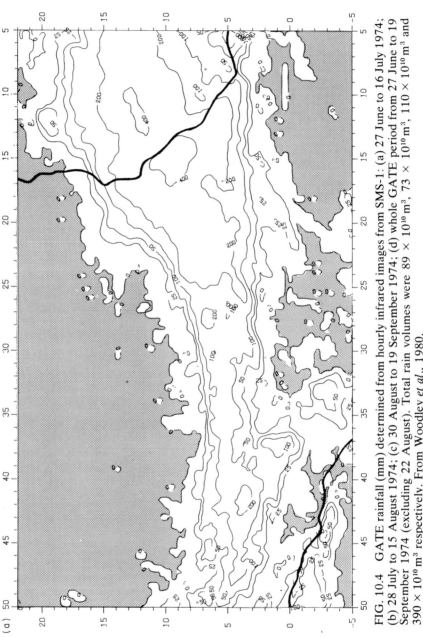

FIG. 10.4 GATE rainfall (mm) determined from hourly infrared images from SMS-1: (a) 27 June to 16 July 1974; (b) 28 July to 15 August 1974; (c) 30 August to 19 September 1974; (d) whole GATE period from 27 June to 19 September 1974 (excluding 22 August). Total rain volumes were $89 \times 10^{10}\,\mathrm{m^3}$, $73 \times 10^{10}\,\mathrm{m^3}$, $110 \times 10^{10}\,\mathrm{m^3}$ and $390 \times 10^{10}\,\mathrm{m^3}$ respectively. From Woodley et al., 1980.

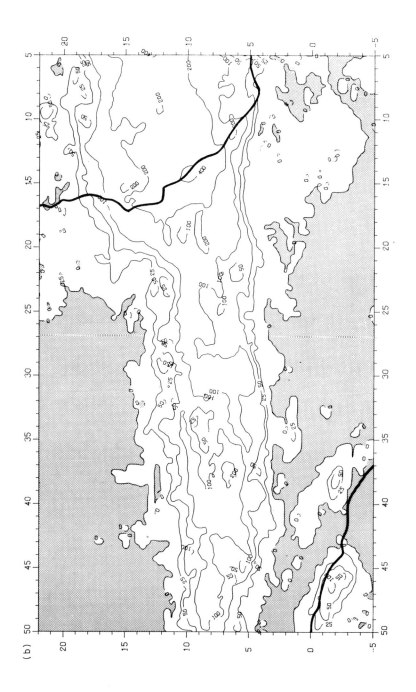

(b)

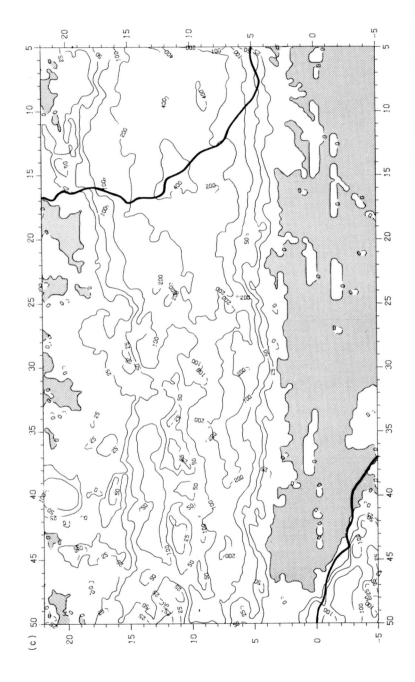

(c)

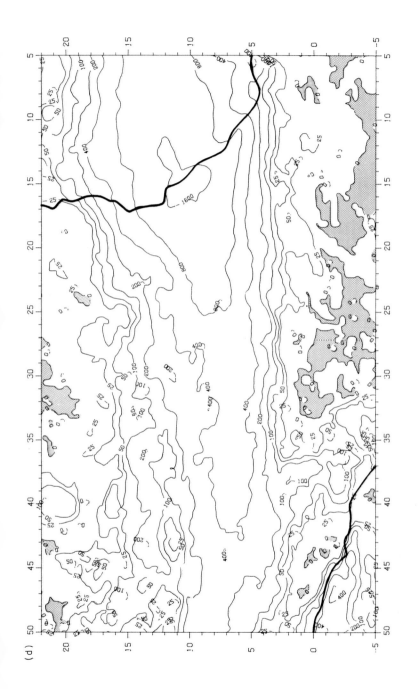

(d)

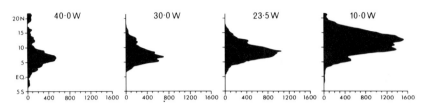

FIG. 10.5   Longitudinal profiles of rainfall (mm) for GATE (27 June to 19 September 1974) from SMS-1 infrared images. After Woodley *et al.*, 1980.

produced by strong surface winds. From similar maps for each rainrate interval (not shown), Kidder and Vonder Haar come to the surprising conclusion that over the tropical oceans the ratios of the frequencies is uniform. They also find little difference between noon and midnight frequencies: only a preference for rain in the very heavy category to occur at local noon.

The second application of Wilheit's model is the "Satellite-derived Global Oceanic Rainfall Atlas" of Rao *et al.* (1976). In this Atlas there are

(a) maps of weekly, monthly, seasonal and annual rainfall rates covering the world's oceans between 72°N and S;
(b) graphs of zonally-average rainfall rates for the Pacific, Atlantic and Indian Oceans separately, and for Global Oceans together, for months, seasons and years; and
(c) some selected sectors from the maps in (a) above, reproduced in colour to illustrate more clearly the utility of the maps for studies of key phenomena such as the Indian monsoon, and the El Niño effect in the Pacific Ocean.

Rainrate was averaged over grid cells 4° in latitude and 5° in longitude and having at least 25% water surface. Weekly and seasonal maps are shown for the period 11 December 1972 to 28 February 1975, and annual maps for 1973 and 1974.

In making their frequency maps Kidder and Vonder Haar (1977) found a systematic variation in brightness temperature between day and night passes; also a variation along scan lines which could not be attributed to limb darkening or angular variations in the emissivity of the sea. Both anomalies were corrected empirically, by requiring 3 month average brightness temperatures over rain-free areas to be independent of scan angle and time of day (see Kidder, 1976).

Rao *et al.* (1976) also encountered problems in the ESMR-5 data, and in the Wilheit model as well. To reduce the sensitivity of the model they chose to represent freezing levels over the globe by three steps of change instead of five. Because of uncertainties in the location of observations it was required

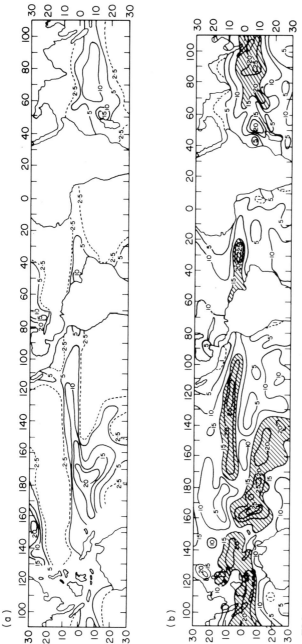

FIG. 10.6 Frequency of precipitation (% of observations), December 1972 to February 1973 (a) derived from Nimbus-5 ESMR data, averaging noon and midnight data; (b) based on ship observations at noon (after McDonald, 1938). From Kidder and Vonder Haar, 1977.

that data be at least one degree from coastlines and significant islands. The variable effects of surface wind, water vapour and clouds were dealt with by imposing thresholds of minimum rainrate—2 mm h^{-1} in the tropics, 1 mm h^{-1} in high latitudes—and the dependence of brightness temperature on scan angles was corrected by subtracting a factor equal to 0·07 of the scan angle measured in degrees.

In the Atlas the authors show for comparison the mostly incomplete January and July rainfall maps of the Deutscher Wetterdienst Seewetteramt, Hamburg, and a map of annual ocean precipitation compiled by R. Geiger. Differences of monthly rainfall range from −30 to +80% in the few places where comparisons are possible. Differences of annual rainfall are comparable.

Biases may be substantial. This is illustrated by latitudinal profiles of rainrate for the Pacific Ocean in 1973 and 1974 (Fig. 10.7). From 20 to 50°N ESMR-5 rainfall is close to the recent estimates of Dorman and Bourke (1979) (which are based on 23 years of ship observations); however, in the major Intertropical Convergence peak between 5° and 10°N, ESMR rainfall is far below the ship estimates (also see Fig. 10.3).

Rao et al. recognized limitations in the Atlas: "No claim is made for reliability in absolute values of rain rate better than a factor of two . . .". They suggested as refinements modifying the model to include liquid water above the freezing level, and better spacecraft ephemeris and attitude. Additional cautions were noted in an appendix: incomplete data during 2 months, and the possibility that observations at high latitudes included sea ice. Regrettably, these cautions were omitted in a subsequent and dependent paper of Rao and Theon (1977). There is now rather widespread agreement that the Atlas should be viewed and used with caution because of

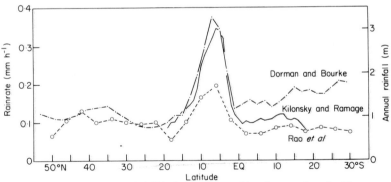

FIG. 10.7   Pacific Ocean rainfall inferred from ship observations by Dorman and Bourke (1979; 23 years of record), and from ESMR-5 observations by Rao et al. (1976; 2 years of record). The rain profile of Kilonsky and Ramage (1976) from visible imagery is also shown.

the details of the methods which were employed in the processing and analysis of the ESMR data.

The Atlas remains, however, a document worthy of discussion, constituting as it does a pioneer example of the scope which ESMR data now afford for the compilation of satellite rainfall climatologies of the major oceans of the world. A selection of the more interesting points of large-scale rainfall climatology which have emerged from an examination of the contents of the Atlas were discussed by Rao and Theon (1977). These exemplify the uses to which such an Atlas can be put in the elucidation not only of major features of global rainfall patterns, but also of the short-period fluctuations to which they may be subject. Indeed, it is in this area of climate dynamics that satellites have potentially most to offer: using conventional data alone it has been difficult enough to establish with confidence a mean annual map for global oceanic rainfall; but it has been extremely difficult to examine inter-annual differences—and virtually impossible to elucidate short-period rainfall fluctuations.

The ESMR Rainfall Atlas has yielded the following types of information.

1. The distribution of major rainfall areas over the world's oceans

Satellite studies have done much to elucidate the dominant cloud structures of the tropical oceans. These involve prominent east-west cloud bands usually just north of the equator (the inter-tropical cloud bands, or ITCBs), and oblique cloud bands extending from the ITCB north-eastwards especially in the north Atlantic and north Pacific, and south-eastwards especially in the south Atlantic and south Pacific (named "intra-tropical cloud bands" by Gruber (1972), but more appropriately and less confusingly named "trans-tropical cloud bands", or TTCBs).

Extensive satellite analyses of tropical cloudiness, for example by Kornfield *et al.* (1967), Hubert *et al.* (1969), Miller (1971), and Sadler *et al.* (1976), have combined to yield substantial knowledge of the short- and longer-term variations in the structures and positions of the ITCBs and TTCBs in the different tropical oceans. ESMR maps of instantaneous rainfall rates correspond very well with the gross rainfall patterns anticipated from the patterns of clouds. For example, it is seen in Fig. 10.8 that, moving eastwards along the equator in the Pacific, the rain belt associated with the ITCZ bifurcates in the region of 170°E: the ITCB branch proceeding eastwards just north of the equator towards Central America, whilst the TTCB branch runs east and south-eastwards before merging with the southern Pacific rainfall zone (mid-latitude depressions) in the vicinity of 160°W. A similar pattern, in mirror-image form, is seen in the north Pacific off the eastern seaboard of Asia.

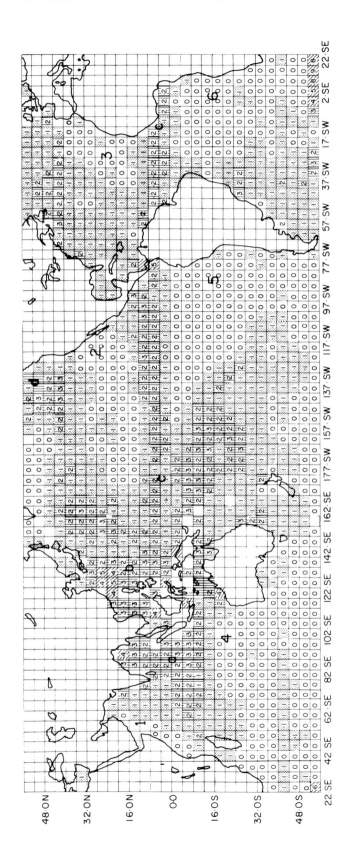
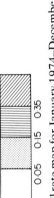

FIG. 10.8 ESMR-derived global oceanic rainfall rate map for January 1974–December 1974. Areas of heavy rain are labelled a through d, and areas of light rain 1 through 6. From Rao and Theon, 1977.

Turning to middle latitude ocean areas, Rao and Theon describe an extensive rainfall area in the south Atlantic to the south-east of South America mainly between latitudes 25–50°S and longitudes 25–50°W. Here, rainfall levels seem to be significantly higher than to the east or west, the average rainrates of 0.1 mm h^{-1} indicating annual rainfall totals of about 900 mm. Rao and Theon say that this rainy region was not known before, and does not appear on any previous map of the global rainfall: there are no islands here, and few ships traverse the region. Recourse to visible and infrared imagery reveals, however, that this is frequently a region of enhanced cloudiness (see e.g. Miller, 1971); also, combined satellite/ conventional data analyses of the southern hemisphere have indicated that it is a preferred region for mid-latitude cyclogenesis (Streten and Troup, 1973). For such reasons, it may well be that this region of enhanced cloud and rain is analogous to the TTCBs of other oceans, especially the north and south Pacific and the north Atlantic, rather than "possibly an extension of the southern Pacific rain zone" as Rao and Theon suggest.

2. The sub-structure of major rainfall areas

An example of the utility of an ESMR Rainfall Atlas at a larger scale is afforded by zonally averaged graphs of rainfall for the Indian Ocean (see Fig. 10.9). In more detail the ITCB is known to be capable of adopting a wide range of different structures, for these have been evidenced in plan form by satellite visible and infrared image time composites (see Barrett, 1974). In these, the zones of most pronounced convectional cloud development appear as brightly reflective bands of cloud, whilst associated (usually parallel or sub-parallel) zones of subsidence within the ITCZ appear as dark bands in which cloud growth has been suppressed.

Of particular interest is the "double ITCB" (see Pike, 1971) which occurs most frequently in the Indian Ocean in the northern hemisphere winter and in the Pacific Ocean near the coasts of Central America in the northern hemisphere spring. Meridional profiles of zonally averaged rainfall reveal that two distinct rain maxima in the Indian Ocean are sometimes evident at certain times of the year between 20°N and S of the equator. Rao and Theon suggest that the northern maximum appears to grow at the expense of that immediately south of the equator as the South Asian summer monsoon advances, the reverse occurring as it retreats. During June–August the amplitude of the northern maximum is three times greater than the southern maximum, whereas during December–February the amplitude ratio is reversed. In some months or seasons, however, the bimodal structure disappears (e.g. December 1974–February 1975): according to Rao and Theon this may be because the 1 mm h^{-1} rainrate cut-off level adopted in the

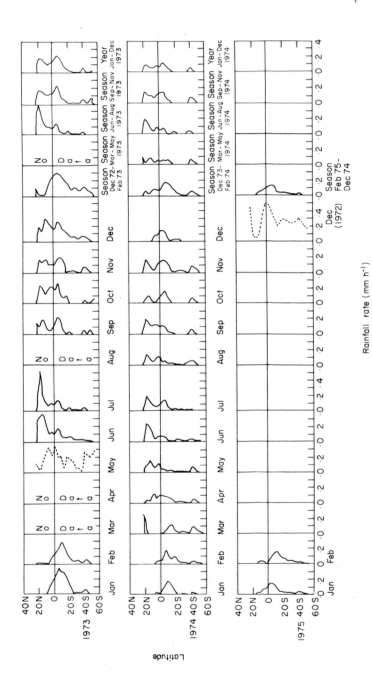

FIG. 10.9 ESMR-derived graphs of zonally averaged rainrate versus latitude in the Indian Ocean. Broken lines represent curves based on inadequate data. From Rao *et al.*, 1976.

preparation of the maps obliterates the lesser maxima in cases like these when the southern maximum is unusually weak.

3. Interannual variations of rainfall

In large-scale studies of climate, attention is often focussed not only on average distributions of elements such as rainfall, or seasonal variations of the same, but also on differences from year to year, and, ultimately, on long-term climatic variations.

Weekly, monthly, seasonally and annually averaged maps generated from ESMR data, like the cloud brightness maps of Kilonsky and Ramage, afford improved opportunities for the study of short-term fluctuations of rainfall intensity and distribution. One example is afforded by the reference in Rao and Theon (1977) to rainfall over the Pacific in January 1973 and January 1974. The first fell in an El Niño year, with the warm current along the equator enhanced as a result of relaxed cold upwellings along the coasts of Peru and Ecuador (see Wyrtki 1975). The second was marked by a much weaker El Niño phenomenon. It appears from the ESMR Atlas maps that, in January 1973, heavy rain extended across the breadth of the Pacific between 0 and 8°N, and spread south of the equator between 170°E and 160°W. In January 1974 the same regions were relatively dry. When rainfall in the equatorial zone between December 1972 and February 1973 was compared with that between December 1973 and February 1974, the ratio was found to be as high as 6:1.

Satellite evaluations of rainfall distributions over the oceans of the world clearly have much to offer in studies where seasonal and interannual, as well as longer-term, anomalies are involved.

IV. Macroscale Rainfall Mapping over the Continents

When we turn to consider rainfall over land we find that satellite studies have been much less integrated, comprehensive and broadly informative than those over the oceans. It was explained in Chapter 9 that passive microwave studies over land have been restricted by added practical problems. One significant effect of this has been that other, less direct, approaches to satellite-assisted rainfall mapping have had to be employed for most continental applications (see Chapters 4–6). Furthermore, since different workers, groups, and agencies have developed their own preferences amongst these approaches, land area results are varied in type. Finally, influenced as they have been by the available data, equipment, funding, and the range, scope and objectives of different projects, these results are piecemeal in their areal coverage and mostly brief and/or fragmentary in their coverage through time. Most of the more extensive studies (e.g. Follansbee's estimates of rain over major portions of the USSR and China (see Section II, Chapter 13)) have been designed to meet near real-time information needs in areas for which climatic records of rainfall are relatively satisfactory. Results from regions less well-supplied with surface rainfall stations (e.g. Barrett's results from north-west Africa (see Section IV, Chapter 12)) have so far been discontinuous or shorter term. For such reasons, land area results must be reported and discussed not here, but in the chapters which follow. It is probable that the most comprehensive work has been undertaken by US military units or US remote sensing contractors, but detailed results are not readily forthcoming from either type of source: it is not the primary function of such bodies to release scientifically interesting findings for general use. However, even in those cases the scales of operation have almost certainly been national or subcontinental rather than continental or global.

It is, therefore, over the continents that there remains the greater need for the implementation of a unified method of improved rainfall mapping for the support of macroscale inventories and models. Whilst the supply of conventional data is far superior to that over the oceans there are still some land areas (e.g. significant regions of Africa and Asia) from which no data may be available from GTS stations in squares as large as those in the ESMR Atlas maps (4° latitude × 5° longitude). This is all the more serious for climate inventories and modelling because more detailed rainfall data are necessary from land areas on account of the higher degree of spatial variability of rainfall over land, influenced as it is by the added factors of exposure, altitude, morphology, nature and composition of the surface, etc.

In that substantial gaps would be seen in every major continent if GTS data were allocated to 1° grid squares, conventional data at present fail to meet many hydrological needs too. Of course, numerous raingauges are deployed in networks supplementary to the GTS system, but in many cases these data are not readily available to the general user community, and from some countries are hard to acquire at all. Satellites would seem capable of making good many of the existing data deficiencies over land, but problems of approach and project co-ordination are much greater here than over the oceans of the world. Consequently, suitable solutions to rainfall inventory problems over continental regions have yet to be found. Possible ways in which the needs for more and better data from land areas may be met are discussed in the following section of this chapter, and in the last chapter of this book.

V. Mesoscale Rainfall Mapping over the Oceans

Applications of satellite rainfall monitoring methods on meso- and small synoptic scales over land have most often served hydrological or flood warning purposes, as discussed in Chapters 10 and 11. Otherwise, such mesoscale applications have been relatively modest, the most significant being related to rainfall mapping over oceans in support of more comprehensive water budget studies. Since applications of this kind are likely to grow in importance, available results from such studies are prefaced here by some remarks on moisture budgets in general.

1. The concept of moisture budgets

A budget in meteorology is simply a statement of sources, sinks and storage over some interval of time. It is entirely analogous to the familiar budget of personal finance involving income, expenditures or losses, and balance or net worth. A single meteorological budget tells of the relative importance of processes enhancing or depleting a substance or property, and it tells of local trends—net gains or losses. Two or more budgets tell in addition of spatial redistribution—relative gains and losses.

Budgets in meteorology are based on the law of conservation of substance or property. They apply to entities or to defined volumes (a hurricane or a cumulus cloud, for example, or the triangular slab of air outlined by ships of the 1969 Atlantic Tropical Experiment (Augstein *et al.*, 1973)).

Because measurements are discrete, in the determination of a budget a distinction usually is made between grid and subgrid scale processes. Some processes will not be resolvable in space or time on the scale of the measurements which can be made. The collective contribution of these unresolved processes, together with the net of errors in measurements of grid scale processes, is carried in the budget equation as a residual. But these terms may be separated if an estimate of probable measurement error can be made. What remains in the residual then is a term representing the effect of subgrid scale processes.

In meteorology, budgets are the main analytical tool whereby concepts for parameterizing cumulus convection are designed and tested. "Parameterization" here means the expression of the net effects of convective scale processes on fields of mass, momentum and moisture, in terms of synoptic scale observations of these properties. Then the period and dimensions of the budget must be scaled to the interactive process under study: large enough to contain the small event, and small enough to be a part of the large event. In addition, budgets are an efficient way of summarizing the vast amounts of data generated by field and numerical experiments.

Moist convection is the subgrid scale process represented in the residual of many recent budget studies (see, for example, Yanai *et al.*, 1973; Cho and Ogura, 1974; Ninomiya, 1974; Nitta, 1977; McNab and Betts, 1978) because in the tropical "boiler" of the atmospheric heat engine, upward transport and release of heat is largely accomplished by deep convective clouds—the hot towers of Riehl and Malkus (1958). Without an accounting of convective heat release in the tropics, accurate weather forecasts of one week and longer are not possible (Miyakoda, 1974; Kuettner *et al.*, 1972). Moisture is a primary variable in the prediction of weather because the 2500 J of solar energy needed to evaporate a gram of water from the sea is released to the atmosphere when that water returns to the sea as rain. Precipitation therefore is a measure of the latent heating of the atmosphere, and (since rain usually falls out of clouds quickly and directly) its distribution and timing as well. As has been shown by Reed and Recker (1971), budgets of moisture complement budgets of mass and energy. Because convective processes must simultaneously satisfy conservation of mass and heat as well as moisture, moisture budgets serve as checks on mass and heat budgets, and they may help to narrow the range of possible modes of behaviour of the atmosphere.

2. Applications of satellite rainfall techniques to moisture budgets

Perhaps the first application of satellite estimates of rainfall to moisture budgets was in BOMEX. Scherer and Hudlow (1971) estimated rainfall beyond the range of the BOMEX radars using the scheme described on pp. 59–60. However, because the BOMEX budgets deliberately avoided disturbed periods of significant rain, the results of this experiment were inconclusive: the main benefit was the knowledge that rainfall could safely be ignored (Holland and Rasmussen, 1973).

Satellite estimates of rainfall contributed in a small way to Betts' (1978) budget analysis of convection on 2 September 1974 of GATE. Ship soundings were dispersed over an area 3·6 times larger than that covered by the GATE radars. However, radar rates agreed closely with ten values of satellite rainrate made using the scheme of Stout *et al.* (see Section I, Chapter 5) for the large area (Fig. 10.10), indicating that rates measured by the radar were representative of the budget area.

The agreement of satellite and radar rainfall is not entirely to their credit, since a part of the radar data was used to calibrate the satellite. Yet, as Betts observed, the average radar rainrate for this day (which coincided closely with the growth and decay of the organized convective disturbance) was within 2% of the rainrate inferred from a budget of heat.

For a variety of reasons some of the work needed to confirm and extend

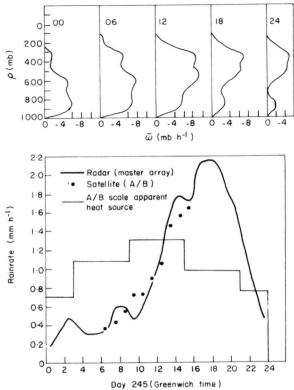

Fig. 10.10   Comparison of radar rainrates (heavy curve) for the "master array" in GATE (a circular area 408 km in diameter), satellite-derived rainrates for the A/B scale ship array (a hexagon 777 km in diameter) and Nitta's (1977) integrated A/B scale apparent heat source (block). Upper curves are A/B scale average vertical motion profiles at 6 h intervals for 2 September 1974. From Betts, 1978.

these and other budget findings will be done over areas beyond the view of the GATE radars. It appears that satellites can make an important contribution here by providing information on rainfall which otherwise would not be available.

On a similar areal scale but for a longer time and for a far more intense disturbance Adler and Rodgers (1977) used ESMR-5 data to determine latent heat release and the distribution of rainrate for a cyclone in the Pacific. This was typhoon Nora, from 1973, which we saw in Chapter 9. After correcting the ESMR data, according to the scheme of Kidder (1976), for day–night differences and scan anomalies, and assuming a 5 km freezing height in the model of Wilheit et al. (1977) (see Chapter 9), Adler and Rodgers calculated rainrates and latent heat release at six times over an area

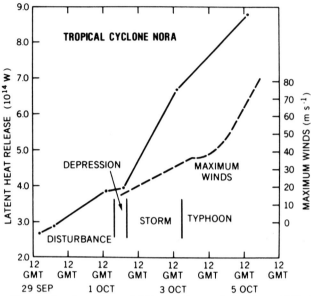

FIG. 10.11   Tropical cyclone Nora, 29 September to 5 October, 1973: latent heat release. Calculations are for a circular area of radius 444 km (4° lat). The maximum $(8.8 \times 10^{14} \text{W})$ corresponds to an area average rainrate of 2.0 mm h^{-1}.

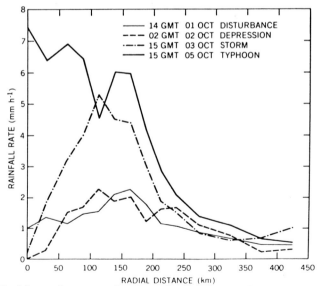

FIG. 10.12   Mean rainrate as a function of radial distance from the storm centre in Fig. 10.11 for four observations as the storm intensified. From Adler and Rodgers, 1977.

of 444 km radius centred on the storm. (Centring on the circulation or on the lowest brightness temperature made little difference.)

Adler and Rodgers discussed six contributions to error in the rain and latent heating rates. Although the magnitudes of the errors were not all well known, they concluded that the ESMR calculations should be reliable at least in a relative sense.

They found that as the storm intensified from a depression to a very strong typhoon latent heat release over the large area tripled, reaching $8 \cdot 8 \times 10^{14}$W (Fig. 10.11). At the typhoon stage for a radius of 111 km ESMR latent heat release was smaller than most earlier results. However at radii of 222, 333, and 444 km it agreed with earlier results, including numerical models. The increase in latent heat release was accomplished both by increases in rain area and increases in average rate. Of particular interest is the radial profile of rainrate (Fig. 10.12). During disturbance and depression stages this peaked between 100 and 150 km. Between depression and storm stages peak rainrate doubled. Between storm and typhoon stages while continuing to increase the peak shifted inward virtually to the centre of the typhoon.

3. Additional applications and outlook

Satellites are contributing also to measurement of rainfall on convective scales in the eastern Pacific Ocean, in the High Plains of the United States and in Florida (Augustine et al., 1979; Griffith et al., 1981). The Pacific work is in support of Equatorial Pacific Ocean Climate Studies (EPOCS), a programme of the Environmental Research Laboratories of NOAA which aims to observe the coupling of atmosphere and ocean in an area of the tropics that has been linked with changes of climate in middle latitudes. The High Plains Experiment (HIPLEX) and Florida Area Cumulus Experiment (FACE) are programmes of the Bureau of Reclamation and National Oceanic and Atmospheric Administration aimed at increasing convective rainfall through cloud seeding. Satellite rainfall estimates will be made for all three experiments by the method of Griffith and Woodley (see Section II, Chapter 6), and for the Pacific experiment by the method of Kilonsky and Ramage (Ramage, personal communication; see Section IV, Chapter 4).

Interest in using satellites to observe rainfall on climatic scales received a big push through the Workshop on Measurement of Precipitation from Space, held on 28 April–1 May 1981 at the Goddard Space Flight Center, Greenbelt, Maryland, USA. Most of the satellite instruments and techniques so far employed to infer rainfall were represented; also a number of new ideas were put forth. The Goddard Space Flight Center will publish proceedings of the workshop.

CHAPTER 11

Applied Hydrology

I. Introduction

As man's population and global economy have grown, so, too, have his needs for living space, food, energy and mineral resources increased. More than ever it is necessary to know the amounts and distributions of such vital resources, to plan for their use and conservation, and even, wherever possible, to work for their enhancement.

In this chapter we will review the more significant ways in which satellites can assist in the assessment of water. In the chapters which follow, consideration will be given to the utility of satellite-improved rainfall data in agroclimatology and agrometeorology with special reference to crop prediction problems, and in the monitoring and alleviation of environmental hazards and associated disasters. Such questions have significance across the entire range of scales from that of the individual, through those of local communities, regions, countries and continents, to that of the international family of nations as a whole.

II. Water Resource Evaluation

Water may be one of the most ubiquitous of the resources of planet Earth, but it is also one of the most widely and heavily exploited minerals. Shortages of water, in terms of either volume or quality of supply, or both, have been traditionally severe in many parts of the world. However, as remarked in Chapter 1, shortages have occurred recently in a number of additional regions, including some which might once have been thought very unlikely sufferers. Examples include parts of north-western Europe (e.g. during the long drought of 1976), and parts of the USA (e.g. in the north-east in the early 1970s). Some of the most striking results of such shortages have been the recent restrictions of water supplies for domestic consumption in some parts of the UK to rations dispensed by standpipes in the streets, and cards on the tables of hotels in the Washington, D.C. area of the USA advising patrons that iced water with meals would be made available only on request. Today the point has been reached at which water supply problems significantly circumscribe or even tightly restrict the lives of men and the economies of nations across by far the greater part of the habitable Earth.

For many purposes water resource evaluation in the field depends more upon assessments of water on or below the surface than that arriving as precipitation. However, a good knowledge of rainfall is often valuable in this context, besides being vital for a broad spectrum of associated applications including some in pure research (e.g. modelling of hydrological cycles) and some in practical fields of great immediacy (e.g. in power management, and flood prediction and control).

Prominent amongst those regions where there is not only a serious water supply problem but also a rapidly increasing requirement for water is the Middle East, where several oil-producing states are attempting to develop industry and agriculture whilst oil revenues last.

The precariousness of natural water supplies in the Middle East may be introduced and exemplified by reference to the small Gulf state of Qatar. Hydrological studies there have suggested that recent levels of extraction of water from the ground have been nearer optimistic than conservative estimates of recharge by rainfall (FAO, 1974). However, even in that tiny nation (11 600 km²) where rainfall is relatively well-monitored by gauges (24 in 1972), the conventional rainfall data have been deemed inadequate for the water budget to be confidently assessed, in part because of the dominance of isolated rain shower activity in the hydrometeorology of the Middle East. In neighbouring states similar problems are found, but they are more serious and difficult to answer by virtue of the greater size of the countries concerned, including Saudi Arabia, the United Arab Emirates, Oman, and Iran. A standard solution employed by hydrologists, for example in Saudi Arabia, involves the identification of "representative basins" which are relatively well-monitored by conventional means, and from which extrapolations may be made into intervening areas (see e.g. WMO, 1974; Henry (ed), 1980). This, however, is recognized to be more of a practical necessity in seeking to improve water resource evaluation and monitoring in poorly documented and relatively inaccessible areas than an ideal answer to such problems. It would seem more than reasonable to believe that satellites might be of significant use in improving knowledge of rainfall distributions and amounts in such areas and situations.

Such a possibility has been tested in the Gulf state of Oman (Barrett, 1977a and 1979), and has been recognized by experienced hydrologists as, perhaps, the only practicable way in which rainfall monitoring might be improved in the foreseeable future in some parts of the Old World arid zone (Thomas, personal communication, 1978).

The Sultanate of Oman occupies the south-east corner of the Arabian peninsula. The only rainfall of note (100–150 mm p.a.) falls over the mountains in the north-east, backing on the Gulf of Oman, and in a smaller area in the south-western Dhofar Province, which is affected by the fringe of the monsoon rains over the north Indian Ocean in the summer season. In any plan to expand agricultural production the larger north-eastern region must figure most prominently. Topographic and soil surveys have revealed areas suitable for agricultural expansion, especially on alluvial fans along the flanks of the Jebel Akhdar mountains, and in restricted valleys and wadis.

This is, however, a much more difficult area than Qatar in which to assess natural water reserves and rates of replenishment: the geology is complex, the terrain often spectacularly rugged, and many areas in the main watershed are almost inaccessible on the ground. These areas are not entirely uninhabited, however, and experience in the field has shown that even rain gauges placed near mountain tops and visited by helicopter for reading are not necessarily secure from human damage or interference. Yet, prior to any agricultural expansion better knowledge of the rainfall and its distribution is required, at least for special evaluation periods, but preferably on a continuous basis.

In 1973–1974 several hydrological consulting firms worked in this region, using conventional techniques to assess rainfall and other elements and phenomena of hydrogeological significance (Ilaco, 1975; Renardet Saute I.C.E., 1975, Gibb, 1975). One of the most interesting conclusions to emerge from a study of part of the area was that "the scarce data of the rain gauges at high altitudes do not point to a rise of precipitation with increasing heights" (Ilaco, 1975, Annex A). This was confirmed by a plot of daily rainfall versus altitude in 1974 for north-east Oman as a whole (Barrett, 1977a). Clearly it could be most unfortunate if isohyets were to be interpolated between the permanent rainfall stations on the basis of an assumed linear, or other comparatively simple, relationship between rainfall and height above sea level.

So a satellite-augmented rainfall mapping exercise was commissioned to test the applicability of a cloud indexing technique to an area in which rainfall is generally infrequent, sporadic and very "spotty" in its distribution: the area was subdivided into about 130 grid squares of 0·2° latitude and

0·2° longitude sides; about one-third of the squares were left with surface rainfall stations after fifteen stations had been reserved for verification purposes. In view of the infrequency of rainfall events in north-eastern Oman, calibration regressions were prepared for the year previous to the study year of 1974 on a subregional basis, taking account of the morphology of the land and available knowledge of climatic variations from place to place. Daily estimates were floated up or down when satellite cells were affected by rain cloud which also prompted events recorded by rainfall stations in gauge cells within the same morphoclimatic region(s). On all other occasions the computed regression relationships between cloud and rain were invoked. No altitude weighting factors were employed because of the apparent absence of a simple relationship between rainfall and altitude; however, it was anticipated that if such a relationship did, in fact, exist it would be evidenced through higher frequencies of rain cloud types over high ground. Although estimates were made where possible on a 24-hour basis, these could be prepared only for the venues of monthly and annual stations; indeed, in the south-eastern half of the study area the available calibration/ verification data were from monthly accumulated gauges, locally precluding daily rainfall estimation.

Figure 11.1 exemplifies the results. The monthly maps involving both raingauge and satellite data suggest that the method coped well with both spatial and temporal variations of rainfall, being logical in terms of the meteorological situations which were noted through the study period. Verification studies (Fig. 11.2) revealed the following:

(a) Satellite-assisted estimates for mean annual rainfall at the 15 reserved stations averaged 97·9% of observed rainfall;
(b) Approximately two-thirds of the estimated annual totals for individual stations were within 20% of observed rainfall; and
(c) About three-fifths of the monthly estimates for individual verification stations were accorded the correct *category* of rainfall in a contingency table plot against observed rainfall.

Three quite significant points emerged from the resulting rainfall maps compiled with and without satellite assistance:

(a) The satellite-assisted maps yielded a total rainfall volume for the area as a whole during 1974 of more than 30% less than a comparable total derived from conventional data alone. This suggests that there may be even less scope for further water resource exploitation in north-eastern Oman than was previously thought;
(b) The satellite-assisted maps suggest stronger gradients between high and low rainfall regions, confining the former to quite small areas; and

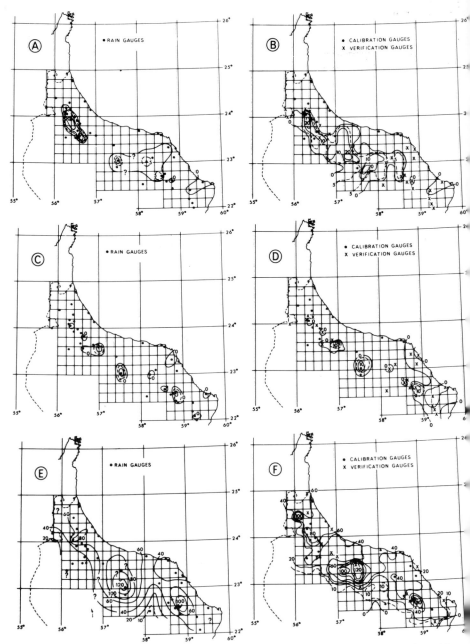

FIG. 11.1 Rainfall maps for north-east Oman prepared from gauge data alone (A, C, E) and gauge plus satellite data (B, D, F). In July 1974 considerable differences are apparent in the two analyses (A, B), whereas in June 1974 (C, D) the differences are slight. The annual analyses for 1974 (E, F) exhibit hydrologically significant differences. From Barrett, 1980a.

(a)

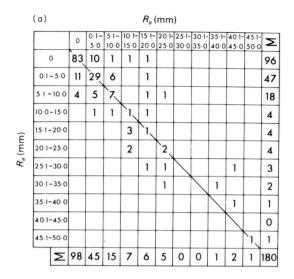

R_e (mm)	0	0.1–5.0	5.1–10.0	10.1–15.0	15.1–20.0	20.1–25.0	25.1–30.0	30.1–35.0	35.1–40.0	40.1–45.0	45.1–50.0	$\geq$
0	83	10	1	1	1							96
0.1–5.0	11	29	6		1							47
5.1–10.0	4	5	7		1	1						18
10.0–15.0	1	1		1	1							4
15.1–20.0				3	1							4
20.1–25.0				2		2						4
25.1–30.0					1	1				1		3
30.1–35.0						1		1				2
35.1–40.0									1			1
40.1–45.0												0
45.1–50.0											1	1
$\geq$	98	45	15	7	6	5	0	0	1	2	1	180

(b)

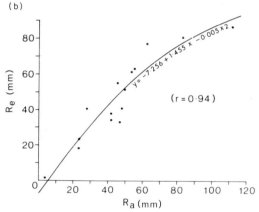

$$y = -7.256 + 1.455x - 0.005x^2$$

$(r = 0.94)$

FIG. 11.2   Verifications of results of satellite-assisted rainfall monitoring in Oman:
(a) a contingency table for monthly estimated rain (R_e) versus observed rain (R_a); (b)
a regression analysis for annual R_e:R_a. Both relate to data from 15 specially reserved
verification rainfall stations. From Barrett, 1977a.

(c) The satellite-assisted maps confirm that there is no simple relationship
between rainfall and altitude over periods no longer than 1 year
because of the significance of the relatively localized heavier rainfall
events.

The conclusions drawn from this study are that the traditional climatic
picture painted of this area (with isohyets and contours running roughly
parallel in the mountainous region) may be generally realistic subject to an

increase of rainfall from south-east to north-west along the main mountain range, but that natural water supplies (e.g. for expanded irrigation agriculture) may be expected to be very variable from one locality to another, and through time.

III. Catchment Monitoring

Interest in hydrology and hydrometeorology often focusses more on individual river catchments than the national or subnational regions discussed in the previous section. In some cases this results from the need to consider representative basins, as indicated earlier. In other cases it stems from the particular significance of a certain basin for water supply, power generation, or some special land use pattern.

Interest in the possibility that satellites might be used to provide rainfall data for individual catchments rather than regular grid squares or intersections has led to a number of practical tests which we will review now.

One obvious type of approach would be to take satellite or satellite-improved isohyetal maps prepared (as on the Oman study) from grid-square estimates and to prepare rainfall depth–area (volume) estimates for the river basins by planimetric methods. This could increase the utility of the data obtained initially for other purposes. However, it seems that such a possibility has not been investigated in any such situation yet (but see Section IV and Fig. 11.5).

A different type of approach, which has been investigated by several groups, entails the treatment of a river basin as a whole at the earlier stage of cloud assessment in terms of rainfall itself. One of the first of these studies was undertaken for three basins in north-western Montana from 15 April to 30 June and 1 October to 31 December, 1970, using imagery from polar orbiters (Davis and Weigman, 1970). Here, sequences of visible and infrared satellite images were used to classify cloud categories, each of which could be expressed in terms of a probable precipitation amount (see Section IV, Chapter 4). Summations of these amounts for the three basins in the study yielded estimates of cumulative basin precipitation. Figure 11.3 illustrates the outcome over a period of 2½ months. It was concluded that the cumulative precipitation for two of the three adjacent basins (A and C) was well-represented by the satellite estimation technique, whilst for one basin (B) the satellite-estimated rainfall was significantly greater than that deduced from raingauge data, but in general agreement with the radar data. This was an encouraging result, and indicated that useful quantitative estimates were possible at the basin level from the satellite evidence, indeed perhaps better estimates than radar provided, for this tended to underesti-

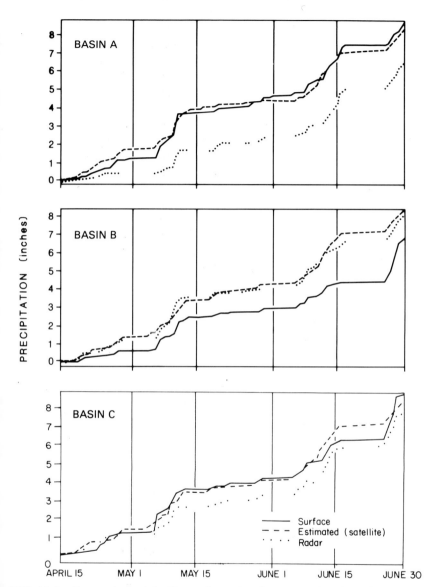

FIG. 11.3   Comparison of satellite-estimated and gauge-analyzed rainfall for three basins in north-western Montana. After Davis and Weigman, 1970.

H

mate cumulative precipitation more, especially as a function of range. A logical next step might have been a more formal integration of raingauge and satellite data to provide an overall assessment of basin rainfall volume.

Other polar-orbiter studies of mean basin rainfall have utilized the NESS cloud-indexing techniques (see Chapter 4). Such studies have been reported for areas in Africa and southern Asia (Follansbee, 1973), and, more recently, for Surinam in South America (Popovich, personal communication, 1978).

In this last case estimation of areal precipitation from weather satellite imagery has been undertaken regularly from 1972 to the present time in support of hydroelectric power plant operations. The Surinam River, which flows northwards to the Atlantic Ocean from the highlands in the south of the country, is controlled by the Afobaka Dam, a major installation about 170 ft high and over 6000 ft long. The drainage area of the river to that point exceeds 5000 sq. miles. A reservoir of about 600 sq. miles is formed behind the dam. Application of the Follansbee method to the estimation of areal rainfall over the drainage basin upstream from the dam has been particularly significant when the accumulated precipitation over a period of time has been large.

Popovich reports that this approach

"has resulted in enhanced operation of the Afobaka development for both flood control and power since 1971. There is absolutely no doubt that, without the use of satellite imagery, the flood flows downstream from the Afobaka Dam would have been substantially greater, and, additionally, a lesser amount of water would have been utilized for power purposes with an attendant reduction in power generation."

He continues that experience indicates that mean basin precipitation has been accurate to within ±6 mm in 60% of all cases, and ±12·5 mm in 90% of all cases. He concludes: "It is my subjective opinion that areal estimates of precipitation derived from satellite imagery are more accurate and reliable than those derived from a low density precipitation gauge network."

Lastly, reference must be made to a unique application of a NESS cloud-indexing method—to areal rainfall estimation over three major lakes in central Africa.

Phase II of a hydrometeorological survey organized by the UN Development Programme in conjunction with the World Meteorological Organization for the catchments of Lakes Victoria, Kyoga and Mobutu Sese Seko has had as its main objective the formulation of a mathematical model to simulate the behaviour of the Upper Nile river system to help in the future development of the water resources of the Upper Nile (Follansbee, personal communication: unpublished WMO report, 1979). Due to the scanty distribution of precipitation gauges over the three lakes, attempts were made from October 1974 to December 1977 to supplement the incomplete and spotty rainfall data through the use of satellite imagery, mostly from polar-orbiters, but subsequently from Meteosat also. Results, compared with climatological gauge data, suggested that:

(a) reasonable monthly and annual estimates were obtained for Lake Victoria and its land catchment area;
(b) estimates for Lake Kyoga were slightly high, by 10–15%;
(c) estimates for Lake Mobotu Sese Seko were definitely high, by 35–40%.

It was recommended that project meteorologists familiar with the region should be able to make suitable adjustments to the rainfall estimates for the synoptic situations in which overestimation is concentrated. Given the facts that the method has operated well over the largest lake (Victoria), and that suitable adjustments are thought practicable for the smaller lakes, there would seem to be a demonstrated potential for practical applications of some existing satellite rainfall monitoring techniques across large inland water bodies, especially in support of river management and control.

The work of Amorocho (1975) may be cited as an example of a river catchment study based on geostationary satellite data alone. In the case of the Upper Sinu river basin in Columbia, South America, the amount of daily precipitation in a headwater area of some 4500 km^2 upstream from a potential damsite in a narrow gorge was inferred from cloud imagery obtained from ATS-3, the second successful geostationary satellite, which was maintained approximately above the equator at 69° W longitude from 1970. Evaluations were made of cloud cover changes over the catchment, taking account of cloud amount, and cloud type with special reference to convectional clouds. Surface data from rain and stream gauges were used to calibrate the observed cloud variations, and thus provide estimates of basin rainfall.

Table 11.1 summarizes the areally integrated results. They confirmed that even raw, unenhanced visible images from the satellite were acceptable for

TABLE 11.1   Comparison between computed and recorded flow volumes and water balance, Upper Sinu River Basin, Colombia, January–December 1973 (after Amorocho, 1975).

Quantity	Volume (m^3 × 10^9)	Depth/area average (mm over 4500 km^2)
(1) Total recorded flow (streamflow data)	9·793	2176
(2) Total computed flow (satellite/raingauge data)	9·908	2202
(3) Difference between (2) and (1)	0·115	26

such work, though, undoubtedly, infrared imagery for night as well as day would have been better for such purposes. Later SMS, and present GOES, satellites have provided such round-the-clock coverage—though such coverage has not been, and is still not regularly available for some parts of the tropics on account of deficiencies in the geostationary satellite network (see Chapter 2).

IV. River Monitoring and Control

Whilst areally integrated rainfall is the key parameter in many hydrological monitoring programmes, especially for ground-water recharge estimation and basin modelling, run-off via rivers or streams is often of at least comparable importance. River gauging records have been used both in the calibra-

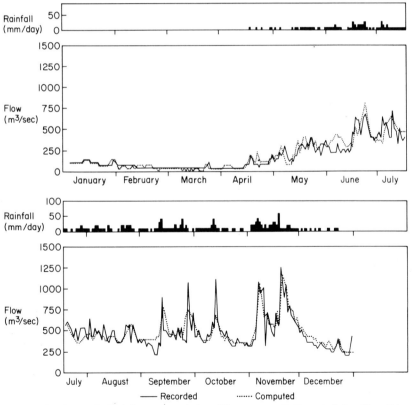

FIG. 11.4   Recorded and computed (satellite assessed) flows of the Sinu River, Colombia, South America, 1973. After Amorocho, 1975.

tion and verification of satellite estimates of rainfall; tentative steps have been taken towards the estimation and even prediction of streamflow itself from satellite evidence.

As noted in the previous section, satellite rainfall estimates were calibrated by streamflow records by Amorocho (1975) for one catchment in Colombia. Figure 11.4 compares recorded and calculated flows for the Sinú River in Colombia in 1973. It is seen that the model reproduced the major floods events quite well, although there was a tendency for it to smooth and subdue the smaller flood peaks. This would be acceptable in any situation in which the chief preoccupation was with extreme events, and in which the ground data alone were inadequate for estimating areal precipitation values owing to the very broken distribution patterns of tropical rain. Although in the case cited the purpose of the study was primarily to verify streamflows and water balances, it is likely that modifications could be made to the method so that acceptable rainfall and run-off data might be obtained from suitable catchments through operational periods with a much lower level of ground instrumentation than in an initial calibration period.

Elsewhere, river gauging records have been used to verify areal rainfall estimates obtained initially by cloud-indexing schemes on a grid square basis. For example, the study by Barrett (1975b) in north-western Sumatra in support of irrigation design entailed an area divisible into 31 grid squares with sides of 10' latitude and longitude (c. 18·5 × 18·5 km), but still containing only five dependable raingauges. In order that the most realistic rainfall maps possible might be prepared, all five raingauges were employed in the mapping programme itself. Some months after the completion of the basic study river gauging records were made available for the two small catchments indicated in Fig. 11.5 (Barrett, 1976). Using a simple mass-balance approach (rainfall − estimated potential evaporation over turf = runoff (river discharge)) the rainfall estimates for the months of July to December in the Krueng Aceh and Krueng Baro catchments were found to be 79·2% and 87·8% respectively of the anticipated values. Since actual evapotranspiration is almost always less than potential evapotranspiration these results were quite pleasing, especially since they depended only on twice-daily images from polar-orbiting (DMSP-Block 5C) satellites, for a period predating the Japanese geostationary satellite. The comparative study was useful also in that it indicated ways in which the particular cloud indexing method employed in that case could be improved by fine-tuning for local conditions.

Turning lastly in this section to rainfall monitoring for the primary purposes of monitoring streamflow and controlling rivers and reservoirs (see also Chapter 12 for discussion of floods) it seems that promising early work in these directions has not been adequately followed up: in the early 1970s

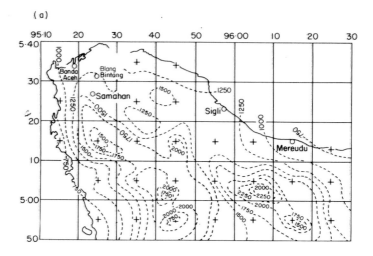

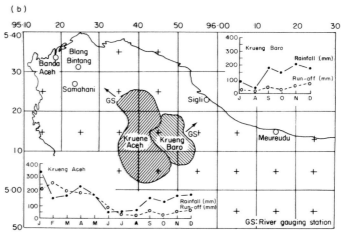

FIG. 11.5   Results of satellite rainfall studies in northern Sumatra. (a) Satellite-assisted annual rainfall assessment, 1974. Rainfall observations were available only for the named locations. Satellite estimates were made for the remaining overland grid squares. (b) Verification of the monthly results obtained prior to the construction of map (a). Shaded areas are two drainage basins for which river flow data became available from river gauges at points marked (GS). Graphs represent observed and satellite-estimated run-off. After Barrett, 1976.

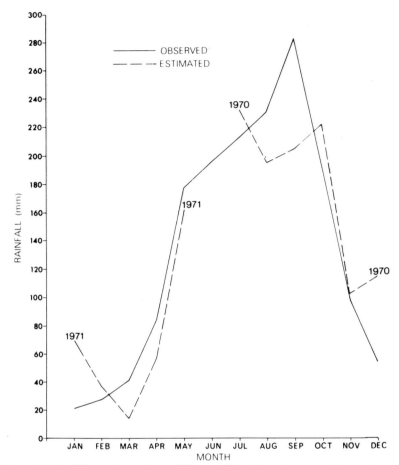

FIG. 11.6   Satellite-estimated monthly rainfall feeding the upper Mekong River, 3 July 1970–1 May 1971, versus monthly mean observed rainfall for 79 stations in Thailand from 1931 to 1960. After Follansbee, 1973.

both cloud indexing and area brightness techniques were demonstrated in such contexts with some success.

One example of the use of the cloud-indexing approach concerned the Mekong River Basin in south-east Asia (Follansbee, 1973). Since no conventional rainfall data were available from the northern half of the basin to the Mekong Committee of the United Nations in Bangkok, a simple cloud indexing method employing a fixed weighting system for different categories of clouds was organized using APT images from NOAA satellites. The resulting estimates of mean areal rainfall (see Fig. 11.6) served as inputs into the computer programme established by the Committee for the forecasting of river discharges and floods. In this way bogus data produced by extrapolation or from climatic means were replaced. Results for the initial trial period in 1970–1971 yielded estimated monthly totals (aggregated from the daily estimates) which were in good agreement with monthly means for 79 rainfall stations in Thailand from 1931–1960.

An example of the area brightness approach is afforded by the research study undertaken by Grosh *et al.* (1973) (see Section IV, Chapter 4). This demonstrated that stream run-off in Surinam in South America, associated with cloud systems observed in the visible by the geostationary satellite ATS-3, was proportional to the area of bright cloud when nocturnal storms were insignificant. It was suggested that hydrologists working in such tropical regions could compile empirical curves relating the volume of direct run-off to bright cloud area for each basin and season; through these, storm area measurement derived from satellite pictures should yield acceptable estimates of direct run-off, and associated potential flood hazards. In this way these data might be provided for remote tropical areas from which it is difficult to obtain conventional measurements of rainfall and river stage in near real-time.

V. Conclusions

There can be little or no doubt that in many areas, inputs of satellite data into operational rainfall assessment programmes could lead to more realistic evaluations not only of rainfall itself, but also a number of dependent variables which have considerable significance for man, especially where information is required quickly. A considerable range of appropriate techniques has been developed and tested with success. Some of these techniques necessitate expensive hardware and specially developed software, whilst others require the services of trained analysts to undertake procedures largely or completely dependent on human skill. Since there are so many potential users of improved rainfall data the question arises as to whether the benefits of such data might be maximized by central processing of the raw inputs from raingauges and satellites, and ready dissemination of the results. This question will be discussed further in Chapter 15.

CHAPTER 12

Floods, Droughts and Plagues

I. Satellite Remote Sensing of Environmental Hazards and Disasters

Most, if not all, of man's activities are subject to the influences of a wide variety of natural hazards. These, in extreme circumstances, prompt effects of disastrous proportions. Without doubt the most important stimuli of natural disasters are atmospheric in origin.

Some atmospheric elements or phenomena pose relatively immediate threats to man and his activities. Table 12.1 lists types of hazardous weather which have been identified as significant for short-term forecasting purposes in one particular region, the eastern USA. Clearly several of these weather hazards involve rainfall as the key atmospheric element. Satellites have been used increasingly in recent years for the recognition and/or prediction of such effects, especially since the advent of geostationary satellites, which are best-suited to the study of dynamic, short-lived phenomena (see Wasserman, 1977).

Other environmental hazards must be seen as medium- or long-term effects, for they result from the abnormal accumulation of different types of conditions through more extended periods of time. Table 12.2 summarizes some of the more important hazards which influence agricultural activity and production, classified on the bases of the time-scales in which they are influential.

In this chapter attention must be focussed on the identification and assessment of rainfall-related hazards, using satellites to augment conventional observations of rainfall amounts and distributions. Related questions,

TABLE 12.1   Types of hazardous weather and weather-related conditions in the eastern USA (after Wasserman, 1977).

Hazardous weather

Heavy rain	Low cloud ceilings
Heavy snow	Clear air and mountain wave turbulence
Freezing rain	Blowing snow
Fog	Frost, freeze, and cold wave
High winds	Heatwave
Strong low level wind-shear	High air pollution levels
Tornadoes and other severe	
thunderstorms (including high	
winds, hail and lightning)	

Related conditions

Flash flooding	High waves
Coastal or lake flooding	Beach erosion potential
River flooding	Fire weather
Aircraft icing	

TABLE 12.2   Examples of types of rural disasters, classified on the basis of the time needed for their impact to be felt on agricultural production (after Howard *et al.*, 1978).

(a) *Short-term impact:* days to months
—severe storms and hurricanes;
—flooding from high intensity, short-period rainfall;
—earthquakes;
—forest fires;
—wind;
—frosts.

(b) *Short-term cumulative impact:* several weeks to several years
—flooding from prolonged rainfall;
—drought;
—seasonal effects caused by snow and ice cover;
—adverse synoptic weather situations;
—continuous heavy cloud cover;
—insect infestations and/or disease.

(c) *Long-term impact:* years to centuries
—climatic changes;
—soil degradation;
—desertification;
—pollution: industrial and agricultural.

involving hazard mitigation and disaster avoidance or relief are beyond the scope of this book. The discussion will for the most part be concerned with approaches and projects with quite narrow and specific objectives. However, it is noteworthy that efforts have been made in some weather forecasting centres to invoke satellite data more generally alongside various conventional data types for the monitoring of a wide range of short-term hazards as part of normal operations. Phenomena involved have included heavy rainfalls, flash flood situations, and severe thunderstorm activity (Wasserman, 1977). In all three cases, "eyeball" techniques for satellite image analysis have been devised, to be applied quickly and simply to the locally available satellite pictures. These techniques contrast rather sharply with the more detailed methods on which most of the results in Sections II–IV depend. Their significance lies in the scope which they reveal for valuable evidence of some acute hazard and disaster situations to be gained comparatively cheaply, without the need for additional staff or equipment.

II. Flooding from Intense Rains

A series of flood disasters has spurred efforts to add satellites to the rain-gauges and radars that form the backbone of modern flood warning systems. Here we discuss applications of satellite observations to the problems of flash floods and tropical storm floods.

1. Flash floods

A flash flood comes so quickly that people in its path have little or no time to react. Thus flash floods may gravely threaten human life, and the magnitude of a flash flood disaster is sometimes out of all proportion to the hydrological flood which produced it. Examples from recent years were mentioned in Chapter 5.

Flash floods result when very heavy rain falls over a watershed which, because of its slope or the nature or condition of its surface, is unable to hold and absorb a significant part of the water mass. Characteristically these heavy rains are convective in nature, and therefore transient, or at least abrupt in onset. Such convection is easily observed and followed in images from geostationary satellites and, as we have seen in earlier chapters, several techniques exist which may be useful in flash flood forecasting. Here we shall examine the performance of the most prominent of these techniques in *post facto* applications to three major flash floods.

(a) *Big Thompson.* The Big Thompson River drains a mountain catchment abutting the continental divide in Colorado. Eastward the basin narrows to a

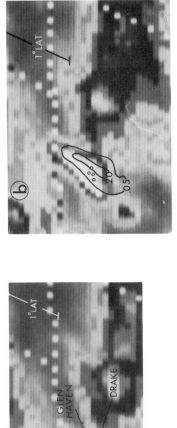

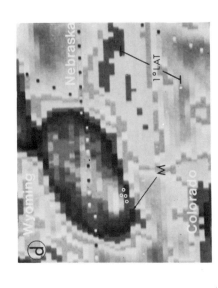

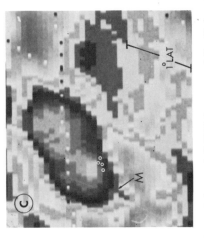

FIG. 12.1   Enhanced GOES imagery of the Colorado-Wyoming-Nebraska region, 1 August 1976: (a) 0030 GMT; (b) 0100 GMT, with satellite rainrate in inches per half hour; (c) 0230 GMT; (d) 0300 GMT. From Scofield, 1978b.

canyon which contains the river in its passage through the Front Range of the Rocky Mountains. The highway paralleling the river on the floor of the canyon is a main approach to the interior mountains. Its flanks have attracted cottages, mobile homes, motels, campgrounds and other enterprises. There are three small towns in the canyon.

During the night of 31 July 1976 the canyon was scoured by a flash flood. One hundred and thirty-nine people died and damage was reckoned at US $35 million. The cause of the flood was a large thunderstorm in the upper watershed, a storm of extraordinary intensity and persistence. Over 300 mm (12 in) of rain fell in parts of the basin, with most of this rain coming in a period of 4 h, and more than half falling at rates exceeding 150 mm (6 in) h^{-1}.

A general account of the Big Thompson flood is given in a Natural Disaster Survey Report (NOAA, 1976). Maddox *et al.* (1978) have identified synoptic- and large meso-scale factors, and Caracena *et al.* (1979) meso-scale and cloud·microphysical factors, which contributed significantly to the flooding. These include:

(i) marked synoptic features: especially a surface front and upper level ridge;

(ii) a very moist, potentially unstable surface airstream impinging from the east at high velocity (15–20 m s^{-1}) upon a north-south mountain barrier;

(iii) light winds in the middle and upper troposphere; and

(iv) a cloud base well below the freezing level.

In the storm which resulted an increasing upward tilt allowed raindrops to fall out of rather than through the updraft, cloud base was at or very close to ground level, warm cloud (coalescence) processes dominated precipitation formation, the movement of cells was slow, and cell development was anchored to an intermediate elevation. For 3 h at and near the storm's peak, according to the calculations of Caracena *et al.* (1979), 85% of inflowing water vapour was converted to rain. The average flux of rain over this period was 9 × 10^6 kg s^{-1} (9000 m^3 s^{-1}). Although flux of rain briefly reached 9000 m^3 s^{-1} over the Big Thompson catchment, most of this rain fell outside. In fact, Big Thompson was only one of several flash floods recorded on eastern slope rivers in northern Colorado and southern Wyoming on the night of 31 July.

Scofield's (1978) use of the Scofield-Oliver technique (described on pp. 75–81) to estimate rainfall from this storm is illustrated in Fig. 12.1. Precipitation began at 0000 GMT (Caracena *et al.*, 1979), shortly before the first picture shown, and abruptly decreased soon after the last picture. A half-hour rate in excess of 50 mm (2 in) was determined for part of the upper

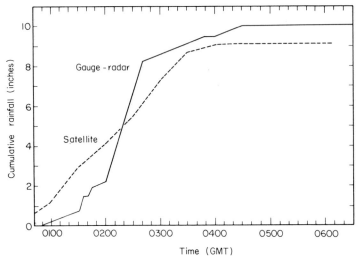

Time (GMT)

FIG. 12.2   Accumulated rainfall curves for Glen Comfort, Colorado, 0040–0630 GMT, 1 August 1976. The gauge/radar curve is from Carcena *et al.*, 1979; the satellite curve from Scofield, 1978.

Big Thompson catchment between 0030 and 0100 GMT during the first burst of heavy storm rainfall (Caracena *et al.*, 1979). This was a period of especially rapid anvil growth (Fig. 12.1a, b). Scofield (1978) draws attention to a feeder band (M in Fig. 12.1c, d) extending south from the anvil, with convective cells forming here and merging with the anvil to the north.

Rainfall was estimated for Glen Comfort, one of a few raingauge sites in the storm area. Cumulative satellite rainfall was only 10% less than gauge rainfall, though the maximum satellite rate was less than half that estimated by Caracena *et al.* (1979) from radar observations (Fig. 12.2). Otherwise agreement is remarkably good.

An independent estimate of rainfall from this storm was made by Woodley *et al.* (1978), using the scheme of Griffith and Woodley (see pp. 69–75). Infrared images of the same GOES/SMS satellite were used, though in this case these were in digital form and were processed by computer. Woodley *et al.* produced two analyses. Sites of reported flash floods fell within or close to the isohyets of 100 mm on the first, a large-area map (Fig. 12.3a). There was no independent analysis of rainfall on this scale; however, Caracena *et al.* (1979) estimated *heavy* rainfall (rates > 136 mm h^{-1}) from gauge-calibrated radar observations for a smaller area containing the Big Thompson catchment. The analysis of Woodley *et al.* encloses this area (Fig. 12.3b). Maxima are comparable, but not coincident, and the satellite estimate is much more broadly dispersed. Time plots of hourly and cumulative rainfall

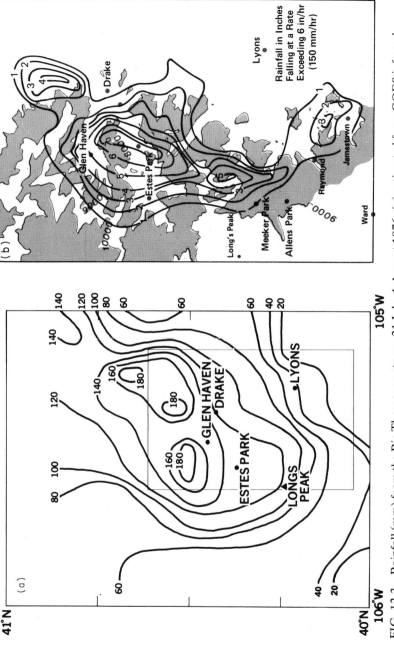

FIG. 12.3 Rainfall (mm) from the Big Thompson storm, 31 July–1 August 1976: (a) estimated from GOES infrared satellite imagery by Woodley *et al.* (1978): (b) estimated from radar for rates over 150 mm h⁻¹ for the boxed area inside (a), from Carcena *et al.*, 1979. After Woodley *et al.*, 1978.

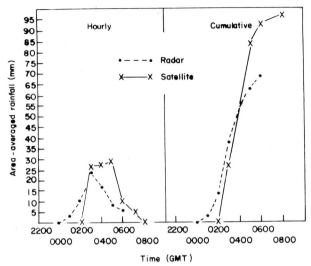

FIG. 12.4 Total rainfall estimates averaged over the Big Thompson basin (843 km²), 31 July–1 August 1976. The radar curve is from Caracena *et al.* (1979), the satellite estimates from Griffith and Woodley. After Woodley *et al.*, 1978.

for the Big Thompson catchment show close tracking of satellite and radar estimates as intensities increase, then satellite overestimates amounting to 25 to 40% by the time rain ended (Fig. 12.4).

Bielicki and Vonder Haar (1978) also estimated rain from this storm. They applied the Scofield-Oliver technique to infrared images displayed on a television monitor. Although their estimate of rainfall for the Big Thompson catchment is close to that inferred from radar, the result is inconclusive: the area of precipitation in their estimate, and its continuity through gaps in the satellite sequence, were adjusted by means of the same radar data.

(b) *Johnstown.* Johnstown lies at the confluence of the Little Conemaugh and Stoney Creek Rivers in Pennsylvania. It is within the watershed of the Ohio River, close to the Appalachian divide. Hills and low mountains crowd the rivers in this part of Pennsylvania, leaving scant room in the valleys for towns and cities. Floods have plagued Johnstown from settlement times, and, in 1889, made its name synonymous with disaster.

The flood of 1889 was caused by rains of 150 to 250 mm (6 to 10 in), falling through one evening and into the next morning over a large part of west central Pennsylvania. Two thousand two hundred and nine lives were lost, most in Johnstown when a wall of water from a collapsed dam on a tributary to Little Conemaugh River swept across the town. Rainfall in the flood of 1977 exceeded 300 mm (12 in), and most of it fell in the space of

10 h. Eleven dams were partly or wholly destroyed. Seventy-seven people are known to have died and damages amounted to $200 million.

McCullough (1968) has documented the flood of 1889, a Natural Disaster Survey Report of NOAA (1977) described the flood of 1977, and Gilbert (1977) summarized the long history of flooding at Johnstown. Hoxit *et al.* (1978) have analysed meteorological conditions attending the 1977 flood, and include as appendices to their study estimates of rainfall from radar (Greene and Saffle, 1978) and from satellite observations (Scofield, 1978c). Though the proportions were different, ingredients in the meteorological mix which brought disaster to Johnstown in 1977 were much like those contributing to the Big Thompson flood, namely, strong surface pressure gradients and an upper shortwave trough providing synoptic forcing, a very moist, potentially unstable surface airstream impinging on an extended barrier, weak shear enhancing precipitation efficiency, slow movement of storms, and repeated cell generation. Perhaps the main differences between the Johnstown and Big Thompson storms were the absence of a front in the former and the augmentation of barrier uplift by surface outflow from earlier storms.

Two waves of thunderstorms moved through Pennsylvania on 19 and 20 July 1977. In the first lower tropospheric moisture content, air mass instability and low level inflow were moderate. Only the second produced significant flooding, and this occurred just inside the western flank of the boundary marking outflow from the first squall line system. Thunderstorms of the second system were very intense here, and came in waves (radar showed that nine raincells crossed Johnstown during the night of 19 July).

The satellite estimates described here suffer from a seven picture gap in coverage across 3½ h late in the storm's life (the kind of lapse which must never occur if satellites become mainstream elements of flash flood forecast systems). Scofield's (1978a, b) estimates did not attempt to bridge the gap, and consequently we show only hourly satellite rainfall (Fig. 12.5), and a single infrared picture (Fig. 12.6) intermediate in time to the estimate in Fig. 12.5. The cloud shown in this picture is the anvil of the second squall line system, here close to the peak of its intensity and organization. Even more so than in the Big Thompson storm, the anvil of the Johnstown storm tended to be symmetric, and in this respect was more like tropical than middle latitude thunderstorms (see, for example, Weickman *et al.*, 1977). The numbers superimposed on the satellite picture represent rain intensity measured by the weather radar at Pittsburg in west-central Pennsylvania as part of a Digital Radar Experiment (D/RADEX; Green and Saffle, 1978). Their corresponding precipitation intensities are given in Table 12.3 The pattern of radar rainfall accumulated through the storm closely matched that of gauge rainfall, but amounts tended to fall below gauge measurements, by as

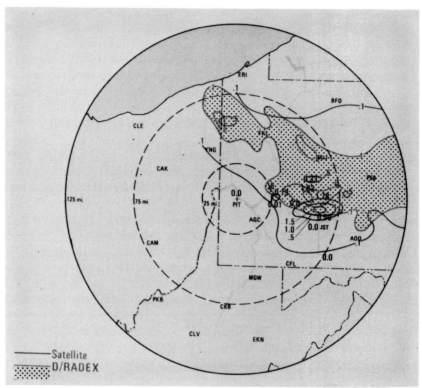

FIG. 12.5 Hourly rainfall analyses for the Pittsburg (PIT) area, 20 July 1977. Contours in inches. Johnstown (JST) is also marked. From Scofield, 1978c.

TABLE 12.3   Rainfall rates for D/RADEX levels.[a]

| D/RADEX level | Rainfall rate (mm h^{-1}) | |
	Range	Midpoint value assigned
1	1·1–2·8	1·8
2	2·8–5·8	4·1
3	5·8–11·4	8·6
4	11·4–28·9	17·3
5	28·9–57·3	40·0
6	57·3–114·3	80·0
7	114·3–180	180
8	180	180
9	180	180

[a]From Greene and Saffle (1978).

much as one-third in the area of heaviest rainfall. In Fig. 12.6 the effect of satellite perspective is an apparent northward shift of the high clouds relative to the map lines and locations shown (and relative to the D/RADEX values as well): cold cloud features actually are about 15 km south of where they appear to be. Allowing for this small displacement, rainfall was confined generally within the area of cold cloud, and tended to be associated with the cold centres. The outstanding exception was rain along the Ohio–Pennsylvania border, which occurred in a small thunderstorm complex close to the north-west flank of the main outflow region.

One consequence of the vastness of the anvil in this storm is a tendency for the satellite rainfall to spread beyond that recorded by the radar (Fig. 12.5). Due to the relative flatness of temperature gradients across the anvil and, as noted by Scofield (1978a), the large textural changes which occurred from picture to picture, there is more uncertainty than usual in the location of heavy rain cells. This may explain the discrepancy in locations of maxima north of Johnstown in Fig. 12.5. Hoxit et al. (1978) note that the radar analyses are smoother than Scofield's satellite analyses. Gauge observations confirm the radar rain maximum in location, and the satellite maximum in amount. Scofield's (1978a) satellite estimate of rainfall at Johnstown is compared with gauge and radar measurements in Fig. 12.7. Allowing for missing radar and satellite data, the curves are remarkably close.

The estimate of Woodley et al. (1978) interpolates across the seven missing images. Satellite and gauge storm rainfall are shown in Fig. 12.8. Woodley et al. note general agreement in the magnitudes and orientations of the main maximum on each map, but there the agreement ends. The satellite algorithm underestimated rainfall in the flood area around Johnstown, but so greatly overestimated rainfall around it that the net effect was a 60% overestimate of rainfall for the map area.

(c) *Kansas City*. Brush Creek enters the Blue River just above the Blue's confluence with the Missouri River at the eastern edge of Kansas City. Through most of its 15 km length it is an urban stream of gentle gradient, draining southern parts of Kansas City and suburbs to the west. Above Kansas City's Country Club Plaza shopping centre, where Brush Creek turns east to meet the Blue River, the catchment is only 38 km². Yet on the night of 12 September 1977 flow on Brush Creek at the Plaza shopping centre reached 500 m³ s⁻¹ (17 600 ft³ s⁻¹). Twenty-five people died in floods in and near Kansas City that night and property was damaged to the extent of $90 million. Most of these losses were along Brush Creek.

The meteorology of the Brush Creek flood is discussed by Hales (1978), from which most of this account is drawn. In fact two storms occurred back-to-back. The first came in the early morning hours of 12 September.

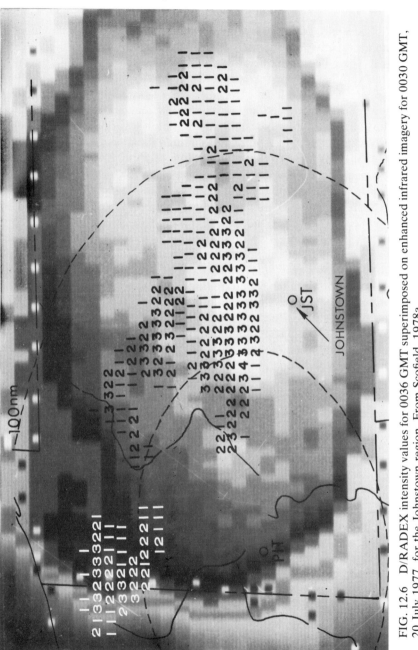

FIG. 12.6 D/RADEX intensity values for 0036 GMT superimposed on enhanced infrared imagery for 0030 GMT, 20 July 1977, for the Johnstown region. From Scofield, 1978a.

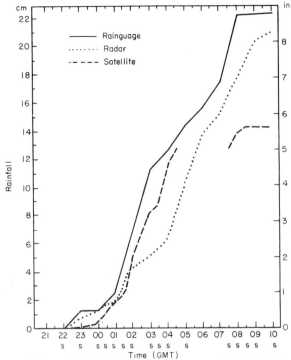

FIG. 12.7   Cumulative rainfall (inches) at Johnstown, 2100–10000 GMT, 19–20 July 1977. S indicates availability of satellite infrared images. After Scofield, 1978a.

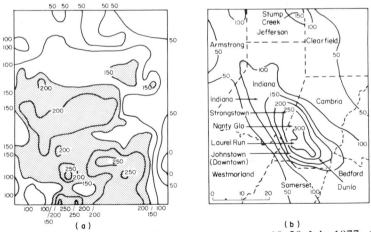

FIG. 12.8   Rainfall (mm) from the Johnstown storm, 19–20 July 1977: (a) as estimated by Woodley et al. (1978) from GOES infrared imagery; (b) as observed by gauges according to the NOAA Disaster Survey Report, (1977). The areas in the two maps are identical. After Woodley et al., 1978.

Although only slightly less intense than the second, this storm's rain fell on dry ground and produced only minor flooding. Rain in the second storm beginning 15 h later exceeded 200 mm (8 in) over much of the upper basin of Brush Creek, and fell at rates as high as 140 mm in 2 h.

The factors contributing to this flood are much the same as in the Johnstown and Big Thompson floods. They include synoptic forcing in the form of a strong low level front and an advancing short wave aloft; a very moist, warm surface airstream impinging on a persisting barrier; large potential instability, and repeated passage of cells across the same catchment. Here the barrier was entirely atmospheric: a pool of cool outflow air left by the first storm and reinforced by the first cells of the second storm. A few hours before heavy rains began along Brush Creek, the southern boundary of this outflow pool was evident in satellite pictures as a line of clouds between regimes clear of small cumulus clouds to the north and dominated by cumulus to the south. The line lay just to the south of Kansas City. Two lines of thunderstorms moved across Kansas City during the evening. These linked up over Brush Creek.

Scofield (1978b) estimated rainfall at Kansas City International Airport (25 km north-west of Brush Creek) for the first and second storms (Fig. 12.9). Times of onset were very close for both storms, but cessation of rain occurred too early and rate was a little high for the second storm. The result was an underestimate of rainfall by the satellite: 25% in the first storm, 6% for both storms. No estimate is available for the scheme of Griffith and Woodley.

Summing up these experiments we may say that both Woodley and Griffith and Scofield and Oliver approaches have demonstrated skill in estimating storm rainfall from measurements made on clouds in satellite images. However, a fairer comparison of the two techniques would be blind testing, because, of course, by the time these experiments were run only the computer did not know the clouds present had produced disastrous flash floods. Even better would be testing under operational conditions.

Two factors in particular require careful attention.

(a) The overriding importance of accurate *Earth location* is apparent if one contemplates the maps of observed rainfall, especially that of the Johnstown flood (Fig. 12.8), where a mere 23 km separated 300 mm rains from 10 mm rains. (The resolution of SMS/GOES infrared imagery at that latitude is about 10 km.) Uncertainties in image navigation may have contributed significantly to the smoothing of isohyets in estimates by the Griffith and Woodley approach.

(b) The importance of continuity in image sequences is apparent in the results of Woodley *et al.* for the Johnstown flood. In such cases a technique

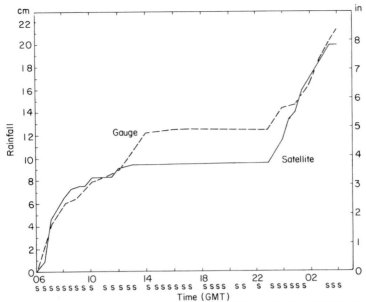

FIG. 12.9   Cumulative curves of observed and satellite-derived rainfall (in inches), Kansas City International Airport, 0600–0400 GMT, 12–13 September, 1977. S indicates availability of satellite images. From Scofield, 1978b.

offering the flexibility of human judgment has advantages. For its part the scheme of Scofield and Oliver tended to produce a hummocky field composed of many uniform cells on a background of (over-extensive) light rainfall. This is a consequence of discreteness in levels of rain rate, and the weight given to domes and towers on top of relatively flat anvils. The Scofield and Oliver technique also tends to underestimate rainfall in the late stages of a storm, to the extent that Oliver and Scofield (personal communication, 1979) have modified their technique as it is applied to tropical storms and cells in large, persistent clusters of thunderstorms by requiring for these situations a 25 mm $(1\frac{1}{2}\,h)^{-1}$ increase in rainrate and a continuance of higher rates 1 h into the warming phase in the evolution of the cells.

A technique in a state of rapid flux cannot be said to be fully operational. Perhaps this, as much as anything else, has slowed the transfer of severe storm monitoring techniques to forecast offices. However, NESS is now furnishing the National Meteorological Center (NMC) with estimates of rainfall derived by the Scofield-Oliver method to support NMC's flash flood forecasting and quantitative precipitation forecast programmes (Follansbee, personal communication, 1979).

The complexity of such an approach may also be a handicap to acceptance by operational forecasters. Stout *et al.* (1979) proposed a simpler method for the operational estimation of convective rainfall, which would employ two measurements only, namely cloud area and area change, and two coefficients. This approach remains to be tested.

2. Tropical storm floods

Flooding is the rule when Atlantic hurricanes, Pacific typhoons, and the cyclones of the Indian Ocean make landfall. It often accompanies weaker versions of these tropical revolving storms, and may pose a graver threat to life than winds or surges of the tides. Examples of severe flooding from recent years include Agnes, from June of 1972 (120 dead in the north-east United States); Fifi, from September of 1974 (3000 known dead in Honduras); David, from August of 1979 (1000 dead in the Dominican Republic); and the "Bangladesh" cyclone of November, 1970 (300 000 to 1 000 000 dead). Tropical storms are so closely monitored now that, at least in advanced countries, assuring public safety in such disasters is less a problem of information than appropriate reaction, especially when floods are extensive. Only mass evacuations, like those organized when hurricanes threaten along the Gulf and Atlantic coasts of the United States, may be effective in protecting people's lives. Then knowledge of the flood threat becomes important. This leads directly to the question of how much rain is falling and has fallen from the storm and the more difficult corollary, how much rain will fall and how rapidly.

The satellite schemes so far described are merely diagnostic inasmuch as no prediction of future rain rate is involved. Both the Scofield and Oliver and Griffith and Woodley techniques have been applied in this mode to tropical storms. (Recall also the application by Adler and Rodgers (1977) and by Allison et al. (1974) of microwave observations to measurement of rainrate in a Pacific typhoon (pp. 203, 153).) However, Griffith et al. (1978) described an adaption of their scheme which gives it true predictive value.

Considering first the main diagnostic applications, the tropical storm for which Scofield and Oliver (1977) experimentally estimated rainfall (see pp. 77–79) produced modest rain amounts. A more interesting test in the present context was hurricane Anita, which struck north-east Mexico 250 km south of Brownsville, Texas on 2 September 1977. Meteorologists at the National Environmental Satellite Service, NOAA, produced operational estimates of storm rainfall which were transmitted to forecasters in Texas at the request of the San Antonio, Texas, Weather Service Forecast Office (McGinnis et al., 1979). The sum of 16·5 h of satellite rainfall was compared with 24 h gauge rainfall. Although some large point differences were apparent, with substantially larger satellite values in the region of heaviest rainfall, there was agreement in the general pattern. Partly on the

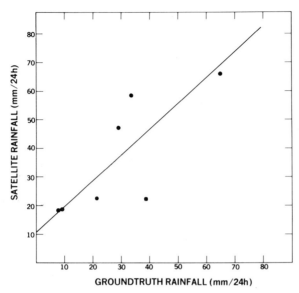

FIG. 12.10   Daily area-average rainfall as estimated from satellite images and as measured by gauges, for three hurricanes over land. After Griffith *et al.*, 1978.

basis of this satellite information, Texas forecasters decided against issuing flash flood warnings for Brownsville and the lower Rio Grande Valley.

As has already been noted (p. 71), Griffith and Woodley modified their technique for use on tropical storms for these may produce copious rain, yet show no growth of the anvil canopy. The modification consists of an upward adjustment of rainfall according to that fraction of the cloud which is above designated secondary cloud temperature or brightness thresholds (Griffith *et al.*, 1978). Skill was demonstrated in estimating average daily hurricane rainfall for three storms (Fig. 12.10).

An estimate of average rainrate, together with information on expected size, speed, and position, raises the possibility of inferring total rainfall for points in the path of a hurricane—a true predictive application of a satellite scheme. Griffith *et al.* (1978) represented total rain potential for a station in the path of a tropical storm as

$$\hat{R} = (\delta/v)<R>$$

δ is the width of the storm along its path relative to the station, v is storm speed, and $<R>$ is storm average rainfall rate. When applied to a dozen hurricanes (Table 12.4), the index indicated high rain potential for the flood hurricanes Agnes and Fifi, and low potential for the dry hurricanes Celia and Edith.

TABLE 12.4   Satellite estimated daily average hurricane rainrate and rain potential.[a,b]

Name of storm	Year	Period of estimate (days)	Total daily rain volume (10^{11} m^3)	Average rain rate (mm day^{-1})	Total storm rain potential (mm)
Agnes	1972	4	0·97	128	393
Fifi	1974	5	1·18	109	232
Caroline	1975	2	0·09	53	150
Anita	1977	2	0·25	46	150
Carmen	1974	4	0·45	116	124
Dottie	1976	2	0·8	48	124
Eloise	1975	5	0·12	60	118
Babe	1977	2	0·29	42	81
Debbie	1969	4	0·24	70	73
Emmie	1976	2	0·31	50	62
Edith	1971	4	0·23	70	57
Celia	1970	4	0·25	39	34

[a]Estimated from visible images through 1975, infrared thereafter.
[b]After Griffith *et al.* (1978).

We may tentatively conclude that satellite visible and infrared techniques can provide useful supplemental information updating gauge and radar observations of tropical storm rainfall over land, and in remote areas perhaps even primary information on the distribution and flux of rainfall. Predictions of storm rainfall also are possible, although best results will probably come through combinations of satellite microwave with visible and infrared image data. Certainly there is merit in further experimental testing with better verification data and tighter controls on *a priori* knowledge of storm rainfall.

III. Monitoring Longer-term Rainfall Anomalies and Extremes

It is convenient for present purposes to differentiate between rainfall *anomalies* (relatively small departures from the climatic norm), and *extremes* (relatively large departures from normal). Whilst on the shorter time-scales extremes are much more important than anomalies, in the longer term there is interest in both, for even quite small departures from the climatic norm may have significant impacts upon the environment, activities, and livelihood of man. Thus, for example, the US Department of Commerce takes an active interest in both anomalies and extremes across substantial sections of the land area of the globe through its Environmental/Resource Assessment and Information programme. Each week a weather report is published by the Environmental Data and Information Service through its Center for Environmental Assessment Services (CEAS). Satellite imagery is used in support of conventional observations in the preparation of these statements, generally but not exclusively through an area estimation technique based on Follansbee (1973) (see pp. 52–54). The CEAS weather reports include:

(a) A weekly energy assessment under the heading "Summary of temperature-related energy consumption".
(b) Agricultural assessments and drought evaluations for a number of major world regions, usually including the USSR, China, Australia, India, South America and Central Africa. "Marginal drought" is defined as 2 or more consecutive months with less than 60% of normal cumulative precipitation; drought as 2 or more consecutive months with less than 40% of normal cumulative precipitation, and severe drought as 2 or more months with less than 20% of normal precipitation.
(c) An outline of the principal weather anomalies under the heading "Major abnormal world weather".

For users with needs more limited to rainfall alone, the Environmental Data and Information Service also publishes some of its rainfall assessments in the form of a weekly "CEAS Weather Summary" for selected areas. The key contents are:

(a) Maps of rainfall as a percentage of the normal, both for the week just ended, and the period since the beginning of the current year (Fig. 12.11).
(b) A tabulation giving the information in (a) in terms of relative rainfall categories (wet, normal, dry, very dry) on a subregional basis.
(c) Summary descriptions of synoptic conditions and associated rainfall through the preceding week and month. The following is an example of such a summary. It appeared in the report for the Caribbean Basin, June 25–July 1, 1979:

"The heaviest rainfalls occurred over Puerto Rico, Nicaragua, eastern Cuba, and Hispaniola. Nicaragua totalled nearly 4·0″; Dominican Republic North, 4·1″; central Haiti, 4·8″; and eastern Cuba, 4·0″. The Lesser Antilles also received abundant rainfall, accumulating up to 6″ by the period's end. Puerto Rico received 2·1″, which is two times normal. The Netherlands Antilles were noticeably dry with virtually no precipitation indicated by satellite or the programmed weather data base. Overall, the Caribbean Basin received above normal amounts of rainfall during this period.

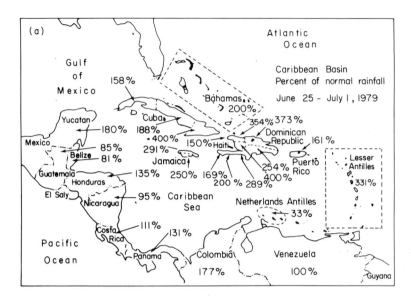

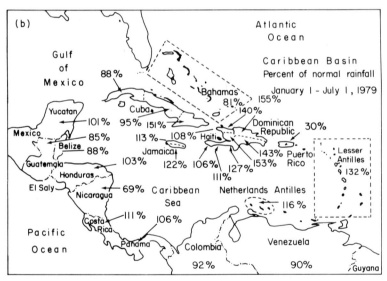

FIG. 12.11   Satellite-assisted rainfall analyses of the Caribbean Basin countries for (a) 25 June–1 July 1979, and (b) 1 January–1 July 1979, from a CEAS Weather Summary, Environmental Data and Information Service, NOAA, 1979.

Rainfall totals for the month also were above normal in most areas, as unusually active precipitation areas moved across the Basin from week to week. Soil moisture should be more than adequate in agricultural areas. Too much precipitation in central Haiti, Puerto Rico, and eastern Cuba may have caused damaging floods to agricultural districts of those countries during this period and throughout the month." (US Dept. of Commerce, CEAS Bulletin)

It would seem that weather reports published by others, with purposes akin to those of the CEAS reports, would also benefit from a consideration of satellite imagery in support of conventional data. One such report is the monthly publication "Foodcrops and Shortages" of the Food and Agriculture Organisation of the United Nations. Since the primary stated purpose of this publication is "To present up-to-date information on general crop weather conditions in over 70 countries round the world so that developing food supply difficulties might be identified", it is obviously concerned with the development and areal distribution of rainfall anomalies and extremes. A specific conclusion of Howard *et al.* (1978) was that satellite-improved rainfall maps prepared primarily for desert locust control trials in north-west Africa (see Section IV) were demonstrably superior to the independent, but conventionally-based, statements for that region in corresponding "Foodcrops and Shortages" bulletins. In this case the fine areal resolution provided by a one-half degree grid-square rainfall mapping programme enabled the interpreters to outline areas of rainfall anomalies and extremes with a much higher degree of areal precision than had been possible using conventional rainfall intelligence alone (see Fig. 12.14). An example cited was the drought in Algeria and Tunisia early in 1977.

One other study which deserves mention here is that of Follansbee (1976). This was concerned with the assessment of precipitation over substantial parts of China and the USSR. One of Follansbee's most interesting products was a map of the number of consecutive days in 1 month for which zero rainfall was estimated for each grid square considered in China and the USSR. This type of map is illustrated here by Fig. 12.12. Produced as a natural by-product of the daily mapping method employed by Follansbee (see Section II, Chapter 4), it has clear potential for the identification and areal assessment of both high and low rainfall anomalies and extremes.

Similar cumulative representations of both rainfall amounts, and rainless days, might be compiled with value in any programme sampling satellite images regularly over land at least once each day.

IV. Pests and Plagues

Many pests and plagues which affect man, his livestock or his crops are weather-sensitive. Satellite remote sensing techniques are being invoked

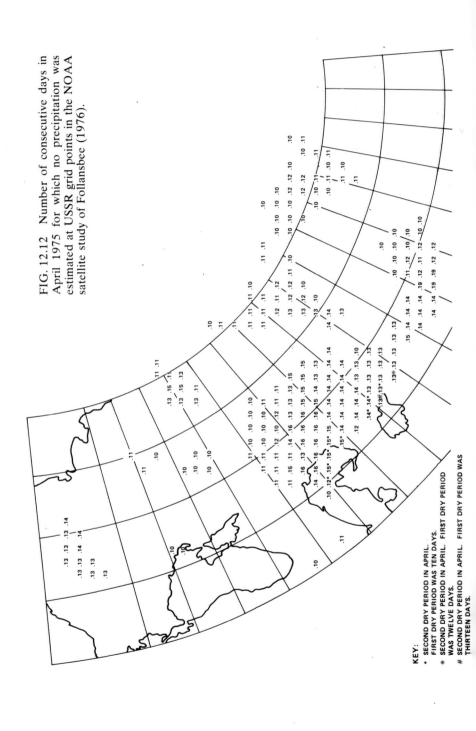

FIG. 12.12 Number of consecutive days in April 1975 for which no precipitation was estimated at USSR grid points in the NOAA satellite study of Follansbee (1976).

KEY:

• SECOND DRY PERIOD IN APRIL.
   FIRST DRY PERIOD WAS TEN DAYS.

⊕ SECOND DRY PERIOD IN APRIL. FIRST DRY PERIOD
   WAS TWELVE DAYS.

SECOND DRY PERIOD IN APRIL. FIRST DRY PERIOD WAS
   THIRTEEN DAYS.

increasingly in programmes whose purposes are the monitoring and control of undesirable insect infestations. An example of such a programme relating to man as the direct object of an insect pest has been the Landsat monitoring of wetlands as mosquito breeding grounds (Barnes, 1978). One programme concerned with a livestock pest involved the Screwworm Eradication Data System (SEDS) summarized by Giddings (1976) and Arp (1976); this entailed the mapping of surface temperatures from NOAA infrared images and an extension and improvement of the coverage provided by ground observing stations (Giddings, 1976). But the project which has most significance in the more narrow context of satellite-assisted rainfall monitoring has been the Desert Locust Satellite Application Project of the Food and Agriculture Organisation (FAO) of the United Nations, undertaken for the international Desert Locust Commission. It is this work, concerned as it is with a major plant pest, which we should discuss in greatest detail.

The Desert Locust (Schistocerca gregaria Forsk.) has historically presented a formidable threat to agriculture in the arid and semi-arid zone of Europe and south-western Asia. Since 1951, when extensive locust swarms and hopper bands threatened many countries, the United Nations, through its agency FAO, has played an increasing role in the coordination of locust survey and control efforts of national plant protection institutes. The Northwest Africa Desert Locust Commission is one of three such bodies established to cooperate with national organizations especially through the provision of coordinative and logistic support.

Until recently, information of both the location of potential desert locust breeding sites, and the assessment of breeding populations, has been gathered by surveys on the ground, complemented occasionally by air searches. However, it has been recognized that such methods are limited in both time and space: often, potential breeding areas cannot be covered to the extent necessary to provide the required information concerning either the scale and distribution of present locust activity, or its potential for future development.

For rapid population expansions, the desert locust is dependent on four principle factors (Anti-Locust Research Centre, 1966). These are:

(a) A desert environment, i.e. a hot and dry climate.
(b) A sandy-silty soil type for egg-layings.
(c) Sufficient soil moisture in the top layer of the soil profile for eggs to hatch.
(d) Green vegetation to feed on during the various stages of development.

Of these four, the first is dictated by rainfall climatology, and the second by desert morphology and pedology. The third and fourth are both heavily dependent on short-term variations in the local availability of water and, therefore, upon rainfall meteorology. It has long been realized that potential breeding grounds for the desert locust are characteristically ephemeral and scattered. They are determined by the spatial and temporal distributions of rainfall events—which, in arid and semi-arid areas are notoriously irregular and "spotty". The feasibility of using satellite data for the detection of potential breeding areas, from which locust swarms of plague proportions can emerge in a matter of only a few weeks given optimum conditions, was first suggested by Pedgley (1974). Pedgley successfully used false-colour renditions of Landsat multispectral images covering parts of the Red Sea coast of the Arabian peninsula for the identification of the locations of areas of active vegetation growth after rains. Proposals were then drawn up for more detailed and extensive research in FAO involving not only the use of Landsat data for vegetation mapping, but, more basically, the use of weather satellite data also for the identification of areas of significant rainfall: within these the search for vegetation would be concentrated. The satellite-assisted rainfall mapping method selected for such purposes was a cloud-indexing technique (see Section II, Chapter 4).

The initial Desert Locust Satellite Application project, organized by the Remote Sensing Centre of FAO, included:

(a) A definition study, in which the structure of a long-term operational programme involving a high degree of international cooperation was proposed by an expert consultant from the UK Centre for Overseas Pest Research (Roffey, 1975) for the area of jurisdiction of one of the regional Desert Locust Control Commissions. This area was north-west Africa, including Algeria, Libya, Morocco and Tunisia. Effective conventional surveillance of this vast, thinly populated region has been virtually impossible.

(b) An initial experimental phase ("Stage I"), in which the remote sensing (weather satellite and Landsat) components of the established project were tested in a selected area of restricted size. The area chosen was dominated by the Ahaggar Massif of southern Algeria, whose flanks are traversed by a number of major wadis in which locust breeding often occurs after significant rains. Stage I was completed in the first half of 1976 (Hielkema and Howard, 1976).

(c) A broader experimental phase ("Stage II") in which the different aspects of the complete programme were tested as nearly as possible in a near real-time context across all four countries of north-west Africa. Stage II was undertaken in 1976/77 (Barrett, 1977b, c; Howard et al., in press).

(d) A staff training phase ("Stage III") in which north-west African nationals were to be trained to staff a fully-operational remote sensing unit. Stage III was commenced in 1979 (Barrett and Lounis, 1979).

Stage 1   Stage 2

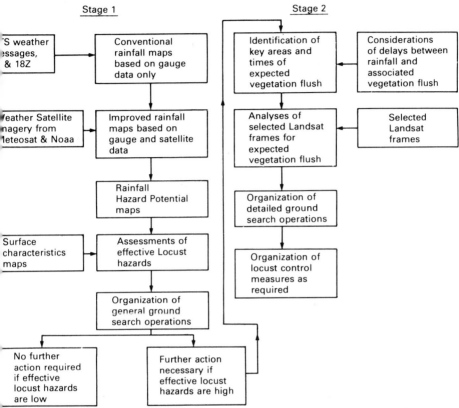

FIG. 12.13   Flow-diagram for satellite-assisted Desert Locust monitoring and control operations in north-west Africa. Rainfall assessment is the first step. From Barrett, 1980b.

The final project objective—the establishment of a fully-operational unit in north-west Africa for satellite-assisted rainfall monitoring, as well as dependent selection and analysis of Landsat imagery—is now being achieved through a new Desert Locust Satellite Applications Development Project (see Barrett, 1980c). Figure 12.13 summarizes the key steps envisaged for routine operational activities.

Figure 12.13 indicates that weather satellite data should be used not only for the improvement of rainfall mapping in a region particularly deficient in GTS weather stations, and for the selection of higher-resolution Landsat images to map areas of subsequent vegetation flush, but also to facilitate route-planning for the existing ground search teams which inspect vegetation, soil moisture, and locust activity in the field. Indeed, early trials have suggested that, through the resulting more economic use of ground inspection teams, economies may be achievable which could at least compensate

J

for the additional expenses of the new remote sensing unit itself. Thus, through the use of satellites for broad-area surveillance, greater efficiency in the locust monitoring and control operation should be achieved at no extra cost (Barrett and Lounis, 1979).

Results of the application of the cloud-indexing method of rainfall in monitoring developed for use in north-west Africa have been obtained twice daily for some 2700 one-half degree grid squares using NOAA visible and infrared images as the satellite data base. The findings have been published in the form of weekly isohyetal maps for north-west Africa, including:

(a) maps compiled from conventional (GTS) data only (Fig. 12.14c) and
(b) maps compiled from conventional plus satellite evidence (Fig. 12.14d).

These results have been discussed in detail by Barrett (1977c and 1980b) and Barrett and Lounis (1979). These authors concluded that

"The satellite-improved maps have undoubted inaccuracies of their own, but it is certain that they are more realistic than those based on gauge data alone . . . They can be viewed with much more confidence in the broad areas not represented by any gauge: whereas the raingauge maps are compiled from, at best, 85 points of data, the satellite maps are based on some type of observational evidence for every one of the 2700 grid squares."

Subsequent analyses of the satellite maps have suggested, too, that they will support valuable new studies of the nature, distribution, behaviour, and movement of rain areas, and the weather systems which prompt them over the north-west Sahara (see Fig. 12.14b). In this way, knowledge and understanding of both weather and climate patterns of north-west Africa will be advanced.

From accumulated results in the springs of 1977 and 1979 maps of likely "hazards" of associated locust development were compiled. Five categories of rainfall-dependent hazard were defined. These were as follows (Barrett 1977c):

(a) "No hazard": in such areas rainfall is deemed to have been zero or negligible;
(b) "Low hazard" (0·1–20 mm rain): conditions are not expected to stimulate vegetation development excepting, perhaps, locally in very favourable sites;
(c) "Moderate hazard" (20·1–40 mm rain): if other factors permit, quite important vegetation development may be found;
(d) "High hazard" (40·1–60 mm rain): such areas should be monitored carefully for vegetation development. This is likely to be widespread and significant excepting where strong inhibiting factors operate; and
(e) "Very high hazard" (≥60 mm rain): widespread, significant, vegetation growth is highly likely unless powerful factors overrule.

Resulting hazard maps were used for the selection of key Landsat frames, as illustrated by Fig. 12.4e. Colour additive analyses of the frames outlined in this example evidenced vegetation flush in central Algeria and north-western Libya. Vegetation was seen to be best-developed along the beds of major wadis, and around and within basins of interior drainage. It was spread much more thinly across plains and plateaux, and then only where surface materials and configurations were suitable.

Such findings suggest that even the improved information on rainfall will, by itself, be less than completely adequate for the most efficient management of ground inspection teams (Fig. 12.14f) and the planning of locust control operations: relationships between rainfall and related vegetation flush in an area of such diverse surface characteristics as north-west Africa are clearly far from simple. The chief variables are thought to include (Barrett and Lounis, 1979):

(a) the length of the lag between rainfall, and vegetation development: it appears that this may be very variable from season to season;
(b) the nature of the surface materials. Heavy rains in some areas are likely to be largely lost by rapid infiltration, whereas in others water is retained longer in the soil; and
(c) the surface topography. In some localities rainwater runs off quickly, and is probably ineffective for promoting much or any vegetation growth. In others it accumulates and even stands for considerable periods of time.

However, the most important conclusion of the work in north-west Africa is that a satellite-assisted rainfall mapping method of the detailed cloud-indexing type can be undertaken for broad areas within the time-constraints of an operational locust control programme. It was only rarely that a single analyst was hard-pressed to complete his 12-hourly rainfall maps in the allotted time. However, it would seem desirable to plan and undertake a parallel research programme to consider the possibilities for at least the partial support of a cheap brightness-contouring system to help pinpoint centres of convectional activity. Indeed, the need for some degree of auto-mation rose sharply with the arrival of Meteosat, the first geostationary satellite to be located permanently over Africa. This type of satellite, with its high frequency of imaging, can help greatly in the assessment of rapidly developing and/or rapidly moving areas of rain. Indeed, it is thought possible that rainfall monitoring for locust control over very broad areas might be effected satisfactorily using Meteosat imagery four times daily for low-resolution (1° square) assessments, and polar-orbiting satellite imagery for higher-resolution (¼ or ½° square) mapping of the more significant rain-clouds as and when necessary (Barrett, 1980c). However, "A continuing basic dependence on polar-orbiting satellites seems necessary, since these

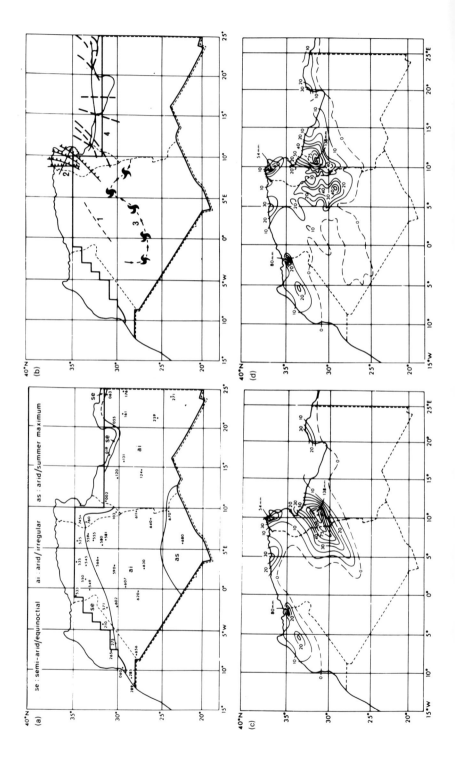

(a)

se = semi-arid/equinoctial   ai = arid/irregular   as = arid/summer maximum

(b)

(c)

(d)

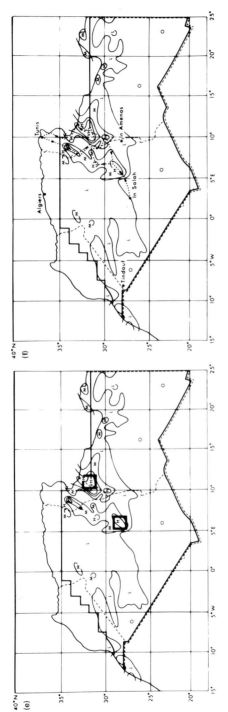

FIG. 12.14   Examples of the use of satellite-improved rainfall monitoring in north-west Africa in support of Desert Locust Control Commission operations: (a) the scatter of GTS rainfall stations, and morphoclimatic regionalization for satellite rainfall estimation: (b) plots of significant rain-cloud systems from satellite evidence: (c) rainfall analysis from gauge data only, and (d) rainfall analysis from gauge plus satellite data, all for 29 March–4 April, 1979: (e) selection of Landsat frames for post-rainfall vegetation assessment, and (f) selection of preferred routes for Ground Inspection Teams, both on the basis of "Hazard Maps" (zero, low, medium, high and very high locust hazard) drawn from satellite-assisted rainfall analyses for Spring 1977. After Howard et al., 1978: Barrett, 1980c.

satellites provide infrared images with a best resolution of 1 km, compared with the 5 km of Meteosat (already reduced to some 7–8 km over the northern most parts of Africa)" (Howard *et al.*, 1978).

More generally, it may be concluded that the north-west African studies have demonstrated a valuable capability of satellite rainfall monitoring techniques in pest control. Such an approach seems worthy of further extension across the vast Old World desert and semi-desert areas which are subject to the desert locust threat. This area extends from the coasts of West Africa across the Middle East as far as north-western India and the semi-arid expanses of south-central Asia. As far as quickly available rainfall observations are concerned, most of this zone is conventional data-remote. Although major locust plagues have not occurred since 1962–1963 recent local infestations, for example in East Africa (probably resulting from the curtailment of spraying operations in the Horn of Africa by the war in Ethiopia), have underlined the need for continued vigilance. Furthermore, major locust plagues seem to have recurred quasi-cyclically over the last century (see Waloff, 1976): it is not impossible that natural controls have been more effective over the last 10 to 15 years than artificial control measures may have been.

Interest is now being shown also in the use of satellite rainfall estimation techniques for an improved understanding of the behaviour of the East African army worm (see Betts, 1971), whose infestations cause substantial damage to crops every year somewhere between southern Mozambique and the Red Sea (Pedgley and Rose, personal communication). It is thought that mesoscale convective rains may help encourage concentrations of breeding and hatching, and that detailed analyses of rain areas evidenced by Meteosat could be invaluable for the direction of control operations especially in areas with few daily reporting rainfall gauges.

On a much broader scale NASA has been conducting research and enquiries into the ability of both aircraft and spacecraft sensors to observe the Earth and collect data considered important in the disciplines of environmental medicine, geographic pathology, ecology and entomology (Barnes, 1978). It is likely that the use of remote sensing for these purposes (including improved rainfall monitoring) will be particularly appealing to agencies responsible for broad area pest and disease control in developing countries having minimal public health facilities. However, Barnes concluded:

> "Our preliminary evaluation of their requirements makes us realize that, while significant progress has been made, we still have an inadequately developed technology (to meet their needs) . . .It appears that there is a basic requisite for continuing remote sensing research within the federal government or at university laboratories."

CHAPTER 13

Crop Growth and Production

I. The Rise of Interest in Agrometeorology and Agroclimatology

We live in an age of rapidly increasing world population. Even if the rate of growth were only 2·5–3·0% the population of 4000 million in 1976 would exceed 7000 million by the year 2000. It has been remarked that

> "The demands of a rapidly increasing world population, with its implications for increasing pollution, an increasing scarcity of renewable resources, and diminishing non-renewable resources, may lead within a few decades to major world-wide traumas historically not experienced by mankind" (Howard, 1977).

Perhaps the most serious single effect of such a massive population increase will be an accelerating threat to agriculture, caused by deforestation, increased soil erosion, loss of agricultural land to housing and industry, and, in some regions, a decline of soil fertility, and increased desertification. However, some major national and international traumas have already resulted from crop failure due to adverse weather and weather-related effects; impact of such effects may be expected to increase as the pressure on world food supplies increases further, and more marginal lands are exploited for food crop production. One result of the recognition of this scenario has been an upsurge of interest in the twin sciences of agrometeorology and agroclimatology, especially insofar as they are concerned with assessments of crop growth and predicted volume of crop production.

Active interest in such matters is now being displayed by a wide range of national, regional and international bodies, including commercial companies vitally concerned with present and future fluctuations in commodity prices on the world market, national governments concerned with the life and health of their subjects, groups of nations strongly affected by tariffs and trade, and international agencies with responsibilities for the provision of advice, aid, and/or relief to countries which are seriously deficient in some related respect.

Through the past two decades extensive research has been undertaken into the nature of the relationships between both semi-natural ("managed") vegetation and food crops, and factors in their environments. So it has become a fundamental tenet of agrometeorology and agroclimatology that plant life is particularly sensitive to atmospheric influences, though the degree and significance of this sensitivity varies considerably from one species to another. Of the many atmospheric elements which affect plant life, precipitation is generally seen to be the one which needs to be monitored in greatest detail on account of both its biological significance and its high variability in space and time. In this chapter we will discuss the utility of satellite-assisted rainfall mapping in programmes designed to improve operational assessments of crop growth and productivity. It seems fair to suggest that, although fewer actual results have been achieved by the time of writing in these spheres of application, it is here that satellite rainfall-monitoring methods may have their greatest humanitarian value in the medium-term future. Therefore more consideration will be given than hitherto in Part III to plans and proposals in addition to project results.

II. The Assessment of Conditions Affecting Vegetation and Crop Growth

1. Crop growth

There are, at present, several important information systems in operation which provide general indications of vegetation and crop performance through summaries of significant weather conditions and elements. The need for such systems was underlined at the World Food Conference held in Rome in 1974. One of the key issues addressed at that Conference involved the capacity of governments to deal with food shortages: this largely depends on the availability of adequate and timely information on the current and prospective crop and food situation. It was recognized that, in view of current information shortfalls, the most urgent action required was to "Improve national and international reporting services on current conditions, and factors affecting crops, stocks, inputs and prices", and that "The use of satellites for special purposes in meteorology and agriculture needs to be studied and assessed in depth".

Some satellite-assisted rainfall monitoring results have been obtained in the pursuit of such goals. These include results from programmes already mentioned in previous chapters—notably the research project of Follansbee (1976), and the operational activities of the US Center for Environmental Assessment Services (see Section III, Chapter 12).

In the first case test results were obtained for rainfall in selected areas of Asia and North America prompted by a need for the estimation of rainfall over vast agricultural regions in autumn, winter and spring, particularly for the improvement of assessments of the performance of spring wheat. Follansbee's "climatological" technique was tested in a quasi-operational mode for China and the USSR in March and April, 1975, and over 24 States in the USA in April 1976. Figure 13.1 illustrates the weekly results for China, aggregated from daily estimates. Verification tests for results from the USSR revealed a mean absolute error of 55% of the mean observed precipitation, with bias virtually zero. Localized convectional activity seems likely to have been the greatest single source of error. In the results of categorical rain, no rain tests on the data for the USA revealed that daily rain was estimated for 80% of observed rain days, and no rain for 70% of the no rain cases.

In the second case, because of the operational nature of the CEAS weather bulletin system, few rainfall reports, maps and tabulations like those illustrated earlier (see Figs 12.11 and 13.2) have been subjected to rigorous verification checks. However, some pertinent points were made in an unpublished paper by LeComte (1977), concerned with an evaluation of satellite rainfall estimates in northern Haiti from January through June 1975. These preliminary findings indicated that useful estimates could be made if the base method (following Follansbee, 1973: see Section III, Chapter 4) was adjusted to increase the rainfall contribution from cumulo-congestus clouds in winter, to obtain a better estimate of the contribution from showery precipitation: monthly errors averaged for four surface stations ranged from -20% to $+26\%$. Today, precipitation estimates for countries in the Caribbean Basin "are based on satellite interpretation" and "Conventional surface and upper air reports are also considered" to take account of LeComte's conclusions (CEAS Weather Summary notes). Figure 13.2 shows recent results for Haiti. LeComte concluded that "This seems to be a very cost-effective method of evaluating the moisture situation if point rainfall estimates are not required."

For the future, two possibilities in particular would seem worth investigating further, irrespective of the type or types of satellite-assisted rainfall mapping methods which might be deemed most appropriate for further use

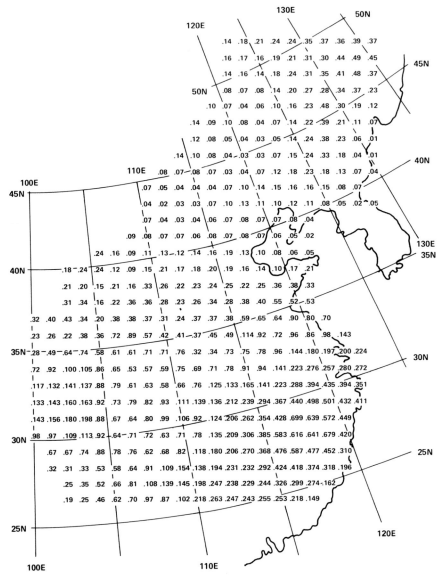

FIG. 13.1   Weekly precipitation estimates (hundredths of an inch) prepared for China, 13–19 April, 1975, by Follansbee (1976).

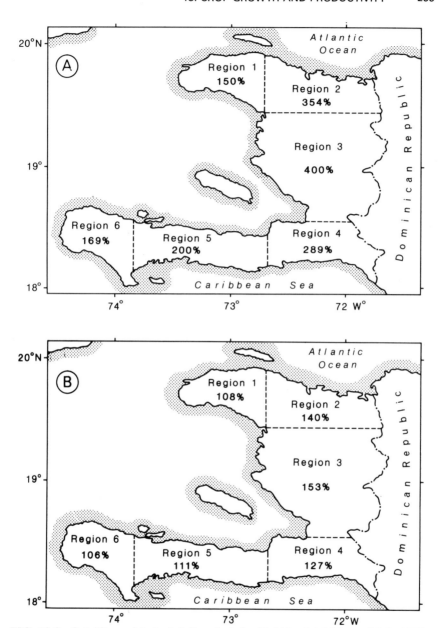

FIG. 13.2   Satellite-assisted rainfall analysis for Haiti for (a) 25 June–1 July 1979; and (b) 1 January–1 July 1979, from a CEAS Weather Summary, Environmental Data and Information Service, NOAA, 1979.

and development. These are:

(a) that rainfall estimates from satellites might be seen to be of value in the assessment of the condition and performance of a wide range of specific crops; and

(b) that such estimates might be used profitably along with ground truth and/or remote sensing data from Earth Resources satellites to provide numerical assessments of vegetation or crop biomass.

One of the biggest challenges seems to be the design and implementation of a broad, comprehensive crop information system, one of whose major purposes would be the issuing of early warnings of human disasters stemming from crop failure and resulting imbalances in the global distribution of food supplies. Proposals have been made for such a system or systems embracing major economic blocs (e.g. the European Economic Community: see Fraysse, 1977), and the world as a whole (Park, 1977).

One detailed study which has been concerned explicitly with the value of satellite rainfall-monitoring techniques for the routine assessment of food crop performances is that of CEAS and the University of Missouri (NOAA, 1979). This study, concerned with the Caribbean Basin, suggested that sufficiently firm climate/crop relationships had been identified for corn, millet/sorghum, beans and rice for satellite monitoring of rainfall to be of potential value in respect of each of them, and that, with further preparatory work, the list might be extended to include some or all of sugarcane, plantain or bananas, cow and pigeon beans and manioc. If some or all of these crops could be assessed using satellite data inputs, the significance of satellite rainfall monitoring would be assured throughout the tropics. A general conclusion reached in the study of the Caribbean Basin was that "Proposed assessment procedures (including the use of GOES imagery for improving regional rainfall estimates) can provide a low-cost, yet potentially useful system . . . for early warning on possible subsistence food shortages due to anomalous climatic conditions in developing countries." General principles suggest that other crops, of great value as commodities of world trade (e.g. rice and cocoa beans), could be assessed through similar means.

2. Biomass estimation

The way ahead for biomass estimation has been prepared already in respect of natural and semi-natural grasslands. Interest in rangeland monitoring and management has risen recently, especially because of widespread droughts in Africa, North America and Australia, and their impacts on forage crop productivity and consumption by animals. Techniques for the assessment of forage crops using Earth Resource observation (Landsat) satellite data in conjunction with surface weather data have been tested with success in a number of natural and semi-natural grassland areas (especially savanna and semi-arid vegetation communities). Results have pointed to a high potential utility for such schemes especially in rangeland management and stock adjustment to varying conditions for grazing. It has been demonstrated that Landsat spectral reflectance curves for grassland areas change through time in harmony with phenological stages of plant growth (see e.g. Carneggie *et al.*, 1975). Graphs of reflectances, or the ratios of reflectances in selected wavebands, can be translated into estimates of green forage production on the basis of comparisons between image characteristics and ground truth, usually presented in regression form. Some workers are of the opinion that, since Landsat bands 5 and 7 are usually the most widely separated during the key (growing) period of the year, these two alone may be adequate for good biomass estimates. However, other authorities prefer bands 5 and 6, arguing that the difference between these two is generally the more sensitive to green biomass differences. In order to minimize the effects of differences in processing data in the chosen wavebands, relationships of the following type are generally preferred to simpler expressions of band differences or ratios:

$$TVI = \frac{\text{Band } 6 - \text{Band } 5}{\text{Band } 6 + \text{Band } 5} + 0 \cdot 5 \tag{13.1}$$

where TVI is a "Transformed Vegetation Index".

Models of this kind have yielded promising results in the Great Plains of the USA (Haas *et al.*, 1975). Further tests, of special relevance here, indicated that the TVI, *plus limited weather data*, may be employed to give acceptable estimates of biomass even in the absence of regular field measurements of vegetation growth (Seevers *et al.*, 1975).

A stepwise multiple regression analysis was performed to select the variables most likely to explain the variation in the Landsat observation:

$$y = \beta_0 + \beta_1 B + \beta_2 P_1 + \beta_3 P_2 + \beta_4 T_{max} \tag{13.2}$$

where y represents a TVI, B represents green biomass (kg ha^{-1}); P_1 represents precipitation since the previous satellite overpass; P_2 is precipitation on the day before the overpass; T_{max} is maximum temperature on the day of the overpass, and the β_i are empirical coefficients.

These variables accounted for more than 90% of the TVI variations in American tests. They were then used in a four-variable model with green biomass as the dependent variable and the other four as the independent variables. The ability to estimate green biomass in increments of 250–300 mg ha^{-1} was indicated, with a 95% probability level.

Although requirements for meteorological data were met from conventional sources in the American studies it has emerged from discussions with many involved with the developing world that those requirements are often not met by surface data especially in tropical and subtropical zones. One example is the Sahelian zone of Africa. It is generally agreed that in such regions temperature variations are relatively unimportant, in the sense that they can be assessed tolerably well from climatic data, being influenced most by latitude, altitude and season of the year. Rainfall, however, is susceptible to a number of less predictable influences, some of which are geographical (e.g. distance from the sea), whilst others are atmospheric (e.g. weather system frequencies and intensities), and still more are hybrid or composite (e.g. the effects of aspect, rain shadow effects of mountain ranges, etc.). It would seem potentially very valuable to attempt, through satellite-improved rainfall monitoring programmes, to establish the total effects of such factors in terms of more detailed rainfall maps than those which can be compiled at present from relatively scant surface data—especially remembering that rain-cloud masses in the rangeland zones are notoriously variable and disjointed.

It should be recognized, of course, that rangeland biomass estimates alone can give misleading indications of the grazing potential of any grassland area: primary productivity is not linearly related to secondary productivity, for nutrition levels of grassland communities are both spatially variable, and subject to variation through time. Thus it may be suggested that the usefulness of any successful satellite-assisted method for the assessment of grassland biomass might be heightened later through suitable ground truth calibration and/or consideration of the likely behaviour of different members of the grassland communities, given the influence of known antecedent weather and grazing conditions.

III. Crop Prediction Programmes

If improved assessments of present vegetation and crop conditions are of commercial, political and humanitarian significance, then improved forecasts of likely conditions and harvests in the future are of even greater potential significance, and constitute some of the most important goals in agrometeorology and agroclimatology. Reference to the famines which may result from adverse growing season or harvest weather has already been made in Chapter 12. Here we will be concerned with the broader question of crop prediction under the whole range of possible weather and climate conditions. Reports at the World Food Conference in Rome in 1974 underlined the need for satellite assistance in this area too. It was reported that

"Despite actions by the WMO, through the World Weather Watch, for the efficient acquisition, exchange, analysis and archiving of weather data on a global basis the existing supply of information at world levels does not permit timely identification of many of the crop shortfalls which it should be possible to prevent. From some countries information is unavailable, from some it is incomplete, and from some it arrives too late to be useful." (Reuter News Report, 1974)

Herein, then, lies one of the biggest challenges yet presented to those involved in satellite improvement of rainfall monitoring systems. And it is in this connection that some of the most extensive, and earliest, operational tests of satellite rainfall methods have been deployed. Unfortunately, it is not possible to present actual findings from some of the more detailed or widespread projects (e.g. those operated by US military and commercial organizations) because results have not been released to the general scientific community. However, we can report enough projects and results to illustrate the types of progress which have been, or could be, made to meet the general needs spelt out in Rome.

At the heart of crop production modelling is the assumption that *crop production* (the volume of the harvest) is the product of *crop area* and *crop yield* (crop productivity per unit area). The object of most crop production programmes is to forecast as accurately as possible the volume of the forthcoming crop harvest. Clearly, for many purposes, there is a premium not only on the accuracy of forecasting but also on the stage by which an accurate forecast may be achieved. In these respects crop area may be expected to pose fewer and lesser problems than crop yield: the area under a particular crop is relatively constant from year to year unless government policies change, and is fixed for a particular year once the crop has been sown; meanwhile the yield is so sensitive to both weather and climate that crop production can vary greatly from one year to the next, and crop condition often changes rapidly within a given growing season. Because of the present difficulties in seasonal weather forecasting, and the propensity for some weather hazards to become acute and prompt agricultural disasters (e.g. through sudden effects such as hailstorms, other severe convectional storms, and unseasonal frosts, as well as through cumulative effects such as floods and droughts) it is natural for the accuracy of crop production estimates to increase as the harvest approaches. Recent research has been directed to the development of crop production models which take more detailed account of weather, and weather variations, than their predecessors. The newer models are amenable to satellite inputs, both of crop area and rainfall estimates. We will concentrate our attention on these.

At the simplest level, crop production models fall into three groups, namely:

(a) statistical models:
(b) agroclimatological models; and
(c) agrometeorological models.

Statistical models exploit the principle that crop area tends to be relatively constant from one year to the next, and/or is reasonably amenable to estimation through statistical returns or samples made in the field, in conjunction with other information such as phenometric measurements of the growing crops (plant height, length of ear, etc.) and/or the sizes of harvests in previous years. Good results may be expected in average years, but not in abnormal years. A large operation has to be organized on the ground, and such methods have therefore been most successful in advanced, literate nations where field sampling on the required scale is both possible and reliable. In the USA crop questionaires are processed annually by the US Department of Agriculture (USDA) from some 150 000 farmers. Farm visits are made by qualified observers to some 16 000 fields (see Wigton, 1976). Through substantial areas of the world (e.g. in most developing countries) such operations cannot be mounted because of low levels of literacy and difficulties of recruiting suitable field survey staff.

Agroclimatological methods of crop production forecasting are preferred in some developed countries (e.g. West Germany), and have the advantage for less developed countries of a lesser dependence on specially collected data from the field. These methods consider broad regional climate characteristics, expressed either as evaluations of climatic trends and/or aggregated weather, through the growing seasons of selected crops (see e.g. Hanus, 1980; Baier, 1977a). However, such methods intrinsically perform little or no better than statistical methods, and their value may ultimately be seen to lie rather in what they reveal of the different influences of given atmospheric conditions on different crop types.

It is in the field of agrometeorological crop production modelling that satellite rainfall data have most to offer. Therefore it is this type of approach which we should consider in more detail in the present context. One of the big advantages of the crop-weather approach is said to be that yield estimates can be updated daily on the basis of conventional weather observations obtained via the international GTS network (Baier, 1977b). Another is surely that, through the use of satellite imagery, crop area estimates can be established even for regions from which suitable field data are unforthcoming for either practical or political reasons: this was one of the basic tenets of LACIE, the Large Area Crop Inventory Experiment, a joint project of the US Department of Agriculture, NOAA and NASA (see e.g. Macdonald and Hall, 1977). LACIE began as a proof-of-concept experiment intended to demonstrate the applicability of remote sensing technology for the global monitoring of the most important single food crop—wheat. It was designed to meet US Department of Agriculture needs in areas from which ground data are not readily forthcoming. The pre-eminence of rainfall amongst the weather variables known to have an important influence on wheat production was emphasized in early LACIE papers (e.g. MacDonald *et al.*, 1976), but subsequent attention was focussed increasingly on problems of crop area assessment from Landsat multispectral signatures, and decreasingly on the potential of weather satellites to improve the inputs of weather data to the crop production equation.

In the meantime, early crop production trials in Iran undertaken by EarthSat Corporation, an American remote sensing consultancy firm using a Follansbee-type rainfall technique, had led to more extensive and detailed tests of a derived, man–machine interactive method (see Section III, Chapter 4), in North America (EarthSat, 1976). Tests of this method in the hard red spring wheat regions of North and South Dakota, Montana, and Minnesota in 1975 resulted in daily precipitation estimates accurate to within 3 mm on 72%, and 7 mm on 90%, of all occasions, despite tendencies to overstate light, and understate heavy, events. Estimates for larger areas than a basic 25 n. mi. grid mesh developed from the numerical prediction net of the US National Meteorology Center (NMC), or for larger intervals of time up to 1 week achieved higher levels of accuracy, as anticipated from a basically statistical cloud-indexing technique. Figure 13.3 exemplifies these results. Under the "Cropcast" trade-mark the applicability of such an approach has been extended to several important crop-producing countries and to a variety of crops. Geostationary satellite imagery are being used routinely in rainfall monitoring over the USA, Canada, Brazil and Argentina. Polar orbiting imagery are sometimes used in rainfall monitoring over the People's Republic of China and parts of Europe. Crops concerned include wheat, corn, barley, oats, soybeans, and sunflower. However, since the results of these operations have been provided for commercial purposes they have not been released for general scrutiny and the performance of the Cropcast system cannot be independently assessed.

It is of further interest to note the broader crop prediction method which the satellite-improved rainfall data serve. The 1975 EarthSat tests utilized conventional meteorological observations from first-order ground stations, plus the satellite rainfall estimates, to derive daily assessments of the weather environment of the wheat plant. Soils, plant physiology and local soil water to plant relationships are also considered in a complex computer model which, in effect, permits a crop to be "grown in the computer" in response to the various environmental data inputs. The daily implementation of the model is driven by a "phenology clock": this activates plant canopy estimates of crop albedo, crop rooting structure, and soil-water extraction coefficients. Yield is derived from accumulations of daily plant response through a regression equation which also includes a technology trend estimate to allow for improvements in crop types, husbandry etc.

Tests of this complex method have given spring wheat yield forecasts

13.3b

Contingency table of OBSERVED PRECIPITATION (mm) (rows) versus ESTIMATED PRECIPITATION (mm) (columns).

OBSERVED \ ESTIMATED (mm)	< 0.01	0.01–1.99	2.00–3.99	4.00–5.99	6.00–7.99	8.00–9.99	10.00–11.99	12.00–13.99	14.00–15.99	16.00–17.99	18.00–19.99	20.00–24.99	25.00–29.99	30.00–34.99	35.00–39.99	40.00–44.99	45.00–49.99	50.00–59.99	60.00–69.99	70.00–79.99	80.00–89.99	90.00–99.99	100.00–109.99	110.00–119.99	≥ 120.00	TOTAL
< 0.01	5739	344	88	45	33	16	7	6	4	2	3	7	3	1	1	0	0	0	0	0	0	0	0	0	0	6299
0.01–1.99	4993	737	313	118	64	46	32	24	11	11	11	25	13	7	2	3	1	0	0	0	0	0	0	0	0	6425
2.00–3.99	1831	522	281	176	120	102	59	40	36	24	17	30	15	10	7	4	4	5	2	2	0	0	0	0	0	3287
4.00–5.99	695	259	155	96	75	64	53	45	21	20	12	25	16	4	8	7	4	3	0	0	0	0	0	0	0	1503
6.00–7.99	511	213	139	96	75	52	34	21	26	18	11	20	18	9	6	5	3	3	1	0	0	0	0	0	0	1282
8.00–9.99	192	63	45	26	29	22	18	9	9	5	3	17	11	5	3	4	3	2	2	0	0	0	0	0	0	508
10.00–11.99	106	50	48	26	15	18	20	12	12	6	8	22	16	7	5	3	0	1	0	0	0	0	0	0	0	414
12.00–13.99	52	20	19	14	5	9	11	6	6	10	4	13	3	3	5	0	0	0	0	0	0	0	0	0	0	253
14.00–15.99	36	9	15	12	5	4	2	2	6	6	3	7	3	7	3	0	5	0	0	0	0	0	0	0	0	160
16.00–17.99	29	17	11	8	13	2	4	8	6	10	3	5	2	3	0	0	1	1	0	0	0	0	0	0	0	151
18.00–19.99	5	12	5	10	11	5	8	4	3	7	0	3	1	0	0	0	0	0	0	0	0	0	0	0	0	104
20.00–24.99	26	12	18	7	9	9	11	7	7	4	1	20	3	3	0	0	1	0	0	0	0	0	0	0	0	164
25.00–29.99	28	6	10	4	4	4	3	1	2	6	3	3	1	1	0	0	0	0	0	0	0	0	0	0	0	83
30.00–34.99	0	1	1	1	1	0	1	1	0	0	0	1	0	0	0	0	0	0	0	0	0	0	0	0	0	8
35.00–39.99	0	0	0	0	0	0	0	0	0	0	0	0	0	0	0	0	0	0	0	0	0	0	0	0	0	0
40.00–44.99	0	0	0	0	0	0	0	0	0	0	0	0	0	0	0	0	0	0	0	1	0	0	0	0	0	1
45.00–49.99	0	0	0	0	0	0	0	0	0	0	0	0	0	0	0	0	0	0	0	0	0	0	0	0	0	0
50.00–59.99	0	0	0	0	0	0	0	0	0	0	0	0	0	0	0	0	0	0	0	0	0	0	0	0	0	0
60.00–69.99	0	0	0	0	0	0	0	0	0	0	0	0	0	0	0	0	0	0	0	0	0	0	0	0	0	0
70.00–79.99	0	0	0	0	0	0	0	0	0	0	0	0	0	0	0	0	0	0	0	0	0	0	0	0	0	0
80.00–89.99	0	0	0	0	0	0	0	0	0	0	0	0	0	0	0	0	0	0	0	0	0	0	0	0	0	0
90.00–99.99	0	0	0	0	0	0	0	0	0	0	0	0	0	0	0	0	0	0	0	0	0	0	0	0	0	0
100.00–109.99	0	0	0	0	0	0	0	0	0	0	0	0	0	0	0	0	0	0	0	0	0	0	0	0	0	0
110.00–119.99	0	0	0	0	0	0	0	0	0	0	0	0	0	0	0	0	0	0	0	0	0	0	0	0	0	0
≥ 120.00	0	0	0	0	0	0	0	0	0	0	0	0	0	0	0	0	0	0	0	0	0	0	0	0	0	0
TOTAL	14,243	2265	1148	639	466	354	289	190	164	145	105	231	137	81	53	42	32	30	10	5	6	3	2	0	2	20,642

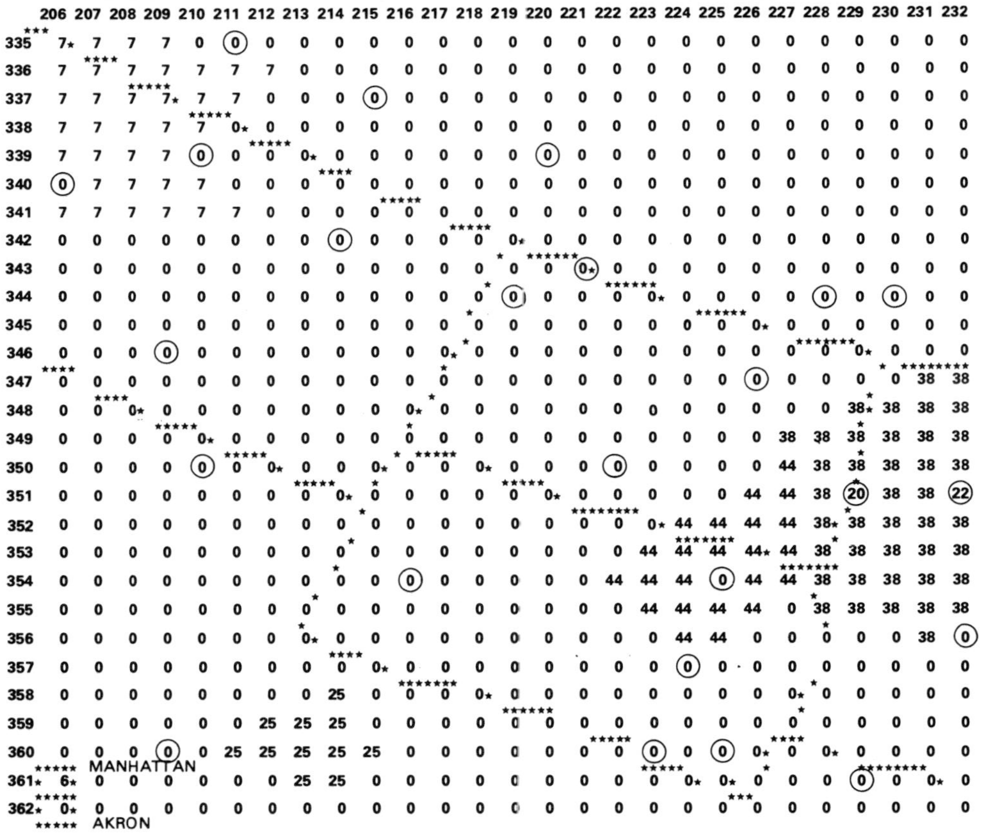

FIG. 13.3   Results of rainfall mapping over the Spring Wheat belt of the USA: (a) sample computer printout maps of rainfall (mm × 10), 1800–2400 GMT, 30 June 1975. Circled values are reported rainfall. Satellite estimates were made from SMS images. (b) Verification table for estimated versus observed rainfall, June–August 1975. From EarthSat, 1976.

13.3a

with an aggregate error of only +2% for the four states in the hard red spring wheat region of the USA. It was concluded that the method "demonstrated an overall capability not previously available to agricultural weather activities". However, it is a very complex system, with heavy requirement for computing facilities. Elsewhere, and for other purposes, simpler methods which could be operated more cheaply, might be preferred.

IV. Towards Future Integrated Crop Prediction and Food Information Systems

The possible way ahead in crop growth and prediction programmes may be clarified through one or both of the following schemes, one of which has national, the other international roots.

The weather satellite shortcomings of LACIE are to be made good in the early 1980s through a new operational research project sponsored by the US government. This is code-named "Agristars" (Agriculture and Resources Inventory Surveys through Aerospace Remote Sensing). This project is seeking to test a wide range of techniques for satellite-improved monitoring of several important meteorological parameters, including rainfall, so that the most appropriate methods might be identified or designed for operational application in NOAA. The improved rainfall data would serve the needs of the US Department of Agriculture for near real-time weather information for most of the world's significant agricultural regions, and the world's most important crops. Areas identified for study include Argentina, Australia, Brazil, Canada, the People's Republic of China, India, Mexico, the USA and the USSR; specified crops include winter wheat, spring wheat, barley, corn, soybeans, sorghum, rice, sunflower, and cotton. Several of the satellite rainfall estimation techniques described in this book are being developed further under the Agristars umbrella. However, Agristars is in a much too early stage of its implementation for further commentary to be either warranted or possible here. Suffice to say that its initial emphases on method testing, vertification, and intercomparison are highly logical in the present situation, and conclusions of Agristars will be awaited with special interest.

Turning finally to the activity of an international aid agency, a last reference may be made to the World Food Conference in Rome. This recognized and concluded that "Measures and policies which governments may decide to undertake, individually or collectively, to promote food security may be ineffectual if the (existing) crop prediction and food information systems are not strengthened and improved". At the end of this Conference, the Food and Agriculture Organisation of the United Nations was directed to design and implement a global "Food Information and Early Warning

System". It was recognized that, amongst the meteorological parameters which are of most significance for crop forecasting (including precipitation, maximum and minimum temperature, wind, relative humidity and cloud cover), precipitation was paramount. Precipitation was also paramount amongst the transformations of such parameters in relation to plant growth and yield (including precipitation, net radiation, potential evapotranspiration, day length and degree days).

In a subsequent Programme Plan for developing the capability of forecasting crop production, Park (1975) stressed that it was "very nearly axiomatic that the distribution and number of synoptic weather stations is inadequate" for such purposes. He reported early work in satellite-improved rainfall monitoring, and indicated how such an approach could be built into a System which woud meet the directives of the World Food Conference to FAO. This involved a crop growth and prediction model of the type outlined at the end of Section III. Unfortunately, although the Programme Plan was approved in principle, and an amended Plan drawn up for acceptance, changes in the leadership of FAO have resulted in the shelving of this project. Whether this will prove to have been only temporary, time alone will tell. There can be no doubt that, despite much experience with models for crop production forecasting, there remains an urgent and overwhelming need for some such approach to be employed on a global scale for the benefit of the whole world community. This could be to the particular benefit of its less fortunate members, through the more speedy and effective organization and mobilization of relief supplies in times of famine.

PART IV

Future Prospects
and Possibilities

CHAPTER 14

Active Microwave Systems

Here we encounter a paradox. Wexler (1954, 1957) and Widger and Touart (1957) envisaged radar as the instrument by which satellites would observe rain. As we noted in Chapter 1, surface radars are so used in routine observations now. They have been adapted for storm surveillance on commercial aircraft. Indeed several recent spacecraft have contained radar systems for the measurement of altitude and surface wind speed. Nevertheless, to date there has been no weather radar on a meteorological satellite.

This chapter addresses two facets of this paradox: first, why, despite their capabilities, radars have not been used to monitor rain from satellites; and secondly, whether and how they may be used in the near future. The heart of the chapter is a summary of proposals for satellite weather radars. These comprise two groups—first generation proposals, which foundered on the question of need, and now a second generation of proposals, which has been stimulated by new needs.

The theory and practice of radar meteorology are of course well documented in books and review articles (e.g. Battan, 1973; Stepanenko, 1973; Atlas, 1964). Stepanenko's book contains a chapter on the use of radars and microwave radiometers on satellites for meteorology and hydrology, and a 1975 NASA report of an Active Microwave Workshop contains a chapter, edited by Bandeen and Katz, on active remote sensing of the atmosphere. In the context of reviewing proposals up to that time, the 1963 report of Dennis offers a general treatment of precipitation detection by satellite radar. Because in recent years new ideas and perspectives have been suggested and developed there is at present no adequate summary of the particular problems attending use of radars on satellites for observations of rain. Thus we preface the discussion of proposals with a review of radar rainfall physics.

I. Radar Rainfall Physics

The physics developed in Chapter 8 as a foundation of rain observations by passive microwave instruments applies equally to weather radars, which usually operate between 1·5 and 30 GHz (20 to 1 cm). Thus the radar signal is characterized by intensity and polarization, and it is affected, through absorption and scattering, by gases, hydrometeors, and the surface of the Earth.

The main difference is of course that the radar provides its own illumination. Radar may also emit an in-phase (coherent) signal (which is the basis for Doppler radars). Usually radars transmit microwave energy in pulses and their emission is beamed. Thus they are able to measure the range of a target and its direction.

The three essential parts of a radar are its transmitter, receiver, and antenna. Of the three, for satellite systems, at least, the antenna is most limiting (Brown and Skolnik, 1975).

Usually the antenna serves a dual function: it beams the transmitted signal and captures the backscattered signal. Two design characteristics are of particular interest. The directional efficiency of an antenna is measured by its *gain* (G). Gain is the ratio (in decibels, dB) of actual power density at a point along the beam to power density at that range if the same signal were radiated isotropically. For a circular paraboloid, gain varies inversely with beamwidth (Probert-Jones, 1962; Battan, 1973). The backscattered power intercepted by an antenna is measured by its *effective cross section* (A_e). For a circular paraboloid, the effective cross section varies directly with the aperture area; it also may be expressed in terms of gain and wavelength, as $A_e = G\lambda^2/4\pi$ (Silver, 1951; Battan, 1973).

1. The reflectivity of raindrops

Following Battan (1973), and ignoring attenuation, the radar equation for average power returned from a volume of scatterers may be stated as

$$\bar{P} = \frac{P_T G^2 \lambda^2 \theta \phi h}{512\pi^2 r^2 2 \ln 2} \sum_{\text{vol}} \sigma_{bsi}. \tag{14.1}$$

P_T, G, λ, θ, ϕ and h—respectively, transmitted power, gain, wavelength, horizontal beamwidth, vertical beamwidth, and pulselength—are parameters of the radar; r is range; σ_{bsi} is the *backscattering cross section* of a single hydrometeor in the volume V determined by $(r\theta/2)(r\phi/2)(rh/2)$.

The backscattering cross section is analogous to the scattering cross section of Chapters 7 and 8, but concerns only the energy returned along the incident beam. It is defined as the cross section of the sphere which intercepts an amount of incident power equal to the power of an isotropic source emitting with the intensity of the backscattered beam. σ_{bsi} is therefore independent of range. The volume sum of backscattering cross sections is the *radar reflectivity*, η, i.e.

$$\eta = \sum_{\text{vol}} \sigma_{bsi}. \tag{14.2}$$

Its units usually are given as $cm^2\,m^{-3}$.

Mie theory gives the general expression for backscattering cross section, this being

$$\sigma_{bsi} = \frac{\pi D_i^2}{4x_i^2} \left| \sum_{l=1}^{\infty} (-1)^l (2l + 1) (a_l - b_l) \right|^2.$$

D is drop diameter and otherwise the symbols (including K, below) are as defined in Chapter 8. It follows that in general the radar reflectivity must be calculated from Mie theory.

In the Rayleigh regime, where drop diameter is much less than wavelength, the backscattering cross section may be approximated as

$$\sigma_{bsi} = \frac{\pi^5}{\lambda^4} \left| K \right|^2 D_i^6. \tag{14.3}$$

Then the radar reflectivity is given by

$$\eta = \sum_{\text{vol}} \sigma_{bsi} = \frac{\pi^5}{\lambda^4} \left| K \right|^2 \sum_{\text{vol}} D_i^6, \tag{14.4}$$

and for a given λ and K the volume sum of backscattering cross sections is a function only of the sum over the pulse volume of the sixth power of the droplet diameters. This summation is the *radar reflectivity factor*, Z, i.e.

$$Z = \sum_{\text{vol}} D_i^6. \tag{14.5}$$

The units of Z usually are given as $mm^6\,m^{-3}$.

In terms of drop size, the Rayleigh regime for scattering is defined by $D \leqslant 0.07\lambda$. Thus for a radar wavelength of 10 cm it embraces the whole spectrum of hydrometeors, but for a wavelength of 1 cm most rain-clouds fall *outside* the Rayleigh regime.

For Rayleigh backscattering the radar equation may be stated

$$\bar{P} = \frac{C\,|K|^2}{r^2} Z, \tag{14.6}$$

where the constant C depends only on the radar. If scattering does not follow the Rayleigh formula.

$$\bar{P} = \frac{C\,|K|^2}{r^2}\,Z_e, \tag{14.7}$$

where Z_e is an *effective radar reflectivity factor* given by

$$Z_e = \frac{\lambda^4}{\pi^5\,|K|^2}\,\eta. \tag{14.8}$$

Since C is a known characteristic of the radar and $|K|^2$ can be calculated (given the temperature and phase of the scattering medium), measurements of $\bar{P}$ and r yield Z or Z_e.

Relationships of radar reflectivity factor Z with rainrate R have been established through measurements of drop size distributions. Rainrates are either measured simultaneously or are calculated (as shown later) from the observed drop size distributions. Drop size distributions also have been modelled by exponential functions of the form

$$n_D = n_0\,e^{-\Lambda D} \tag{14.9}$$

(Marshall and Palmer, 1948), where $\Lambda = c_1 R^{c_2}$, and n_D is the number of drops per increment of diameter per unit of volume. The coefficients n_0, c_1, and c_2 are determined by direct measurements of a number of drop size distributions. Marshall and Palmer found values of $n_0 = 0{\cdot}08\,\mathrm{cm}^{-4}$ and $\Lambda = 41\,R^{-0{\cdot}21}\,\mathrm{cm}^{-1}$ to be good over a wide range of rainrates, but other values also have been reported, and Battan (1973) remarks on the large variability sometimes observed between samples.

Whatever the form of the relationship, Z is calculated from $\sum_{\mathrm{vol}} D_i^6$, and Z and R are related as

$$Z = aR^A. \tag{14.10}$$

The coefficients a and A are determined by least squares fit of Eq. 14.10 to large samples of observations of Z and R. Battan (1973) lists 69 Z–R relations for rain.

If Rayleigh scattering does not apply, Z_e is calculated from Mie theory or is inferred from simultaneous measurements of P and R. The relation of Z and R still is expressed in power law form, but the coefficients vary with wavelength and rainrate.

Received power P must be averaged because the power returned from any volume of particles is a function of the sum of both amplitudes and phases of backscattered waves. Phase relationships depend on the relative positions of the scatterers. The distribution of raindrops within a volume is random (thus the backscattered signal is incoherent). It also is constantly and rapidly changing, thus a measurement made after 10^{-2} s will be essentially independent of the one preceding. Independence may also be achieved by a 50%

displacement of the pulse volume either for the same pulse along the range axis or for the next pulse across the range axis (Battan, 1973; Dennis, 1963). The actual backscatter intensity for a uniform, steady state rain-cloud is approached as an exponential function of the number of independent measurements of returned power. About ten independent measurements are needed to obtain acceptable accuracy in $\bar{P}$ (Dennis, 1963).

2. The reflectivity of the Earth

The Earth for an ordinary weather radar is a platform: it is seldom seen and of no particular interest. But for a satellite radar the Earth always will be in view. Its echo will form a continuous backdrop to echoes from rain and therefore is a compelling concern.

Over the oceans the role of satellite radars in the measurement of rainrate probably will be secondary to that of passive sensors (Atlas *et al.,* 1978). In any case interest in the use of radar scatterometers for measurements of surface wind has stimulated much theoretical and empirical research on microwave reflection from the surface of the sea (see, for example, Barrick, 1972, Vonbun and Sherman, 1975). We summarize pertinent findings in these remarks from Vonbun and Sherman:

> "At microwave frequencies, the ocean produces two types of scattering processes. The larger ocean waves essentially behave as an ensemble of specular reflectors such that the strength of the scatter is proportional to the tilts (or slopes) of the gravity waves. Because the length of the ocean wave is much greater than the height, quasi-specular scattering occurs only at angles close to the nadir direction (usually within 25°). Therefore, near nadir, the active microwave system is closely linked with the physics that controls the gravity wave slope. For angles beyond 20°, resonant (Bragg) scattering occurs from those waves that are comparable to the wavelength of the incident electromagnetic wave. At microwave frequencies, this type of scattering is controlled by the capillary wave structure. Because the capillaries have a short time constant for growth and decay, the scattering strength is linked to the local wind fields and the surface tension of the ocean surface."

These two regimes of ocean backscatter are illustrated in Fig. 14.1. The dependent variable is differential backscatter cross section of the surface, $\sigma°$ ($\sigma° = 10 \log_{10}\sigma$, where σ is the backscatter cross section per unit area; its units are decibels). Backscatter intensity depends strongly on nadir angle and the surface wind speed. As a consequence values of $\sigma°$ range over three to four decades.

The reflectivity of land is neither as well understood nor as easily expressed. In 1960 Keigler and Krawitz reported a large range in $\sigma°$, from -27 dB for deserts to -2 dB for cities, but practically no dependence of $\sigma°$ on wavelength and nadir angle. Observations since then, including some from a

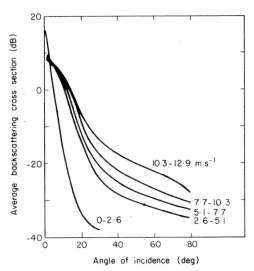

FIG. 14.1   Typical average backscattering cross section of the sea as a function of incident angle for various surface wind speeds. $\lambda = 1\cdot25$ cm. From Vonbun and Sherman, 1975.

space platform (Skylab), were summarized by Eckerman *et al.* (1978). According to these authors the factors important to the reflectivity of land are soil wetness, roughness and type; vegetative cover; and relief. Roughness is gauged through the wavelength of the incident beam: a surface is rough if the size of the scatterer D is greater than one-fifth the wavelength. Thus $D > 0\cdot6$ cm (a conservative value) implies the surface is rough for all $\lambda < 3$ cm. The expectation that reflection is diffuse at large nadir viewing angles (θ_N) was confirmed in coarse resolution Skylab scatterometer observations at $13\cdot9$ GHz ($2\cdot2$ cm) over large areas of the United States (Fig. 14.2). σ° decreased as θ_N increased, and the range $\Delta\sigma^\circ$ became small (4–5 dB) for $\theta_N > 10^\circ$. $\Delta\sigma^\circ$ was much smaller yet for adjacent footprints: $0\cdot5$ dB in the mean, with a standard deviation of $0\cdot4$ dB. The effect of soil wetness and roughness and vegetation on σ° were assessed from University of Kansas ground experiments. At large nadir angles σ° was insensitive to wetness of bare soil and (at vertical polarization) also insensitive to roughness. In general, vegetation attenuates microwave radiation in proportion to water content. Differences for four field crops were small and decreased with increasing θ_N. Soil moisture differences were small at $\theta_N > 40^\circ$ and $\lambda < 3$ cm.

Normally ground return could be expected to be orders of magnitude greater than rain return, therefore, if discrimination is possible, it would make a useful reference for determining the altitude of hydrometeors

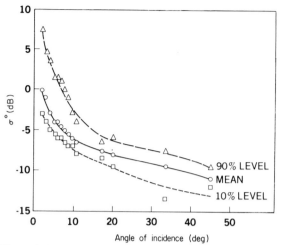

FIG. 14.2   Differential scattering coefficient $\sigma°$ measured at various sites in the United States with the Skylab S-193 scatterometer during summer 1973. Polarization is vertical. From Eckerman *et al.*, 1978, after Moore, 1974.

(Keigler and Krawitz, 1960). As will be seen later there is in addition potential for use of the land surface as a reference target in measurements of attenuation.

3. Attenuation

In analogy with optical attenuation (see Section III, Chapter 7), the depletion of a radar signal is given by $d\bar{P} = -2k_L \bar{P}\, dr$, where k_L is the linear attenuation coefficient. Then the radar equation becomes

$$\bar{P} = \frac{C\,|K|^2}{r^2} Z_e e^{-2\int_0^r k_L dr} \qquad (14.11)$$

If the attenuation coefficient is expressed (as k) in logarithmic rather than linear form and integrated, we have the two way total attenuation A_2 measured in decibels. Thus

$$A_2 = -10\log_{10}\bar{P}/\bar{P}_0 = 2\int_0^r 10(\log_{10}e)k_L dr = 2\int_0^r k\,dr, \qquad (14.12)$$

where $\bar{P}_0$ is power returned in the absence of attenuation and overbar indicates an average over many samples. Outside the $1·35$ cm water vapour absorption line and above the $0·5$ cm oxygen absorption line, absorption by gases is inconsequential. However, depletion of a radar signal by hydrometeors may be substantial, and thereof springs the potential for measuring rainrate from attenuation.

As in the extinction of light by cloud droplets, attenuation is a loss of energy from a beam due to scattering (redirection) and absorption (conversion). The effectiveness of drops in scattering and absorbing is given by scattering, absorption, and extinction cross sections, σ_{sc}, σ_{ab}, and σ_{ex}, where, as before, $\sigma_{ex} = \sigma_{sc} + \sigma_{ab}$. Unit volume coefficients are given by

$$\sum_{\text{vol}} \sigma = k_L, \text{ in units of area per volume, or } \beta \sum_{\text{vol}} \sigma = \log_{10} e \sum_{\text{vol}} \sigma = 0.4343$$

$$\sum_{\text{vol}} \sigma = k, \text{ in units of decibels per kilometre. Thus}$$

$$k_{ex} = k_{sc} + k_{ab} = \beta \sum_{\text{vol}} (\sigma_{sc} + \sigma_{ab}),$$

where attenuation is logarithmic and the subscripts represent scattering or absorption or the sum of both. A useful alternative expression for the attenuation coefficient (Goldhirsh and Katz, 1974) is

$$k = \beta \int_0^\infty \sigma(D)n(D)dD, \tag{14.13}$$

where, again, k and σ may represent scattering or absorption or the sum of both.

In general σ_{sc} and σ_{ab} are determined from Mie theory (see Section III, Chapter 7), but the special case of Rayleigh scattering, where $\sigma_{sc} \ll \sigma_{ab}$, has a simpler solution. There $\sigma_{ex} \cong \sigma_{ab}$, and

$$k_{ex} \cong 0.4343 \sum_{\text{vol}} \sigma_{ab} = 0.4343 \left[\frac{\pi^2}{\lambda} \left(\sum_{\text{vol}} D^3 \right) \text{Im}(-K) \right] \tag{14.14}$$

Since $\sum_{\text{vol}} D^3$ is simply the volume of water per unit volume of scattering medium, Eq. 14.14. can be written

$$k = K_1 M, \tag{14.15}$$

where the coefficient K_1 depends only on wavelength and temperature, M is the liquid water content and we have dropped the subscript on k.

Rayleigh and Mie solutions for attenuation by individual drops are compared for a series of radar wavelengths by Gunn and East (1954). If we tolerate errors as large as 25%, the Rayleigh regime for attenuation extends to drops of 0.4 mm diameter at a wavelength of 0.9 cm, and 2.2 mm diameter at a wavelength of 10 cm. Thus Rayleigh attenuation (Eq. 14.15) is valid for *clouds* at all radar wavelengths, but is generally valid for *rain* only at wavelengths longer than 10 cm.

Gunn and East also give a table of values of the coefficient K_1 for water and ice at various temperatures and wavelengths. Because the imaginary part of the complex index of refraction of ice is much less than that of water, for a given M attenuation by ice is two orders less than attenuation by water, and for present purposes can be ignored.

TABLE 14.1   Attenuation k/R (dB km^{-1}/mm h^{-1}).[a]

Wavelength (cm)	Marshall-Palmer (at 0°C)	Gunn and East (18°C)
0·62	0·50–0·37	
0·86	0·27	
1·24	$0·117R^{0·07}$	$0·12R^{0·05}$
1·8		$0·045R^{0·11}$
1·87	$0·045R^{0·10}$	
3·21	$0·011R^{0·15}$	$0·0074R^{0·31}$
4·67	0·005–0·007	
5·5	0·003–0·004	
5·7		$0·0022R^{0·17}$
10	0·0009–0·0007	0·0003

[a]After Wexler and Atlas (1963).

To account for attenuation by rain, an expression between k and R is needed. The most direct approach assumes a power law relation:

$$k = K_2R^\gamma = K_2R^{\gamma-1} R = K'R. \tag{14.16}$$

K' is calculated from Mie theory for assumed or measured drop size distributions. Values from Wexler and Atlas (1963) and Gunn and East (1954) for several distributions and a range of wavelengths are presented in Table 14.1. Except near 3 cm, K' is constant. Then k is linear in R, and as Gunn and East (1954) point out the total rain attenuation (in decibels) is a function of the path integrated rainrate. Attenuation in rain at 10 cm clearly is small and for present purposes may be neglected.

A more analytic approach to the determination of attenuation by rain depends upon the drop size distribution. Combining a power-law relationship between fall speed and drop diameter with an expression for the contribution of a distribution of drops to rainrate, Atlas and Ulbrich (1977) found

$$R = 10·6 \pi \int_0^\infty n(D)D^{3·67}dD \tag{14.17}$$

Here D is expressed in units of cm and $n(D)$ in units of m^{-3} cm^{-1}; the units of R are mm h^{-1}. But the cross section for extinction, σ_{ex}, also can be expressed as a power law relation in D. Then, still following Atlas and Ulbrich, Eq. 14.13 for the attenuation coefficients becomes

$$k = 0·4343 Q \int_0^\infty n(D)D^q dD, \tag{14.18}$$

where Q and q are, respectively, the power law coefficient and exponent. In the special case of $q = 3·67$, the integrand of Eq. 14.18 is identical to the

K

integrand of Eq. 14.17, and *attenuation is a function only of rainrate*. This "crossover" occurs at a wavelength of 0·88 cm. There attenuation in decibels per kilometre is expressed by the simple relation,

$$k = 0.228R. \tag{14.19}$$

Attenuation as a gradient property requires for its measurement at least two independent observations. These may be made with the same radar if precipitation is uniform along the path or if a suitable target is present behind the rain. They may also be made by adding a second radar, matched in pulse volume and path to the first, but operating at a less attenuating or non-attenuating (longer) wavelength. In either case the measurement of precipitation is independent of the calibration of the radar.

In the one-way or bistatic mode (see Bandeen and Katz, 1975) the receiver of the single wavelength radar is distant from the transmitter. This configuration, with the transmitter on a geostationary satellite, has been used successfully to deduce precipitation characteristics (Tanaka, 1979), but coverage is limited to the few points where receivers can be installed. Usually in a single wavelength systems the radar is operated over a two-way path with a target of rain or a reference reflector. With rain as the target and assuming that loss of the signal (beyond the range effect) is due to rain, the average attenuation coefficient over a distance Δ at least as large as the pulse width is proportional to $1/\Delta \log[r^2 P(r)/(r + \Delta)^2 P(r + \Delta]$ (Goldhirsh and Katz, 1974). If rainfall is uniform over the range increment Δ, R may be estimated from an appropriate k–R relation. But rarely is R uniform, even for small Δ, and this technique has very limited utility.

More often the attenuating system is operated with a reference target (see Atlas, 1964; Battan, 1973; Atlas and Ulbrich, 1977). Then the total two-way attenuation is given (from Eq. 14.12) by

$$A_2 = 2 \int_{r_0}^{r_N} k\,\mathrm{d}r = -10 \log P'/P \tag{14.20}$$

where r_0 and r_N are the slant ranges to the top of the rain and to the target, P' is power returned in the presence of rain and P is power returned in the absence of rain. If k is linear in r there exists a $\hat{k}_{AV}$ such that $\int_{r_0}^{r_N} k\,\mathrm{d}r = r_e \hat{k}_{AV}$. where $r_e = r_N - r_0$. Expressing $\hat{k}_{AV}$ as a power law relation (see Eq. 14.16) and solving for the estimated path average rainrate $\hat{R}_{AV}$ yields

$$\hat{R}_{AV} = \left[\frac{5}{K_2 r_e} \log P'/P \right]^{1/\gamma}. \tag{14.21}$$

k is not linear in R for wavelengths less than ~ 0.8 cm (Atlas, 1964). Furthermore, according to Atlas and Ulbrich (1977) and Eckerman *et al.* (1978). k–R relations become increasingly sensitive to drop size distributions as wavelengths increase from ~ 0.88 cm, and near 0.88 cm attenuation is large. Thus selecting a wavelength for this technique involves a balance between accuracy, range of rainrates, and areal coverage, with the practical consequence that measurements may be needed at more than one wavelength.

The dual wavelength approach employs two radars with rain as the target. If scattering is governed by the Rayleigh formula (so that the difference in reflectivity depends only on wavelength) and the attenuation coefficients at each wavelength can be expressed as power law functions in rainrate, rainrate averaged over the range interval Δ is proportional to the difference between attenuation (over Δ) at each wavelength (Atlas, 1964). Assuming equal exponents in the power law we have

$$\hat{R}_{AV} = \left[\frac{10}{2\Delta(K_1 - K_2)} \log \left(\frac{P_1}{P_1'} \middle/ \frac{P_2}{P_2'} \right) \right]^{1/\gamma}, \qquad (14.22)$$

where the subscripts here represent values at the two radar wavelengths.

The larger the differences in attenuation, the more accurate is the determination of $\hat{R}_{AV}$. Thus it is advantageous for the longer wavelength to be 10 cm, where $K_2 = 0$. A relation of this sort was proposed by Eccles and Mucller (1971) for calculating liquid water content.

Goldhirsh and Katz (1974) carried the dual wavelength concept one step further by proposing the measurement of drop size distribution. Assuming drop size distributions can be modelled in relations of two parameters, as for example, by the Marshall-Palmer relation in $n(D)$ and Λ (see Eq. 14.9), they showed how the distribution parameters could be calculated from observations at three combinations of wavelengths: 10 and 3 cm, 10 and 1 cm, and 3 and 1 cm (with 10 cm being used for surveillance). In each of the first two modes the long wavelength (10 cm) observation yields reflectivity, the short wavelength observation (assuming uniform rain) yields attenuation, and the distribution parameters are calculated from relations of the form k/Z. The advantage of 1 cm over 3 cm is a larger attenuation; the penalty is decreased range. In the third mode the distribution parameters are determined from attenuation alone, in relations of the form k_1/k_3. Calibrations of the radars are not important, however rain must be uniform (hence surveillance at 10 cm). Goldhirsh and Katz concluded these techniques may be impractical because demanding performance standards for the radars, large numbers of samples, uniformity of rain, and large attenuation are required for useful accuracies. In the context of inferring rainrates Bandeen and Katz (1975) reexamined the original error analysis and concluded that errors generally less than $\pm 50\%$ are possible with the second and third configurations. However, their analysis also indicated measurement of rain over a useful range of rates would require several wavelengths.

4. Polarization

Larger raindrops tend to flatten as they fall. The departure from sphericity in a medium of raindrops canted similarly will induce polarization in waves passing through. According to Bandeen and Katz (1975), at small wavelengths, where differences in dipole moment across the flattened drop and along its axis are large, the induced polarization difference can be measured. Measurements at one wavelength provide information on mean drop size and phase, at two wavelengths on drop size distribution. A limited test of these principles was reported by Tanaka (1979). By means of the bistatic system described earlier Tanaka inferred the presence of ice and snow crystals above the freezing level in precipitation over Japan. Whether useful measurements of rainrate can be made remains to be seen.

5. Summary

The average power received by a radar antenna from a volume of cloudy atmosphere depends directly on the radar constant and the radar reflectivity and inversely on the attenuation and the square of range. Of the parameters affecting the radar constant, gain and beamwidth are governed by antenna design. Radar reflectivity represents the average area of reflecting target per unit of pulse volume. Hydrometeors contribute to radar reflectivity according to the sum of their backscatter cross sections, which is a function of phase, temperature, wavelength and drop size. The reflectivity of the Earth is much greater than that of rain. For water reflectivity is a known function of nadir angle and roughness, hence wave spectrum and surface wind speed. For land it depends on soil roughness, wetness and type plus vegetative cover and relief, but tends to become constant at large nadir angles, especially for adjacent fields of view. Attenuation measures the power lost from the beam due to scattering and absorption between the antenna and the target. Attenuation by gases and by snow and ice crystals is small. Attenuation by water drops is proportional to the sum of volume scattering and absorption cross sections.

In the special case of long wavelengths (compared with drop diameters), backscattering and attenuation follow the Rayleigh formula. Then the radar reflectivity is proportional to the sum of the sixth powers of drop diameter. Attenuation is due mainly to absorption, which is proportional to the sum of the third powers of drop diameter, or liquid water content. Over the Rayleigh regime, reflectivity varies inversely with the fourth power of wavelength; attenuation varies inversely with the first power of wavelength. The Rayleigh relation for backscatter embraces heavier rainrates than does the Rayleigh relation for attenuation. Neither relation is appropriate to heavy rain at wavelengths less than ~5 cm.

Because they are directly related to the drop size distribution, reflectivity and attenuation offer two independent measures of rainrate. The relationships are expressed in power law functions of the form $X = aR^A$, where the coefficients a and A are determined by empirical methods. Coefficients for backscatter in the Rayleigh regime are independent of wavelength and rainrate. At a wavelength of $0\cdot88$ cm attenuation is a linear function of rainrate and at wavelengths of 10 cm and longer attenuation by rain is negligibly small. Radar reflectivity requires one set of measurements for its determination, but the radar must be calibrated. Attenuation is independent of calibration but requires two sets of measurements. These are made by measuring gradients in range with rain or a reflector as targets. Accuracy improves if two radars of different wavelength are operated together, for then reflectivity and attenuation may be determined independently, or attenuation may be determined independently of linear gradients in rainrate.

II. Early Proposals for Satellite Weather Radar

1. A visionary view

The papers of Wexler (1954, 1957) and Widger and Touart (1957) mentioned at the beginning of this chapter and a 1959 paper of Mook and Johnson comprise what may be called the visionary view of satellite radar as an instrument for monitoring rain. Their perspective is mainly meteorological, yet there also is in these papers a progression towards detail and commitment.

Whereas Wexler merely mentioned the detection of precipitation areas, Widger and Touart suggested design characteristics for a satellite radar and simulated an echo map. Their radar would operate between 3 and 5 cm, scanning crosstrack, with a nadir resolution of 55 km. Widger and Touart anticipated three big problems: weight limits, power requirements, and the separation of echo and ground return. The 3 cm radar proposed by Mook and Johnson (1959) would achieve 8 km horizontal (300 m vertical) nadir resolution through the deployment of a 1° pencil beam antenna at 480 km altitude. Every second the antenna would scan crosstrack through 90°, distributing 100 pulses along the scan. Average power output for the radar stage alone was estimated to be 25 W. Mook and Johnson mentioned nuclear generators as a possible source.

Then came a series of papers, by Katzenstein and Sullivan (1960), Keigler and Krawitz (1960) and the US Weather Bureau (1961), reflecting a growing appreciation for the technical problems inherent in weather radar for satellites.

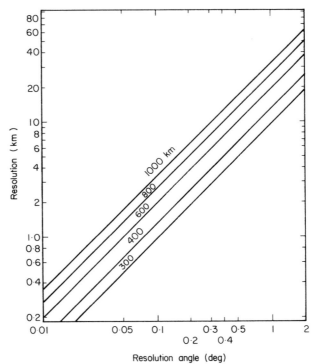

Resolution angle (deg)

FIG. 14.3   Nadir ground resolution as a function of altitude and resolution angle. From Keigler and Krawitz, 1960, copyrighted by the American Geophysical Union.

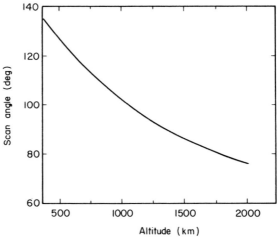

FIG. 14.4   Scan angle for complete coverage at the equator. From Keigler and Krawitz, 1960, copyrighted by the American Geophysical Union.

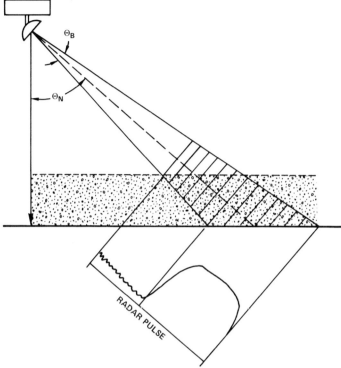

FIG. 14.5   Geometry of a satellite radar. The plane of the orbit is normal to the page. From Eckerman *et al.*, 1978.

A radical solution, offered by Katzenstein and Sullivan, deployed a fan beam pulsed Doppler radar to detect targets and measure altitude by means of Doppler phase shift relative to the Earth.

The systematic analysis of Keigler and Krawitz includes plots of ground resolution at nadir (Fig. 14.3) and the scan angle required for complete orbit-to-orbit coverage at the equator (Fig. 14.4). Aside from weight and power consumption, they found the main problems to be balancing reflectivity and attenuation in the selection of wavelength, distinguishing ground and precipitation returns, and coverage in area and height.

The problem of coverage involves several interrelated factors. Increasing coverage by scanning across the satellite track would degrade height resolution at the ends of the scan (Fig. 14.5). The limit of useful height resolution, 1·5 to 2·0 km, is reached very quickly for beamwidths larger than 0·1° (Fig. 14.6). Scan raises a second problem, illustrated in Fig. 14.7: during echo delay (the time required for a pulse to reach a target and return) the antenna rotates through an angle proportional to the scan rate. If antenna dwell on a

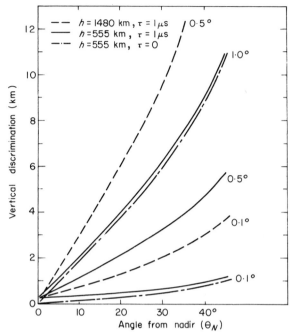

FIG. 14.6   Vertical discrimination as a function of nadir angle for various beam widths, altitudes and pulse lengths. From Keigler and Krawitz, 1960, copyrighted by the American Geophysical Union.

unit area of target is less than echo delay, the antenna will not be positioned to receive the return signal. Increasing angular resolution to improve height resolution doubly compounds the dwell problem, for beamwidth is made smaller and (if coverage is to be maintained) scan is made faster.

With these problems in mind Keigler and Krawitz considered as possible designs for a satellite weather radar, conventional (mechanical) scan, electronic and hybrid scan, correlation monopulse radar and Doppler radar. But the design they proposed is much more elementary than any of these: their precipitation sounding radar would simply view the Earth at nadir. The antenna beamwidth would be 1·75°, giving a ground resolution of 14 km from 480 km altitude. Keigler and Krawitz acknowledged that the precipitation sounding radar would be much more valuable for engineering than for meteorological studies: its main advantage was a design well within the current technology.

The apex of these early proposals was reached in the 1961 design study of the US Weather Bureau's Meteorological Satellite Laboratory. This presumed a requirement for satellite radar to provide radar observations including "information on intensity, vertical extent, and geographic distribution of

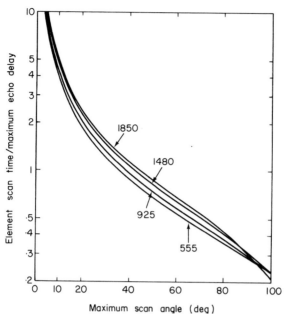

FIG. 14.7   Ratio of dwell time per unit of angle to echo delay time for adjacent sweeps of a 1° beam (multiply by ϕ for a $\phi°$ beam). Altitudes on each curve are in km. From Keigler and Krawitz, 1960, copyrighted by the American Geophysical Union.

precipitation" in uninhabited areas. Power and sensitivity were found to be the main engineering problems. A condition of design was detection of rainrates as low as 3 mm h^{-1}. Measurements would be useful even if differentiated only as to light, moderate and heavy. Global coverage was believed to be important, but impractical at the outset because of geometry and limited power. Three types of radar were considered: fixed narrow-beam pointing down; narrow beam scanning crosstrack; and fan beam.

The satellite was assumed to be at 970 km altitude, moving at a speed of 6 km s^{-1}. On it was a 6·1 m parabolic antenna and a source for up to 50 W continuous power. The preferred wavelength was ~3 cm and nadir resolution was ~7 km. For sensitivity testing the critical weather situations were detection of light rain and detection of the ground in heavy rain. Given these parameters and probable system noise, and estimating ground and rain reflectivity and rain attenuation, the study concludes that "some thunderstorms will prevent detection of the ground by any feasible satellite weather radar".

A singular disadvantage of the fixed beam radar is unrepresentative sampling. The fan beam system smears the signal over too large an area and in addition is less sensitive because of reduced gain. But a scanning narrow beam radar might observe a swath up to 37 km wide under the satellite, achieve resolution as good as 7 km, measure echo height up to ±300 to 600 m, and still detect light rainfall. This design was recommended as "the best chance of proving the feasibility of obtaining weather radar data from a satellite".

2. The pragmatic view

The slow awakening to intractable problems which is apparent in these papers was completed by Dennis in 1963. His premise was, match the components to the needs. In remote areas at least, Dennis argued these needs were less than had been assumed: much existing data, for example from radars on weather ships, saw no systematic use. Nor was satellite radar likely to contribute much to forecast analysis, for in spite of the widely accepted precipitation models for extratropical and tropical cyclones, there is little correlation of instantaneous precipitation patterns with synoptic features of pressure and temperature. In fact the precipitation eddy spectrum tends to peak between 100 and 200 km. Furthermore, in original analyses of marine rain echo coverage within 37 by 37 km blocks, Dennis found correlations dropping below 0·8 at distances as small as 75 to 150 km and at intervals no longer than 1 h.

To determine design requirements for a precipitation radar, Dennis simulated satellite scans of warm season rain over land and sea, correlated echo height and diameter, and counted echoes by size. Considering also distributions of rainrate over the ocean, Dennis concluded that detecting a reasonable fraction of rain required a sensitivity to rates as light as 1 mm h^{-1}, resolutions of 7 km horizontal and 3·7 km vertical, and broad scan. No system so far proposed had any chance of meeting these standards.

Dennis' exhaustive analysis of engineering requirements for such performance identified the most critical limiting factors as power, clutter-signal discrimination, and beam geometry. According to Dennis, "The basic problem facing a satellite radar designer is that of detecting a signal returned from a thin layer of precipitation against a background of circuit noise, antenna noise, and clutter signals from the earth's surface." The detection problem is coupled to the scan problem; furthermore, the reflected signal is incoherent and fluctuating, whereas rainfall is related to an average (in space or time) of reflectivity.

The ratio of ground to rain signals may be as large as −50 dB. Distinguishing the two on the basis of differences such as variability is not practical because the motion of the satellite will induce variation in the ground echo. Nor is multi-frequency discrimination practical, because resolution would have to be matched through the use of two antennas, and the character of the clutter reference might depend on frequency. Differences due to polarization would be small compared with those of ground and rain returns, and Doppler shift could not distinguish the effects of satellite motion from those of fallspeed and wind. A sharp beam is the only way to make this discrimination.

If height resolution is held to 3·7 km for scans 900 km long, beam width must be no more than 0·2°. At 3 cm wavelength this implies an antenna nearly 12 m in diameter; larger if, as seems probable, sidelobes contribute ground signal to the rain signal.

Dennis' analysis pointed to a nexus of engineering problems—power, pulse repetition frequency, sensitivity, wavelength, scan, and antenna size—which alone would have sobered the most enthusiastic advocate. He noted that errors inherent in inferring rain rate from reflectivity were 25 to 50%. His most telling point, however, was phenomenological: because rain is highly variable in time and space "precipitation data from a single satellite radar would be of negligible meteorological significance".

III. Recent Proposals for Satellite Weather Radars

The hiatus that followed Dennis' paper was interrupted once, by a 1969 NASA report which mentioned radars in geostationary orbit. But by the middle 1970s changing needs and advancing technology were urging a new look at the whole question of rain measurement by satellite radar.

1. Present needs

The Atmospheric Panel of an Active Microwave Working Group told a 1974 workshop convened by NASA of possible uses in large scale analysis, numerical weather prediction and climatology (Bandeen and Katz, 1975). Accepting Dennis' contention that a precipitation system should be sampled at least ten times, the Panel encouraged observations on large scales, particularly observations of long lived tropical and extratropical cyclones. For tropical storms the minimum useful standard was observations of semiquantitative reflectivity twice per day at 20 km resolution; the desirable standard was observations to ±3 dB accuracy twice per day at 2 km horizontal and 1 km vertical resolution. The value of these observations would lie in records of precipitation history, especially during genesis. They also would support model studies (through verification) and cloud seeding modification experiments. Other uses suggested by the Panel were providing input data to numerical weather prediction models (especially in relation to cyclogenesis), a global mapping of rain intensity by height, and a global climatology of precipitation echoes. Echo climatology is an apt example of new needs: because attenuation of signals and interference due to rain are serious problems in terrestrial and Earth-satellite microwave communication (Konrad, 1978; Tanaka, 1979) echo climatologies are of particular interest to designers of microwave communication systems.

A second group composed mostly of meteorologists and hydrologists was convened by NASA a year later to discuss and define meteorological requirements of a radar to be flown on the Space Shuttle (Eckerman and Wolff, 1975). The Shuttle group confirmed most of the Panel applications and added to them verification of precipitation parameterization schemes and latent heating in numerical prediction models. They also cautioned that the requirement of operations for continuous coverage in time and space would limit the usefulness of a Shuttle radar.

2. Proposed satellite radar systems

Satellite radar had also caught the attention of Soviet meteorologists. Part of the chapter on passive and active satellite microwave radiometers in Stepanenko's (1973) book, *Radar in Meteorology,* appears in translation in the workshop report (Bandeen and Katz, 1975). Also four designs proposed by Stepanenko are summarized. In the first a pencil beam, $3\cdot5°$ wide along the track and $0\cdot5°$ wide across the track, is scanned through $78°$. Rainrates as light as $1\cdot5$ mm h^{-1} can be detected. The second design employs two fixed fan beams oriented perpendicular to the satellite track. Each subtends $30°$ cross-track and $0\cdot23°$ along-track. Because ground and precipitation returns are smeared together, the detection threshold for this design is $3\cdot1$ mm h^{-1} over water and $8\cdot0$ mm h^{-1} over land. Both of these radars would operate at 3 cm and for both global coverage would be achieved by four satellites. In the third design two frequencies, $0\cdot8$ and 3 cm, improve ground and precipitation discrimination and lower the detection threshold over water to $0\cdot7$ mm h^{-1}. The last design aims only at precipitation detection, by means of a $0\cdot43°$ beam scanning radar in geosynchronous orbit. Resolution is 80 to 90 km.

Bandeen and Katz' Atmospheric Panel also considered the scanning pencil-beam and fixed side-looking fan-beam antennae, and a continuously rotating conical scan configuration as well. The fan beam, they concluded, is a poor antenna for observations of precipitation because unknown beam filling will compromise intensity measurements. The most complete three dimensional mapping of precipitation would be achieved by a scanning pencil-beam antenna, the configuration the Panel chose for its design proposal.

Required performance standards were that signal to noise ratio be sufficient to detect a "reasonable" fraction of precipitation, resolution cells (in vertical and horizontal dimensions) be small enough to exclude ground return, and vertical resolution be $1\cdot5$ km. These yielded a radar of $0\cdot2°$ beam width operating at $0\cdot86$ cm. Average transmitter power was 16 W. The $2\cdot6$ m antenna scanned an arc of 90° through nadir. From an orbit altitude of 400 km, this gives a swath of 800 km, thus a revisit time of 12 h or less except near the equator. Even with but one pulse per beam position a very high pulse repetition frequency—2300 Hz—is needed to assure continuity of coverage across scans. With this scan geometry separate transmitting and receiving beams displaced in angle are needed, for the antenna is looking $3\cdot3$ to $4\cdot7$ beamwidths downscan by the time the pulse is returned. Thus, "the most demanding component of the radar is the antenna".

Because it offers sampling frequencies appropriate to the detection of severe storms, Bandeen and Katz (1975) also considered radar on a geostationary satellite. With low power consumption and a small antenna such a radar must achieve 10–20 and $1\cdot5$ km horizontal and vertical resolution, therefore they suggest a wavelength of 3 to 9 mm. But vertical resolution degrades rapidly away from the subpoint, and even at 3 mm achieving 30–40 km horizontal resolution requires a beam width of $\sim0\cdot05°$, "which is beyond the reach of technology in the near future".

Because practical antenna sizes could not provide adequate angular resolution nor would uncertainties in Z–R relations and in radar calibration allow adequate accuracy, Eckerman et al. (1978) ruled out radars at nonattenuating wavelengths. They proposed instead to measure rain over land by means of attenuation at a single wavelength, with the surface of the Earth just outside each rain area as the reference target.

According to Eckerman et al. P' in Eq. 14.21 is the sum of powers from rain and from ground in the surface cell $[P_R(r_N)$ and $P'_G(r_N)]$ plus receiver noise (P_{N_1}); P is the sum of power from ground in the surface cell [outside the rain area, $P_G(r_N)]$ plus receiver noise (P_{N_2}). Then Eq. 14.21 may be written

$$\hat{R}_{AV} = \left[\frac{-5}{K_2 r_e} \log \frac{P_R(r_N) + P'_G(r_N) + P_{N_1}}{P_G(r_N) + P_{N_2}} \right]^{1/\gamma} \tag{14.23}$$

and it is apparent that the main sources for measurement error in this technique are slant range rain depth and the various components of received power. The estimate is the true average rainrate if powers from rain and from receiver noise are negligible, if the differential scattering cross section is constant, if $K_2 r_e$ is the true $K_2 r_e$, and either $\gamma = 1$ or rainrate is constant in range. Eckerman et al. makes an exhaustive and instructive error analysis of this equation; here we summarize their results first in terms of the dependence of signal-to-noise ratios $(P'_G/P_N, P'_G/P_R)$ on radar design parameters and then in terms of statistical models of various errors.

Since in theory $A_2 \sim P_G'/P_G$, Eq. 14.23 implies that an accurate estimate of R requires large signal-to-noise ratios, P_G'/P_R and P_G'/P_N, and small fluctuations in P_G' and P_G. Then the nadir viewing angle θ_N and wavelength λ should be chosen to keep $\sigma°$ large and (more important) insensitive to the character of the surface, particularly wetness. Factors affecting signal strength also bear significantly on the selection of the surface cell, nadir viewing angle, pulse length and beamwidth. These are summarized in Table 14.2.

TABLE 14.2   Performance factors determining the design of a reference target satellite radar.[a]

Parameter	Performance factor
nadir viewing angle, θ_N, and wavelength, λ	constant $\sigma°$; also (secondarily) large $\sigma°$
θ_N	small θ_N gives larger P_G'
surface cell	minimize P_R
pulse length, h	larger h increases P_G'/P_N, but (for large θ) decreases P_G'/P_R and compromises r_e
beamwidth, θ	small θ increases $P_G'/(P_N + P_R)$ but means fewer samples per footprint, thus larger variance in $\hat{R}$

[a]After Eckerman *et al* (1978).

Interrelationships and the cumulative effects of various errors are seen more clearly through the statistical error analysis. Errors in ground power and total power were modelled as independent random variables. $\sigma°$ was a function of wavelength only. Its mean was taken as $-10\,\mathrm{dB}$ at 2 cm; its coefficient of variation (standard deviation/mean) was 0·2. The rain slant depth r_e was modelled as a random variable in range resolution; K_2 as a random variable of empirical characteristics, dependent on wavelength; and noise power as an inverse function of pulse duration. To complete the model, rain storm dimensions followed the composite model of Konrad (1978); the altitude of the satellite was taken to be 600 km and its speed 7 km s^{-1}. Then rainrate observations were simulated by means of the Monte Carlo technique for true rates between 1 and 100 mm h^{-1}. Performance was gauged through sample standard deviation (S) and sample normalized offset (O), calculated, plotted and tabulated from true R in increments of 5 mm h^{-1} (Fig. 14.8). The standard for allowable error was $S < 0·25$ and $|O| < 0·2$. Because variance is critical for light rain (P_G' large), the condition $S < 0·25$ defines a minimum rainrate R_{MIN}; offset on the other hand is critical to heavy rain (P_G' small), thus the condition $|O| < 0·2$ defines a maximum rainrate R_{MAX}.

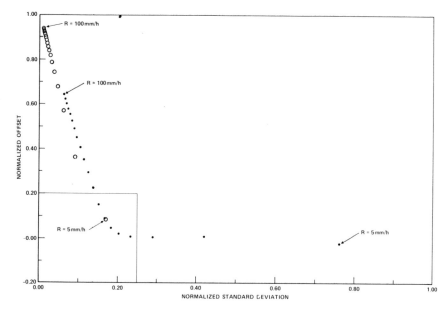

FIG. 14.8   Normalized offset versus normalized standard deviation for rainrates from 5 mm h^{-1} to 100 mm h^{-1} in 5 mm h^{-1} increments with $\theta_B = 0 \cdot 25°$, $P_t = 100$ W. Open circles: $\lambda = 0 \cdot 86$ cm; closed circles: $\lambda = 2 \cdot 4$ cm. From Eckerman *et al.*, 1978.

This behaviour is reflected in Fig. 14.8: for a given wavelength increasing R increases offset but decreases variance. If λ is increased with R constant, the pattern is reversed. Thus measurement of low rainrates always will contain large variance and measurements of high rainrates always will contain a negative offset; yet the interval over which measurements are acceptable can be controlled through radar wavelength. Other design parameters influencing R_{MIN} and R_{MAX} are beamwidth, transmitted power, and pulse width. Multiple beams would decrease R_{MIN} (by reducing variance), but correcting for the contribution of rain in the surface cell to power returned by subtracting P_R from the cell above does not necessarily lead to improvements in the estimated rainrates.

Clearly no single wavelength will cover both light and heavy rainrates. Eckerman *et al.* suggest the best compromise is 2 cm; better still would be a radar of two wavelengths, perhaps 0·86 and 1·87 cm.

Whether single or dual wavelength such a radar system operating alone would have no hope of achieving complete ground coverage or timely revisit. But Atlas *et al.* (1978) foresee eight such radars in four equally spaced polar orbits with satellite pairs opposite each other. Then ground coverage is complete and the revisit interval is reduced to 3 h at most.

If we accept the performance criteria stated by Dennis for operations, eight radar satellites each operating at two wavelengths would seem to be necessary. Taking meteorological satellites of 1980 as a measure of likely commitments of resources, and considering that the first geophysical satellite with active sensors (Seasat) was lost due to power failure, such a system is remote. But proposals for precipitation radar on satellites have advanced enormously over the last few years. At least for research purposes the prospect of directly measuring rainrates with weather radar on satellite platforms now is better than ever before.

CHAPTER 15

Integrated and International Programmes for Rainfall Monitoring

I. Introduction

This book opened with an outline of the needs for improved rainfall data. It has reviewed broadly the range of ways in which satellite data, past, present and future, has been and is being used to improve the chances that such needs will be met. In conclusion, however, it is necessary to discuss the organizational problems which have yet to be solved if progress is to be made with improved monitoring of rainfall on a significant scale.

Conventional rainfall measurements are made by a wide variety of different means. Of these, the commonest by far are the relatively simple collecting raingauges. But even these come in a variety of shapes, forms and sizes, and stand at various heights above the ground. Resulting problems of comparability of measurements from different types of gauges are well-known, and have been documented both generally (see e.g. Rodda, 1971) and particularly (see e.g. Andersson, 1965) on numerous occasions. The World Meteorological Organization has responded to the need for greater standardization in rainfall measurement through its specification of a standard collecting raingauge, the International Reference Precipitation Gauge (IRPG). Unfortunately, though, the rate of adaption of this gauge for synoptic weather station use has been very slow, chiefly because the substitution of this gauge type for existing gauges inevitably introduces an inhomogeneity into a rainfall record, posing problems for the evaluation of long-term climatic normals, and climatic fluctuations and trends.

Consequently conventional raingauge data from different countries are rarely perfectly compatible, and such data from within individual countries are often mixed in the sense that different authorities (meteorological, hydrological etc.) often operate networks equipped with different types of gauges.

In the realm of satellite-assisted rainfall monitoring there is, at present, a similar situation. It will have become abundantly clear from even a casual reading of the chapters in Parts II and III that there is no general agreement concerning the "best" method of satellite monitoring of rainfall, nor could there be: different methods are preferred by different groups, bodies and agencies, because the requirements for rainfall data are wide and various. In the same way in which different consumers of conventional rainfall data find that different types of instruments are best suited to their particular needs, so, too, different consumers of satellite-improved rainfall data recognize that one approach is better suited to their situation than the rest. For this reason, if for no other, it is difficult, if not impossible, to identify a satellite method for rainfall monitoring which is clearly the "best".

II. Potential Users of Satellite-assisted Rainfall Monitoring Methods

Having said this, however, it should be possible to group present and potential satellite rainfall data consumers into a small number of categories, and to suggest procedures and arrangements which might particularly suit them. Viewed in the context of scale of operation, the public users include:

(a) international organizations and groups whose sphere of interest covers the whole world (e.g. the WMO);

(b) international organizations and agencies whose areas of concern are more localized, involving geographic regions (e.g. FAO, with its emphasis on assistance for less developed countries) and/or ethnic blocs (e.g. the Arab League);

(c) national and regional undertakings, usually with relatively narrow terms of reference (e.g. water authorities).

Research users are found in each of these groupings.

In addition to such public users, there are others, prominent among which are commercial firms who have interests generally falling into categories (b) and (c) above, but often with very specific needs. The most common requirement for rainfall data by commercial users today is in connection with production estimates for different crops.

Since the needs of private companies are rarely publicized we may focus our attention here upon the needs of public bodies, and how these might best be met.

The history of the development of satellite remote sensing techniques in other fields, and their adoption for operational use, suggests that there is often much more resistance to the implementation of broader-scale operations than those which are more limited in scope and application. This certainly seems to have been so in the use of satellites for monitoring rainfall too. The most truly operational uses of satellite techniques reside today in national institutions such as NOAA and regional international agencies such as the North-west African Desert Locust Commission. These have arisen principally because of the interest of particular individuals in the potential of the satellite approach, and result in data not generally available outside the institutions involved in the data analysis and interpretation.

In the case of rainfall monitoring there is a particularly significant fact which has hindered the possibility of broader-scale, even global efforts, namely that there has been only one satellite sensing system dedicated to this application—the experimental ESMR—and no satellite or satellite system whose primary purpose has been to supply data for such use. Consequently, although rainfall has been a parameter detailed within schemes such as the World Weather Watch for elucidation by satellites, there has been no clearly organized and/or concerted effort within the auspices of these schemes to ensure that reasonable progress is made towards the stated goals. For example, Table 15.1 alludes to satellite objectives within WWW as identified as long ago as 1967. Most of the other objectives listed in this table have been achieved already. Although the rainfall objectives are comparatively modest, they have not. Nor is it likely that they will be met in the foreseeable future, for there is still no firm policy to that end, despite the types of detailed proposals outlined in Chapter 2. Given such a situation, the best that can be expected at the broadest scale in the new few years is unsteady progress towards a global scheme which seems most likely to be sponsored by a body under no obligation to make the results fully available to any interested party. Even Agristars (see Section IV, Chapter 13), the new research programme directed to that end in the USA, may be seen in that light. At the same time, we may expect some further progress to be made with the development and utilization of microwave sensors, both active and passive, for rainfall monitoring, although technological problems and funding limits may become serious. There is also the real possibility that progress with microwave systems will be slow and fragmentary because of the absence of a suitably defined programme plan.

Of particular interest at this juncture are the needs which have been recognized for rainfall data in major programmes of climate monitoring and research. Two such programmes may be cited as examples.

TABLE 15.1   Weather satellite objectives set up to be achieved by 1971 (from WMO, 1967).

Parameter	Accuracy/resolution	Frequency	Mode
Cloud pictures	2 km at sub-point by day, 8–15 km at sub-point by night	At least 12-hourly, quasi-continuous pictures being the ultimate aim	CDA and APT
Ice- and snow-cover	As for cloud pictures	24-hourly	CDA and APT
Surface temperature and cloud-top temperature	Appr. ±1°C, but with reduced accuracy at sub-zero temperatures	12-hourly	CDA
Radiation- and heat-balance data	About 100 km at sub-point (±2% to 5%)	12-hourly	CDA
Vertical temperature profile	±4°C or better	12-hourly	CDA
Vertical humidity	±10%	12-hourly	CDA
Precipitation intensity	Light, moderate or heavy	12-hourly	CDA and APT

1. The Global Atmospheric Research Programme (GARP) of WMO

This is currently centred on the First GARP Global Experiment (FGGE) through its observational phase in 1979 and the subsequent period of data processing and analysis.

The FGGE requirements for improved rainfall monitoring are not as readily apparent in the benchmark report entitled "Physical basis of climate and climate modelling" (WMO/ICSU, 1975) as might have been expected, largely, it seems, because of the way in which it subdivides the "climatic system". Five physical components are identified, namely the atmosphere, oceans, cryosphere, land surface and biomass. One of the consequences of this subdivision is that the cryosphere receives the kind of attention which, it could be argued, the hydrosphere might equally deserve. It is true that the significance of precipitation is acknowledged, but such references are only:

(a) implicit, e.g. the brief reference to the need to model the hydrologic cycle more realistically;

(b) indirect, e.g. in a section concerned with short-term climatic extremes, where mention is made, *inter alia,* of precipitation anomalies including the south Asian summer monsoon, droughts (lack of rainfall) and floods (excessive rainfall), and their influences in either direct or feedback mechanism roles; and

(c) oblique, e.g. through a mention of the need to understand and observe the fluxes of heat, momentum, and water vapour at the ocean surface, specifically the latent heat fluxes involved in evaporation and precipitation processes.

However, precipitation data requirements for FGGE are listed with the other parameters of significance (see Table 15.2), although land and sea areas are subsequently treated differently from one another. FGGE depends primarily on a rather coarse data-collection grid with cells measuring 500×500 km. It is recognized, though, that the desired resolution of global precipitation and associated hydrologic parameters is much higher, at 100×100 km; even this represents a much larger scale than the basic processes themselves. Over both land and sea surfaces it was reckoned that conventional methods of measuring precipitation (i.e. by gauges) are less than adequate because discrete measurements are seldom representative of mean rainfall over cells as large as the elements in the FGGE grid (WMO/ICSU, 1975). Neither is it considered practical to consider establishing a global network of weather radars.

The proposed solutions for FGGE involved geostationary satellite infrared cloud brightness temperature data and ESMR-type data from Nimbus satellites over the oceans, and, perhaps rather curiously, microwave and Earth resources satellite imagery in support of existing GTS rainfall data for assessments of stored water in the soil and runoff from river basins. However, it should be stressed that FGGE is a programme concerned with a relatively limited span of time. Furthermore, ESMR observations have not yet been planned for operational satellite systems. Consequently FGGE must be seen as the experiment which it is, and we must look elsewhere for specifications of climate programmes with more continuous objectives.

2. The United States Climate Program

The goals of this program, as identified by the US Inter-departmental Committee for Atmospheric Sciences (ICAS), are "to anticipate climatic fluctuations and their domestic and international impacts, and to identify man's impact and potential influence on regional and global climate". Something of the scope of this Program with respect to rainfall/precipitation may be indicated by reference to proposals by NASA for their possible contributions to it (NASA, 1977). For scientific and organizational reasons NASA divided the total climate spectrum into four separate but interrelated portions, each having its own long-term goal. These may be summarized as follows:

(a) Climate A: the current state of the climate. The goal is to provide

TABLE 15.2 Tentative specifications for global observation of precipitation and hydrology proposed for FGGE in 1975 (from WMO/ICSU, 1975).

Variable	Class	Space resolution	Time resolution	Accuracy (1 σ) of determination	Period	Additional remarks, observing technique, etc.
1. Precipitation over oceans	(II)	(100 km)	(5 days)	(1 mm/day-4 levels of discrimination)	–	
Precipitation over oceans	II	500 km	5-15 days	4 levels of discrimination	FGGE	Passive microwave (1·55 and 0·86 cm) radiometry from NIMBUS-F and -G also geo-stationary satellite IR and visible cloud images[a]
2. Soil moisture	II	(100 km)	(5 days)	(10% of field capacity—2 levels of discrimination)	–	
Soil moisture	II	500 km	15 days	2 levels of discrimination	FGGE— limited regions	Detection of vegetation wilting from multi-spectral visible radio-metry and passive microwave brightness mapping[b]
3. Water run-off	(II)	(major river basin)	(15-30 days)	(10%)	–	
Water run-off	II	major river basin	15-30 days	To be estimated	FGGE	River gauges IHD programme; augmentation required

[a]Requires a moderate (but continuous) data processing effort plus further calibration studies.
[b]Method for extracting an unambiguous signature of stored water depletion from colorimetric and microwave brightness data needs to be developed.

timely information in usable form on the current state of the climate system, and to provide analyses of the same.

(b) Climate B: the regional climate that occurs on time scales longer than one month and shorter than one decade. The goal is to predict natural fluctuations over the globe in individual regions on a scale of several hundred km.

(c) Climate C: the climate that occurs on time scales of one decade and longer. The goal is to understand the causes of natural changes of global climate.

(d) Climate X: the climate produced by man's activities on all time and space scales. The goal is to understand and anticipate man's climatic impacts so that appropriate related actions may be taken.

Table 15.3 summarizes NASA's view of the 1977 status of monitoring those important climate parameters for which satellite support is considered necessary. Two of these 40 parameters are distinguished from the others in that:

(a) they are considered of significance for all four "climate types" (A, B, C and X); and

(b) their associated requirements cannot be met by current observing systems, approved future systems, or even a specially designed "Climate Observing System" for the 1980s.

One of these especially elusive parameters is soil moisture in the root zone of vegetation; the other is precipitation.

NASA state that, in the case of the requirements for Climate A, the accuracy, and spatial and temporal resolutions for all parameters have yet to be defined. But they confirm that basic measurements at FGGE levels will not meet the likely Climate A requirements for rainfall, and that daily raingauge measurements at surface weather stations will be totally inadequate for global coverage at the spatial resolution which will be set. In the case of Climates B, C and X, the required resolutions of rainfall data have been set at 500 km every 12–24 h, with a desired accuracy of 10% and a base requirement of 25%. Three tasks are identified in order to progress towards such goals, and the more ultimate objective of finer tuning and verification of climate models. They are:

(a) The improvement of the accuracy of satellite-determined rainfall over ocean areas. At first this should involve further development of the present microwave techniques to increase their accuracy and to adjust for sampling errors, especially aliasing errors arising from the 12-h imaging cycle. Later, research is intended using a cluster of instruments on Spacelab in the early 1980s. This may include 19·37 GHz

TABLE 15.3   Status of climate parameter monitoring, according to NASA (1977).

No.	Parameter	Climate type A	B	C&X	Current systems?	Approved future systems?	Climate observing system (1980's)
						Can requirement be met by . . .	
Weather Variables							
1.	Temperature Profile	↑	↑	↑	[Standard	Yes	Yes
2.	Surface Pressure		Basic		Weather	(No remote technique)	
3.	Wind Velocity		FGGE		station	(from cloud motion)	
4.	Sea Surface Temperature		Measurement		obsns.]	Yes	Yes
5.	Humidity	↓	↓	↓		Yes	Yes
6.	Precipitation	✓	✓	✓		No	No
7.	Cloud Cover	✓	✓	✓		Yes	Yes
8.	Boundary Layer Stability	✓					
Ocean Parameters							
4a	Sea Surface Temperature	✓	✓	✓	No	Yes	Yes
9.	Evaporation		✓	✓	No	No	No
10.	Surface Sensible Heat Flux		✓	✓	No	No	No
11.	Wind Stress		✓	✓	No	No	No
12.	Sea Surface Elevation			✓	No	Maybe	Maybe
13.	Upper Ocean Heat Storage			✓	No	No	No
14.	Temperature Profile		.	✓	Partially	Partially	Partially
15.	Velocity Profile			✓	Partially	Partially	Partially
Radiation Budget							
7a	Clouds (Effect on Radiation)		✓	✓	Almost	Yes	Yes
16.	Regional Net Radiation Components		✓	✓	No	No	Yes
17.	Equator-Pole Gradient		✓ ·	✓	No	No	Yes
18.	Surface Albedo		✓	✓	Yes	Yes	Yes
19.	Surface Radiation Budget	✓	✓	✓	Partially	Partially	Partially
20.	Solar Constant		✓	✓	No	Yes	Yes
21.	Solar Ultraviolet Flux		✓	✓	No	Yes	Yes
Land, Hydrology, and Vegetation							
6a	Precipitation	✓	✓	✓	No	No	No
18a	Surface Albedo		✓	✓	Yes	Yes	Yes
22.	Surface Soil Moisture	✓	✓	✓	No	Partially	Yes
23.	Soil Moisture (Root Zone)	✓	✓	✓	No	No	No
24.	Vegetation Cover (Non-Forest)	✓	✓	✓	Almost	Yes	Yes
25.	Evapotranspiration	✓	✓		No	No	No
26.	Plant Water Stress	✓	✓		No	No	No
Cryosphere Parameters							
27.	Sea Ice (% Open Water)	✓	✓	✓	Yes	Yes	Yes
28.	Snow (% Coverage)	✓	✓	✓	Yes	Yes	Yes
29.	Snow (Water Content)	✓	✓	✓	No	Yes	Yes
30.	Ice Sheet Surface Elevation			✓	No	Maybe	Yes
31.	Ice Sheet Horizontal Velocity			✓	Yes	Yes	Yes
32.	Ice Sheet Boundary	✓		✓	Yes	Yes	Yes
Atmospheric Composition							
21a	Solar Ultraviolet Flux			✓	No	Yes	Yes
33.	Stratospheric Aerosol Optical Depth	✓			No	Almost	Almost
34.	Tropospheric Aerosol Optical Depth	✓		✓	No	No	No
35.	Ozone	✓		✓	No	Yes	Yes
36.	Stratospheric H_2O			✓	No	Yes	Yes
37.	H_2O, NO_x	✓		✓	Yes	Yes	Yes
38.	CO_2			✓	Yes	Yes	Yes
39.	CFM's			✓	Yes	Yes	Yes
40.	CH_4			✓	Yes	Yes	Yes

and 94 GHz microwave radiometers, plus visible and infrared radiometers, to be calibrated by 3 and 10 cm ground radars and high-density gauge networks.

(b) The development of visible and infrared techniques for (more uniform) satellite rainfall observation over land.

(c) The assembling of conventional data sets to permit adequate verification of both land and ocean rainfall mapping techniques.

We must conclude that there is a pressing need for satellite assistance in rainfall monitoring at the macroscale, and it is still not entirely clear how in practice this need may best be met.

III. The Present Choice of Satellite-assisted Rainfall Monitoring Methods

Let us examine the different approaches which might be followed if, instead of the more restricted schemes discussed in this book, a global scheme or global schemes were to be sought for implementation by an appropriate public body. There can be no doubt that both the need, and the potential, for such schemes exist. Indeed, the need is probably far greater than is apparent at first sight, for, in the absence of a global scheme from which data could be made available in near real-time to much of the user community, there will inevitably grow substantial overlaps and duplication of effort if regional and national programmes occupy the operational field alone. This is wasteful of both manpower and resources, and should be avoided wherever possible, especially in the less developed countries where the need for supplementary data for local use is most acute.

In terms of the selection of satellite sensors for rainfall monitoring, the following would seem to be the range of choices:

1. Single sensor approach

In the light of the experience summarized in this book this could involve:

(a) visible/infrared radiometer data;
(b) passive microwave radiometer data;
(c) active microwave (radar) data.

Implementation of a programme based on (a) could be envisaged in the short-term, or on (b) in the mid-term only, because of unresolved problems in the analysis of passive microwave data over land, and on (c) in the long-term on account of our lack of experience with such data from orbital altitudes.

2. Multi-sensor approach

At present this type of approach might be better than the single-sensor approach in the sense that, arguably, passive microwave data are already better than visible and/or infrared data over oceans, whilst visible and/or infrared give better results over land.

In terms of the selection of satellite *platforms* for rainfall monitoring the following choices are available:

1. Single satellite systems

These include:
  (a) Polar orbiters, of which the new Tiros-N family are the most appropriate for fairly immediate use. In this system two satellites operate in opposition to each other to provide a 6-hourly coverage of the globe.
  (b) Geostationary satellites. These seem less promising than (a) in that they give foreshortened views of middle latitudes north and south of about 45°, and cannot view high latitudes, because of their equatorial orbits and the curvature of the Earth. Although the WWW planned continuous circumglobal coverage for the First Garp Global Experiment (FGGE) in 1979 through five geostationary satellites fairly evenly spaced around the equator problems arose with several of the satellites, and their coverage was incomplete. A more serious break in coverage arose at the end of 1979 with the movement of GOES-IO back to the Americas, and the failure of Meteosat-1. Gaps in data coverage may be expected to continue especially over the Indian Ocean. As a result of such considerations, dependence on geostationary satellite data alone for rainfall assessment even in low to middle latitudes would not seem to be a practical possibility yet. However, the fixed views obtained from geostationary satellites carry very significant practical advantages through the relative ease of their image navigation and data normalization: where geostationary data are available, they may be used in preference to data from polar orbiters.

2. Multi-satellite systems

If data more frequent than the 6-hourly data from the Tiros-N system were thought desirable where and when available, it should be possible to employ geostationary satellite data, subject to availability, for rainfall monitoring in low and middle latitudes, and polar-orbiter data in all remaining regions. In that it is in the tropics that rapid convectional cloud growth is most important, it is fortunate that such areas are those monitored from geostationary altitudes. In middle to high latitudes a 6-hourly data coverage by satellite should, perhaps, be usually adequate for the cloud/precipitation processes which are dominant there.

Given present data supplies, both from satellite and conventional sources, it has been argued (see Barrett, 1977e) that the best solution to the global problem in the short-term future might involve the following strategy, at least for near-real time rainfall mapping operations monitoring over land:

(a) Identify those existing rainfall stations from which dependable data are, or could reasonably be, expected.

(b) Augment the network in (a) by as many additional rainfall stations as possible within the scope of the GTS circuit, ensuring that observational methods are standardized as far as possible, and that quality control is rigorously applied.

(c) Map those areas which could not be represented adequately by the data even from the augmented network: these are the areas for which satellite estimates of rainfall would be most usefully obtained.

(d) Select the satellite technique most appropriate for rainfall estimation in each geographical region. The choice for the short- to medium-term future might involve:

   (i) Infrared data from available geostationary satellites, for use in the estimation of daily rainfall over land areas in the tropics and middle latitudes.

   (ii) Microwave data from ESMR-type systems on polar orbiting satellites such as the Tiros-N series (although at present, there is no firm plan to fly such a sensor on Tiros-N). Using these data precipitation might be mapped over the oceans in low and middle latitudes, and possibly deduced also over frozen land and sea surfaces in high latitudes.

   (iii) Visible and infrared data from polar-orbiting satellites such as Tiros-N. From these, rainfall might be assessed over land areas in middle latitudes, and over oceans where (i) and (ii) are absent.

(e) Develop a system capable of receiving, storing and processing the vast and diverse data sets which would be generated, establishing and maintaining a system calibration for the integrated conventional and satellite rainfall estimates, and communicating the information in a timely and dependable manner, without degradation.

Clearly such an approach is complex and pragmatic, rather than simple and elegant. However, if there is—as there seems to be—a pressing need for relatively homogeneous and regular rainfall data around the globe as a

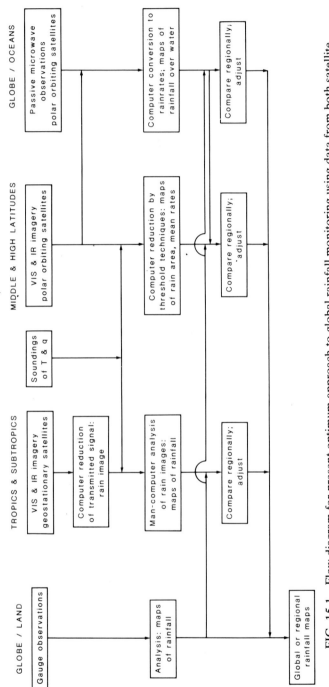

FIG. 15.1 Flow diagram for present optimum approach to global rainfall monitoring using data from both satellite and conventional sources.

whole there seems little doubt that they should be more readily forthcoming from such a satellite/raingauge mix than from conventional sources alone, and that they would be more cheaply provided by the satellite/raingauge approach than the logical alternative of a global surface radar/raingauge combination.

The outstanding problems which would have to be solved before the type of solution summarized in Fig. 15.1 could be implemented include:

(a) Interfacing the results obtained in different regions from the different techniques implicit in (d) above;
(b) Incorporating suitable models of cloud behaviour into the rainfall estimation algorithms to take care of all significant types and organizations of clouds, and to make suitable allowances for cloud movement and/or development between imaging times in every circumstance. At present such models are available for some types of tropical convective cloud activity, but comparable models have yet to be researched for others.
(c) Developing a high degree of automation in the programme, initially in the data handling areas, and, as soon as possible, in the rainfall estimation procedures themselves. This would necessitate further research on the relations between the appearance of clouds from satellite altitudes and associated rain, and the behaviour of clouds and cloud systems between imaging/calibration times.

For such a programme to be undertaken on the global scale envisaged here, a major sponsor would be required. Economy and efficiency would further require that the programme be undertaken in a single facility. This would not preclude, however, the implementation of similar programmes at a regional or national level where more detailed rainfall might be required, or more specialized products for use in hydrology, agrometeorology, civil engineering etc. In these cases it would seem most appropriate to integrate the smaller-scale rainfall operations with other data-analysis procedures based on the same satellite data, and with access to the full GTS weather messages. A wide range of operations might be grouped together in this way. It would seem that many international and national bodies, concerned with a wide range of environmental phenomena, could be provided with useful data from a smaller number of suitably organized satellite centres. Proposals to this effect have been made with respect to more data-sparse regions of the world, either under the auspices of agencies like FAO (see Barrett, 1980a), or groups of nation states, for example in the Sahel region of west Africa (Barrett, 1979: unpublished notes for the Canada Center of Remote Sensing). Such suggestions have been accepted as logical developments in the rise of satellite remote sensing of the environment; however, sponsorship and funding are not readily forthcoming.

IV. The Future of Satellite-assisted Rainfall Monitoring Methods

In the course of summarizing for this book the state of the science of monitoring rainfall by satellites opinions have formed and gelled with respect to several topics. These include the following:

(a) Research in Methods of Rainfall Estimation. The information contents of present weather satellite data are almost certainly much richer in terms of indicators of rainfall than has yet been demonstrated. Research must continue to explore this likelihood, especially through new combinations of data and data derivatives alongside conventional observations. Research must also continue to explore new designs of instruments for observing rainfall, at higher resolutions and in various spectral intervals.

(b) Development of Methods for Rainfall Forecasting. There seems to be valuable scope for the development of satellite-assisted methods for rainfall *forecasting*, to augment and increase the practical benefits derived from the rainfall estimation methods and the better knowledge and understanding we have gained through them. The best approach might well combine a satellite rainfall estimation scheme (providing an improved view of rainfall at the beginning of the forecast period) with an existing scheme for synoptic weather forecasting. The most advanced project of this type is Frontiers (Forecasting Rain Optimized using New Techniques of Interactively Enhanced Radar and Satellite), described in detail by Browning (1979). This programme primarily aims to analyse and forecast detailed patterns of precipitation over the British Isles and adjacent waters for periods from 0 to 6 h ahead, using raingauge, radar and geostationary satellite data as the inputs.

(c) Standardized Verification Procedures. We believe that some common standard(s) should be identified for the evaluation of the performance of any satellite rainfall monitoring method. These standards might include reference to the following basic characteristics of any method:

  (i) How often was it right in predicting rain?
  (ii) How often was it right in predicting no rain?
  (iii) How well does it reproduce the distribution of rainfall?

(iv) How well does it estimate rain intensity?

Statistics answering these questions include (partly after Lovejoy and Austin, 1979):

> (i) and (ii) The averages over many cases of correctly estimated rain f_R and correctly estimated no rain f_N, defined as

$$f_R = R_R/(R + N)$$
$$f_N = N_N/(R + N)$$

> where $R = R_R + R_N$ and $N = N_N + N_R$; also the range and variance of f_R and f_N;

(iii) The map correlation coefficient ρ, defined as

$$\rho = (R_R N_N - R_N N_R)/RN;$$

the areal bias, defined as

$$\beta = \frac{1}{n}\sum_n \hat{A}_R/A_R;$$

the error factor, defined as

$$\varepsilon_F = \frac{1}{n}\sum_n (\hat{A}_R/A_R \text{ if } \hat{A}_R \geq A_R, A_R/\hat{A}_R \text{ if } A_R > \hat{A}_R);$$

and the root mean square error, defined as

$$\varepsilon_{RMS} = \left[\overline{(\hat{A}_R - A_R)^2}\right]^{1/2}/\bar{A}_R;$$

> (iv) Regression analysis, including slope, offset, and linear correlation, and bias, error factor, and root mean square error, as defined above, for intensity.

(d) A Satellite Rainfall Monitoring Paradigm. Since different rainfall monitoring methods appear more suitable at present for different applications, due most fundamentally to the different problems encountered across the range of time and space scales which have been considered, it seems necessary to search for a satellite rainfall estimation paradigm which could lead to a comprehensive, physically consistent, and unified approach to rainfall monitoring by satellites. Through such a single scheme results might be provided for a wide range of applications with the required precision, frequency, and spatial resolution.

(e) New Satellites and Satellite Systems.

TABLE 15.4   Quantities and measures in hydrology which purpose-built hydrologic satellites might help to assess (after Gilg *et al.*, 1975).

1.	*Mapping*
1.1	Systems of waters
1.2	Terrain formation
1.3	Vegetation
1.4	Run-offs
1.5	Flood zones

2.	*Precipitation, evaporation, percolation*
2.1	Precipitation
2.2	Evaporation
	2.2.1 Interception
	2.2.2 Evaporation
	2.2.3 Transpiration
	2.2.4 Evapotranspiration
2.3	Percolation

3.	*Surface hydrology*
3.1	Surface run-off of spring discharges
3.2	Surface run-off of groundwater
3.3	Total surface run-off
3.4	Relation between precipitation and surface run-off (shape, kind, size of the zone)
3.5	Exploitable offer of water

4.	*Subsurface hydrology*
4.1	Subsurface groundwater systems
4.2	Subsurface run-off
4.3	Exploitable offer of groundwater

5.	*Water management*
5.1	Planning
	5.1.1 Determination of the exploitable offer
	5.1.2 Determination of the demand
	5.1.3 Balancing
	5.1.4 Planning of measures
5.2	Surveillance
	5.2.1 Flood forecast
	5.2.2 Water quality
5.3	Run-off control
	5.3.1 Storage management
	5.3.2 Withdrawal of water
	5.3.3 Addition of water
	5.3.4 Modification of precipitation, evaporation, and percolation
5.4	Hydraulic engineering
	5.4.1 Water storage
	5.4.2 River works
	5.4.3 Irrigation
	5.4.4 Drainage

Looking somewhat further into the future, it may be expected that operational analyses of satellite data for different purposes will increase, heightening the need for an integrated approach to both the space and ground segments of environmental satellite systems (Barrett, 1977d). Viewed very broadly, the needs of the operational satellite data user community could be met by a multi-disciplinary satellite system comprised of the following parts:

(i) A general-purpose satellite/sensor system feeding basic data into central reception/processing facilities for a variety of different applications. Now that the electromagnetic spectrum has been rather thoroughly and systematically explored it seems that the value of data from multispectral systems does not increase in proportion to the number of channels used. Given, too, the fact that the ideal choices of wavebands for a wide variety of potential data applications coincide, a family of multi-disciplinary satellites for operational monitoring of the environment need not be over large or complex.

(ii) Restricted-purpose satellite/sensor systems involving relatively cheap and simple satellites dedicated to applications not adequately met by (i) above. Perhaps a hydrological satellite might be one of these. A detailed Mission Study for an operational remote sensing satellite system for hydrology observation in developing countries was undertaken recently as part of the space programme of the Federal Republic of Germany (Gilg *et al.*, 1975). The "partial fields" identified in that study for investigation by a hydrologic satellite are listed in Table 15.4. Possible sensors capable of providing data to meet the associated needs are set out in Table 15.5. Since the data requirements in the visible and infrared regions are shared with the needs of many other data users, it is in the microwave region that a dedicated hydrologic satellite might best operate, although, as indicated in Chapter 14, advances in *active microwave* systems design would seem to warrant a combination of active and passive microwave sensors in such a satellite.

Of special interest is the proposal of Atlas *et al.* (1978) to meet the operational needs of American meteorology in the 1990s with a system comprised of three geostationary and eight polar orbiting satellites. The polar orbiters would occupy four equally spaced, circular orbits, each with two satellites opposed along the orbit. On each satellite would be a multi-channel passive microwave radiometer for measurement of precipitation over water, and a one or two channel attenuating radar for the measurement of precipitation over land. This system would measure global rainfall every 3 h.

A realization of this vision could have as big an impact on global rainfall monitoring as all the earlier weather satellites put together—though an

L

TABLE 15.5 Possible payloads for a specialized hydrological satellite (after Gilg et al., 1975).

Sensor / Payload	Multi-spectral scanners				Radar systems		Spectral regions (number of channels)				Resolution (orbit height 920 km) (m)	Days to complete Earth coverage (days)
	Multi-spectral scanner (MSS)	Thematic mapper (TM)	High resolution pointable imager (HRPI)	Filter wheel in addition to MSS	Synthetic aperture radar (SAR)	Coherent imaging radar (CIR)	Visible (μm)	near IR (μm)	thermal IR (μm)	micro-waves (GHz)		
1	X						0·5-0·8 (3)	0·8-1·1 (1)	10·4-12·6 (1)		30/240^a	18
2	X	X	X				0·5-0·8 (2×3)	0·8-1·1 (2×1)	10·4-12·6 (1)		30/240^a 10	18
3		X					0·5-0·8 (3)	0·8-1·1 1·55-1·75 (3)	10·4-12·6 (1)		50-30/ 90-150^a	18
4	X		X				0·5-0·8 (2×3)	0·8-1·1 (2) 1·55-1·75 2·08-2·35 (2)	10·4-12·6 (1)		50-30/ 90-150^a 10	18
5	X			X		X	0·5-0·8 (3)	0·8-1·1 (1)	10·4-12·6 (1)	1·37	30/240^a 25	18/ 33
6		X			X		0·5-0·8 (3)	0·8-1·1 1·55-1·75 2·08-2·35 (3)	10·4-12·6 (1)	1-2[L] 8-12[X]	50-30/ 90-150/ 30	18/ 80
7		X	X		X		0·5-0·8 (2×3)	0·8-1·1 (2) 1·55-1·75 2·08-2·35 (2)	10·4-12·6 (1)	1-2 [L] 8-12 [X]	50-30/ 90-150/ 10 30	18/ 80

^{a}Ground resolution of thermal IR channel.

effective integration of the new microwave rainfall observations with conventional rainfall data might take several years to achieve.

Until such time, the user community has no choice but to depend on the physically less direct visible, infrared and/or microwave evidences of rain. Faced with the multiplicity of approaches summarized in this book, the would-be routine (as distinct from research) user of satellite-improved rainfall data could be forgiven for asking "Which technique(s) might best meet my need?" In detail this question can be answered only in the context of an individual user's need-to-know, and the resources which he could expend to meet that need. However, some general conclusions can be reached on the types of methods from which his choice might be made:

(a) Mesoscale convective rainfall assessment. For real-time flood-warning purposes, each of the life-history methods offers certain advantages; however only the Scofield-Oliver method was designed for use overland in temperature latitudes. In maritime situations in the tropics, other life history methods (see Chapter 5) are the most appropriate, though bi-spectral methods (Chapter 6) seem promising.

(b) Global or broad-scale rainfall monitoring over land. The best-developed and most widely applicable methods ready for operational usage are of the cloud-indexing types (see Chapter 4). Where rain is predominantly convective in origin, the Griffith-Woodley life-history method may also be useful.

(c) Global or broad-scale rainfall monitoring over oceans and seas. Passive microwave methods are the most promising, but require high-technology facilities—and a regular supply of ESMR-type data. In the absence of one or both of these, the Kilonsky-Ramage type approach has most to recommend it, within the tropics; elsewhere another cloud-indexing approach might prove more suitable.

Once again, as Fig. 15.1 implies, *the urgent and overriding needs are for agreement on the types of systems to be deployed through a specified period of time for operational use, and for international organization and sponsorship of both space and ground segments of the operational systems.* Progress in both directions has been slow. It is our conviction that many valuable opportunities for satellites to be put to good practical use in the service of mankind have been lost, and others continue to be delayed as a result. If this book results in increased interest in these problems, both small and large, and leads to an acceleration of programmes in this and related fields, we shall be well pleased.

References

Adler, R. F. and Fenn, D. D. (1979). Thunderstorm vertical velocities estimated from satellite data. *J. Atmos. Sci.* **36**, 1747–1754.

Adler, R. F. and Rodgers, E. B. (1977). Satellite-observed latent heat release in a tropical cyclone. *Mon. Wea. Rev.* **105**, 956–963.

Aida, M. (1977). Reflection of solar radiation from an array of cumuli. *J. Meteorol. Soc. Japan* **55**, 174–181.

Allison, L., Rodgers, E. Wilheit, T. and Wexler, R. (1974a). *A Multisensor Analysis of Nimbus 5 Data on 22 January 1973.* NASA-TN-D-7911, 43 pp.

Allison, L., Rodgers, E., Wilheit, T. and Fett, R. (1974b). Tropical cyclone rainfall as measured by the Nimbus 5 electrically scanning microwave radiometer. *Bull. Amer. Meteorol. Soc.* **55**, 1074–1089.

Amorocho, J. (1975). *An Application of Satellite Imagery to Hydrologic Modeling the Upper Sinu River Basin, Colombia.* Presented to the Symposium on the Application of Mathematical Models in Hydrology (Bratislava, 8–13 Sept., 1975), 9 pp.

Anderson, R. K. and Veltischev, N. F. (1973). *The Use of Satellite Pictures in Weather Analysis and Forecasting.* WMO Technical Note No. 124, WMO, Geneva, p. 275.

Andersson, T. (1965). Further studies on the accuracy of rain measurement. *Ark. Geofys.* **4**, 359–393.

Anti-Locust Research Centre (1966). *The Locust Handbook.* Ministry of Overseas Development, London, 276 pp.

Arkin, P. A. (1979). The relationship between fractional coverage of high cloud and rainfall accumulations during GATE over the B-scale array. *Mon. Wea. Rev.* **107**, 1382–1387.

Arp, G. *et al.* (1976). *System Development of the Screwworm Eradication Data System (SEDS) Algorithm,* Technical Memorandum, LEC-7646, Lockheed Corporation, Houston, Texas.

Atlas, D. (1964). Advances in radar meteorology. *Advances in Geophysics,* Vol. 10, Academic Press, London and New York, 318–478.

Atlas, D. and Ulbrich, C. W. (1977). Path and area-integrated rainfall measurement by microwave attenuation in the 1–3 cm band. *J. Appl. Meteorol.* **16**, 1322–1331.

Atlas, D., Bandeen, W. R., Shenk, W., Gatlin, J. A. and Maxwell, M. (1978). Visions of the future operational meteorological satellite system. In *EASCON '78 Record:*

Electronics and Aerospace Systems Convention (Arlington, Va., 25–27 Sept., 1978), pp. 576–591.

Augstein, E., Riehl, H., Ostapoff, R. and Wagner, V. (1973). Mass and energy transports in an undisturbed Atlantic trade-wind flow. *Mon. Wea. Rev.* **101**, 101–111.

Augustine, J. A., Woodley, W. L., Griffith, C. G. and Browner, S. P. (1979). Rain estimation from geosynchronous satellite imagery for tropical South America. *Preprints, 3rd Conference on Hydrometeorology* (Bogota, 20–24 Aug., 1979), American Meteorological Society, Boston.

Augustine, J. A., Griffith, C. G., Woodley, W. L. and Meitín, J. G. (1981). Insights into errors of SMS-inferred GATE convective rainfall. *J. Appl. Meteorol.* **20**.

Austin, P. M. and Geotis, S. G. (1978). *Evaluation of the Quality of Precipitation Data from a Satellite-borne Radiometer.* Final Report under NASA Grant NSG 5024, Massachusetts Institute of Technology, Cambridge, 30 pp.

Austin, P. M. and Houze, R. A., Jr. (1973). A technique for computing vertical transports by precipitating cumuli. *J. Atmos. Sci.* **30**, 1100–1111.

Baier, W. (1977a). *Crop-Weather Models and their Use in Yield Assessments.* Technical Note No. 151, WMO-458, World Meteorological Organization, Geneva, 48 pp.

Baier, W. (1977b). Information requirements for regional and global operational systems in agricultural meteorology. In *Earth Observation Systems for Resource Management and Environmental Control* (D. J. Clough and L. W. Morley, eds.), Plenum Press, New York and London, 123–144.

Bandeen, W. R. and Katz, I. (1975). Active microwave sensing of the atmosphere. In *Active Microwave Workshop Report* (R. E. Matthews, ed.) NASA SP-376, Washington, D.C., 287–367.

Barkstrom, B. R. and Arduini, R. F. (1977). The effect of finite horizontal size of clouds upon the visual albedo of the earth. In *Radiation in the Atmosphere* (H. J. Bolle, ed.), Science Press, Princeton, N.J., 188–190.

Barnes, C. M., and Cibula, W. G. (1978). Some implications of remote sensing technology in insect control programs including mosquitoes. *Mosquito News* **39**, 271–282.

Barrett, E. C. (1970). The estimation of monthly rainfall from satellite data. *Mon. Wea. Rev.* **98**, 322–327.

Barrett, E. C. (1971). The tropical Far East: ESSA satellite evaluations of high season climatic patterns. *Geog. J.* **137**, 535–555.

Barrett, E. C. (1973). Forecasting daily rainfall from satellite data. *Mon. Wea. Rev.* **101**, 215–222.

Barrett, E. C. (1974). *Climatology from Satellites.* Methuen, London, 418 pp.

Barrett, E. C. (1975a). Analysis of image data from meteorological satellites. In *Processes in Physical and Human Geography* (R. F. Peel, M. Chisholm and P. Haggett, eds.), Heinemann, London, 169–196.

Barrett, E. C. (1975b). *Rainfall in Northern Sumatra: Analyses of Conventional and Satellite Data for the Planning and Implementation of the Krueng Baro-Krueng Jreue Irrigation Schemes.* Final Report, Binnie and Partners (Consulting Engineers), London, 50 pp.

Barrett, E. C. (1975c). *The Scope for Remote Sensing in the Search for Solutions to Water Supply Problems in Countries Bordering The Gulf.* Consultant's Report, W/2413, FAO, Rome, 18 pp.

Barrett, E. C. (1976). *Rainfall in Northern Sumatra: Verification of Rainfall Estimates.* Final Report, Supplement 1, Binnie and Partners (Consulting Engineers), London, 7 pp.

Barrett, E. C. (1977a). *The Assessment of Rainfall in North-eastern Oman through the Integration of Observations from Raingauges and Meteorological Satellites.* Consultants Report, W/KT629, FAO, Rome, 55 pp.

Barrett, E. C. (1977b). *Mapping Rainfall from Conventional Data and Weather Satellite Imagery across Algeria, Libya, Morocco and Tunisia.* Consultant's Report, W/K4647, FAO, Rome, 13 pp.

Barrett, E. C. (1977c). *Rainfall Monitoring in the Region of the North-west African Desert Locust Commission.* Consultant's Report, W/K8207, FAO, Rome, 43 pp.

Barrett, E. C. (1977d). Monitoring precipitation: a global strategy for the 1980's. In *Monitoring Environment Change by Remote Sensing* (J. L. van Genderen and W. G. Collins, eds.), Remote Sensing Society (UK), 53–58.

Barrett, E. C. (1977e). Identifying the optimum system for observing Earth. In *Earth Observation Systems for Resource Management and Environmental Control* (D. J. Clough and L. W. Morley, eds.), Plenum Press, New York and London, 145–163.

Barrett, E. C. (1979). The use of weather satellite data in the evaluation of national water resources, with special reference to the Sultanate of Oman. In *Space Research XIX* (M. J. Rycroft, ed.), Pergamon Press, Oxford and New York, 41–46.

Barrett, E. C. (1980a). The use of satellite imagery in operational rainfall monitoring in developing countries. In *The Contribution of Space Observations to Water Resource Management* (V. V. Salomonson and P. D. Bhavsar, eds.), Pergamon Press, Oxford and New York, 163–178.

Barrett, E. C. (1980b). Satellite-improved rainfall monitoring by cloud-indexing methods: operational experience in support of desert locust survey and control. In *Proceedings of the AWRA Symposium on Satellite Hydrology,* Sioux Falls, S.D., June 1979.

Barrett, E. C. (1980c). *An Operational Method for Rainfall Monitoring in North-west Africa.* Consultant's Report, FAO, Rome, 43 pp.

Barrett, E. C. and Lounis, M. (1979). *Estimating and Monitoring Rainfall in North-west Africa, 1978–79.* Training Report, W/N4601, FAO, Rome, 39 pp.

Barrick, D. E. (1972). First-order theory and analysis of MF/HF/VHF scatter from the sea. *IEEE Trans. Antennas Propagat.* AP-20, 2–10.

Basharinov, A. E., Gurvich, A. S., Yegorov, S. T., Kurskaya, A. A., Matvyev, D. T. and Shutko, A. M. (1971). Results of microwave sounding of earth surface according to experimental data from the satellite Cosmos 243. *Space Res.* **11**, 713–716.

Battan, L. J. (1953). Observations on the formation and spread of precipitation in convective clouds. *J. Meteorol.* **10**, 311–324.

Battan, L. J. (1973). *Radar Observation of the Atmosphere.* The Univ. of Chicago Press, Chicago and London, 324 pp.

Bergeron, T. (1960). Problems and method in rainfall investigation. In "Physics of Precipitation" (H. K. Weickman, ed.), *Geophys. Monograph No. 5,* Am. Geophys. Union, Washington, D.C., 5–30.

Berlyand, T. G. and Strokina, L. A. (1975). Seasonal cycle of tropospheric cloudiness. In *Physical Climatology* (M. I. Budyko, ed.), Gidrometeoizdat, The A. I. Boekova Main Geophysical Observatory, Transactions, 338.

Betts, E. (1971). Forecasting infestations of tropical migrant pests: The Desert Locust and African Armyworm. In *Insect Flight* (R. C. Rainey, ed.), vol. 7, Blackwell, Oxford, 113–134.

Betts, A. K. (1978). Convection in the tropics. In *Meteorology over the Tropical Oceans* (D. B. Shaw, ed.), Royal Meteorological Society, Bracknell, Berkshire, 105–132.

Bielicki, D. E. and Vonder Haar, T. H. (1979). *Estimation of Big Thompson Flood*

Rainfall Using Infrared Satellite Imagery. Atmospheric Science Paper No. 300, Colorado State University, Ft. Collins, 62 pp.

Blackmer, R. H., Jr. (1975). *Correlation of Cloud Brightness and Radiance with Precipitation Intensity.* Final Report on Contract N66314-74-C-2350, Stanford Research Institute, Menlo Park, Cal., 121 pp.

Blackmer, R. H., Jr. and Serebreny, S. M. (1968). Analysis of maritime precipitation using radar data and satellite cloud photographs. *J. Appl. Meteorol.* **7**, 122–31.

Braham, R. R., Reynolds, S. E. and Harrell, J. H. (1951). Possibilities of cloud seeding as determined by a study of cloud height versus precipitation. *J. Meteorol.* **8**, 416–418.

Brooks, C. E. P. (1925). The distribution of thunderstorms over the globe. *Geophys. Mem. Lond.* **24**, 264–274.

Bristor, C. L. and Ruzecki, M. A. (1960). TIROS 1 photographs of the midwest storm of April 1, 1960. *Mon. Wea. Rev.* **88**, 315–326.

Brown, W. E. and Skolnik, M. L. (1975). Active microwave sensor technology. In *Active Microwave Workshop Report* (R. E. Matthews, ed.), NASA SP-376, Washington, D.C., 369–497.

Browning, K. A. (1979). The FRONTIERS plan: a strategy for using radar and satellite imagery for very-short-range precipitation forecasting. *Meteorol. Mag.* **108**, 161–184.

Browning, K. A. and Harrold, T. W. (1969). Air motion and precipitation growth in a wave depression. *Q.J.R. Meteorol. Soc.* **95**, 288–309.

Busygin, V. P., Yevstratov, N. A. and Feigel'son, Ye. M. (1973). Optical properties of cumulus clouds, and radiant fluxes for cumulus cloud cover. *Izv. Atmos. Oceanic Phys.* **9**, 1142–1151.

Byers, H. R. and Hall, R. K. (1955). A census of cumulus-cloud height versus precipitation in the vicinity of Puerto Rico during the winter and spring of 1953–1954. *J. Meteorol.* **12**, 176–178.

Caracena, F., Maddox, R., Hoxit, L. R. and Chappell, D. F. (1979). Mesoanalysis of the Big Thompson storm. *Mon. Wea. Rev.* **107**, 1–17.

Carneggie, D. M. (1975). Usefulness of Landsat data for monitoring plant development and range conditions in California's annual grassland. In *Proceedings of the NASA Earth Resources Survey Symposium,* Houston, Texas, NASA TM X-58168, 19–41.

Cekirda, A. Z. and Jakovleva, T. A. (1967). The use of satellite information on outgoing radiation for the purpose of determining precipitation zones. *Meteorol. Hydrol.* No. 6.

Chandrasekhar, S. (1950). *Radiative Transfer.* Oxford University Press, Oxford, 393 pp.

Cheng, N. and Rodenhuis, D. (1978). An intercomparison of satellite images and radar rainfall rates. *Papers, 11th Technical Conference on Hurricanes and Tropical Meteorology* (Miami Beach, 13—16 Dec., 1977), American Meteorological Society, Boston, 224–226.

Cho, H. R. and Ogura, Y. (1974). A relationship between cloud activity and the low level convergence as observed in Reed-Recker's composite easterly waves. *J. Atmos. Sci.* **31**, 2058–2065.

Clapp, P. F. and Posey, J. (1970). *Estimating Environmental Parameters from Macro-scale Video Brightness Data of ESSA Satellites.* Extended Forecast Division, NMC, Washington, D.C., 27 pp.

Committee on Space Research (COSPAR) Working Group VI (1967). *Status Report on the Applications of Space Technology to the World Weather Watch.* COSPAR Trans. No. 3, Xth Assembly of COSPAR, London, 47–49.

Conover, J. H. (1962). *Cloud Interpretation from Satellite Altitudes.* Res. Note 81, Air Force Cambridge Research Laboratories, L. G. Hanscomb Field, Mass., 77 pp.

Conover, J. H. (1963). *Cloud Interpretation from Satellite Altitudes.* Suppl. 1 to Res. Note 81, Air Force Cambridge Research Laboratories, L. G. Hanscomb Field, Mass., 17 pp.

Cunningham, R. M. (1952). *Distribution and Growth of Hydrometeors around a Deep Cyclone.* M.I.T. Weather Radar Research Report, No. 18.

Davies, R. (1978). The effect of finite geometry on the three-dimensional transfer of solar irradiance in clouds. *J. Atmos. Sci.* **35**, 1712–1725.

Davis, P. A. and Wiegman, E. J. (1973). *Application of Satellite Imagery to Estimates of Precipitation over Northwestern Montana.* Technical and Semiannual Report No. 1 on Contract 14-08-0001-13271, Stanford Research Institute, Menlo Park, Cal., 39 pp.

Davis, P. A., Wiegman, E. J. and Serebreny, S. M. (1971). *Estimation of Precipitation over Flathead Drainage Basin Using Meteorological Satellite Photographs.* Final Report, Contract 14-06-D-7047, Stanford Research Institute, Menlo Park, Cal.

Debye, P. (1929). *Polar Molecules.* Chemical Catalogue Co., New York.

Deirmendjian, D. (1969). *Electromagnetic Scattering on Spherical Polydispersions.* American Elsevier, New York, 290 pp.

Dennis, A. S. (1963). *Rainfall Determinations by Meteorological Satellite Radar.* Final Report to NASA, Stanford Research Institute, Menlo Park, Cal., 105 pp.

Dismachek, D. C. (ed.) (1977). *National Environmental Satellite Service Catalog of Products.* NOAA Technical Memorandum NESS 88, Washington, D.C., 102 pp.

Dittberner, G. J. and Vonder Haar, T. H. (1973). Large scale precipitation estimates using satellite data; application to the Indian Monsoon. *Arch. Met. Geoph. Biokl.* Ser. B, **21**, 317–334.

Dorman, C. E. and Bourke, R. H. (1979). Precipitation over the Pacific Ocean, 30°S to 60°N. *Mon. Wea. Rev.* **107**, 896–910.

EarthSat Corporation (1975). *EarthSat Spring Wheat Yield System Test, 1975.* Final Report, NASA Contract NAS 9-14655, EarthSat Corporation, Bethesda, Md., 249 pp.

Eccles, P. J. and Mueller, E. A. (1971). X-band attenuation and liquid water content estimation by a dual-wavelength radar. *J. Appl. Meteorol.* **10**, 1252–1259.

Eckerman, J. and Wolff, E. A. (1975). *Spaceborne Meteorological Radar Measurement Requirements Meeting.* X-900-75-198, NASA Goddard Space Flight Center, Greenbelt, Md., 57 pp. plus 2 appendices.

Eckerman, J., Meneghini, R. and Atlas, D. (1978). *Average Rainfall Determination from a Scanning Beam Spaceborne Meteorological Radar.* NASA Technical Memorandum 79664, Goddard Space Flight Center, 35 pp. plus appendices.

Elliot, R. D. (1977). Final Report on Methods for Estimating Areal Precipitation in Mountainous Areas. Contract 6-35358, North American Weather Consultants, Santa Barbara, Cal. for NWS, NOAA, Silver Spring, Md., 179 pp.

Erickson, C. O. (1964). Satellite photographs of convective clouds and their relation to the vertical wind shear. *Mon. Wea. Rev.* **92**, 283–296.

Erickson, C. O. and Hubert, L. F. (1961). *Identification of Cloud-Forms from TIROS I Pictures.* Meteorological Satellite Laboratory Report No. 7, US Weather Bureau, 68 pp.

FAO (1974). *Hydro-agricultural Resources Survey: Qatar, Water Resources and Use.*

AGL: DP/QAT/71/501, Technical Report 2, Food and Agriculture Organization, Rome, 131 pp.

Findeisen, W. (1939). Das Verdampfen der Wolken-und-Rogentropfen (The evaporation of cloud- and raindrops). *Meteorol. Z.* **56**, 453–460.

Follansbee, W. A. (1973). *Estimation of Average Daily Rainfall from Satellite Cloud Photographs.* NOAA Technical Memorandum NESS 44, Washington, D.C., 39 pp.

Follansbee, W. A. (1976). *Estimation of Daily Precipitation over China and the USSR using Satellite Imagery.* NOAA Technical Memorandum NESS 81, Washington, D.C., 30 pp.

Follansbee, W. A. and Oliver, V. J. (1975). *A Comparison of Infrared Imagery and Video Pictures in the Estimation of Daily Rainfall from Satellite Data.* NOAA Technical Memorandum NESS 62, Washington, D.C., 14 pp.

Fraser, R. S. (1975). Interaction mechanisms—within the atmosphere. Chapter 5 in *Manual of Remote Sensing, Vol. 1: Theory, Instruments and Techniques* (F. J. Janza, ed.), American Society of Photogrammetry, Falls Church, Va., 181–233.

Fraysse, G. (1978). Agroclimatological methods involving remote sensing techniques. In *Satellite Remote Sensing Applications in Agroclimatology and Agrometeorology* (E. C. Barrett, ed.), ESA SP-1020. European Space Agency, Paris, 84–115.

Fritz, S. (1954). Scattering of solar energy by clouds of "large drops". *J. Meteorol.* **11**, 291–300.

Fritz, S. (1961). Satellite cloud pictures of a cyclone over the Atlantic Ocean. *Q.J.R. Meteorol. Soc.* **87**, 314–321

Fritz, S. (1963). Research with satellite cloud pictures. *Astronaut. Aerosp. Eng.* **1**, 70–74.

Fritz, S. (1964). Pictures from meteorological satellites and their interpretation. *Space Sci. Rev.* **3**, 541–580.

Fritz, S. (convener), Johnson, A. W., Oliver, V. J. and Pyle, R. L. (1965). *Synoptic Use of Meteorological Satellite Data and Prospects for the Future.* National Weather Satellite Center, US Weather Bureau, Washington, D.C., 62 pp.

Fritz, S. and Winston, J. (1962). Synoptic use of radiation measurements from satellite TIROS II. *Mon. Wea. Rev.* **90**, 1–9.

Fusco, L., Lunnon, R., Mason, B. and Tomassini, C. (1980). Operational production of sea-surface temperatures from Meteosat image data. *ESA Bulletin* **21**, 38–43.

Garcia, O. (1981). A comparison of two satellite rainfall estimates for GATE. *J. Appl. Meteorol.* **20**.

Garstang, M. (1972). A review of hurricane and tropical meteorology. *Bull. Amer. Meteorol. Soc.* **53**, 612–630.

GATE Data Catalogue, 1975, plus supplements. Available from GATE Archive, World Data Center-A, Asheville, N.C.

Gibb, Sir Alexander (1975). *Water Resources Survey of Northern Oman.* Final Report to the Government of the Sultanate of Oman on Phase 1, Soil and Agricultural Studies.

Gilbert, B. (1977). Floodtrap city: Johnstown, Pa. *Audubon* **79**, 2–9.

Giddings, L. E. (1976). *Extension of Surface Data by Use of Meteorological Satellites.* Technical Memorandum LEC-8377, Lockheed Corporation, Houston, Texas, 32 pp.

Gilg, W., Francois, H., Schussler, H., Tiemon, A., Trogus, W., Velten, E. and Werner, R. (1975). *Mission Study for an Operational Remote Sensing Satellite*

System for Hydrology Observation in Developing Countries. Final Report under Contract No. RV 21-V64/74-KA-50, Dornier System GMBH, Friedrichshafen, 132 pp.

Gloerson, P. and Hardis, L. (1978). The scanning multichannel microwave radiometer (SMMR) experiment. In *The Nimbus 7 Users' Guide,* Goddard Space Flight Center, Greenbelt, Md., 213–245.

Golden, J. H. (1967). *The Life Cycle of Convective Cloud Systems as Portrayed by Radar and Tiros Photographs.* Florida State University Technical Note No. 67-73, 44 pp.

Goldhirsh, J. and Katz, I. (1974). Estimation of raindrop size distribution using multiple wavelength radar systems. *Radio Sci.* **9**, 439–446.

Gorelik, A. G. and Kalashnikov, V. V. (1971). Determination of integrated water content of rain clouds and of rain layer height by a superhigh-frequency radiometric method. *Advances in Satellite Meteorology,* Wiley and Sons, New York.

Gorelik, A. G., Kalashnikov, V. V., Kutuza, B. G. and Semiletov, V. I. (1971). Measurements of space distribution of brightness temperatures of clouds and rain at 0·8 and 1·35 centimeter wavelengths. *Advances in Satellite Meteorology,* Wiley and Sons, New York.

Greene, D. and Saffle, R. (1978). D/RADEX rainfall estimates. Appendix B in *Meteorological Analysis of the Johnstown Pennsylvania Flash Flood, 19–20 July 1977* (L. Hoxit *et al.,* NOAA Technical Report ERL 401-APCL 43), pp. 61–74.

Griffith, C. G. and Woodley, W. L. (1973). On the variation with height of the top brightness of precipitating convective clouds, *J. Appl. Meteorol.* **12**, 1086–1089.

Griffith, C. G., Woodley, W. L., Browner, S., Teijeiro, J., Maier, M., Martin, D. W., Stout, J. and Sikdar, D. N. (1976). *Rainfall Estimation from Geosynchronous Satellite Imagery during Daylight Hours.* NOAA Tech. Rep. ERL 356-WMPO 7, Boulder, Col., 106 pp.

Griffith, C. G., Woodley, W. L., Grube, P. G., Martin, D. W., Stout, J. and Sikdar, D. N. (1978). Rain estimation from geosynchronous satellite imagery—visible and infrared studies. *Mon. Wea. Rev.* **106**, 1153–1171.

Griffith, C. G., Woodley, W. L., Griffin, J. S. and Stromatt, S. C. (1980). *Satellite-Derived Precipitation Atlas for the GARP Atlantic Tropical Experiment.* Division of Public Documents, U.S. Government Printing Office, Washington, D.C., 284 pp.

Griffith, C. G., Augustine, J. A. and Woodley, W. L. (1981). Satellite rain estimation in the U.S. High Plains. *J. Appl. Meteorol.* **20**, 53–56.

Grosh, R. A., Weinman, J. A. and Van Scherpenzeel, C. W. (1973). Cloud photographs from satellites as a hydrological tool in remote equatorial regions. *J. Hydrol.* **18**, 147–161.

Gruber, A. (1972). Fluctuations in the position of the ITCZ in the Atlantic and Pacific Oceans. *J. Atmos. Sci.* **29**, 193–197.

Gruber, A. (1973a). *An Examination of Tropical Cloud Clusters Using Simultaneously Observed Brightness and High Resolution Infrared Data from Satellites.* NOAA Technical Memorandum NESS 50, U.S. Dept. of Commerce, 22 pp.

Gruber, A. (1973b). Estimating rainfall in regions of active convection. *J. Appl. Meteorol.* **12**, 110–118.

Gunn, K. L. S. and East, T. W. R. (1954). The microwave properties of precipitation particles. *Q. J. Roy. Meteorol. Soc.* **80**, 522–545.

Gurvich, A. S. and Demin, V. V. (1970). Determinaton of the total moisture content in the atmosphere from measurements on the Cosmos 243 satellite. *Atmos. Oceanic Phys.* **6**, 453–457.

Haas, R. H. (1975). Monitoring vegetation conditions from Landsat for use in range management. In *Proceedings of the Earth Resources Survey Symposium*, Houston, Texas, NASA TM X-58168, 43–52.

Hales, J. E., Jr. (1978). The Kansas City flash flood of 12 September 1977. *Bull. Amer. Meteorol. Soc.* **59**, 706–710.

Hall, C. D., Davies, R. and Weinman, J. A. (1978). *The Distribution of Precipitation derived from Nimbus 6 Data*, unpublished ms. (also presented at the 18th Conference on Radar Meteorology, Atlanta, 28–31 Mar. 1979, American Meteorological Society), 16 pp. plus tables and figures.

Hansen, J. E. (1971). Multiple scattering of polarized light in planetary atmospheres. Part II. Sunlight reflected by terrestrial water clouds. *J. Atmos. Sci.* **20**, 1400–1426.

Hansen, J. E. and Travis, L. D. (1974). Light scattering in planetary atmospheres. *Spa. Sci. Rev.* **16**, 527–610.

Hanus, H. (1980). Regression agromet yield forecasting models. In *Remote Sensing Applications in Agriculture and Hydrology* (G. Fraysse, ed.), Balkema, Rotterdam, 111–135.

Harris, R. and Barrett, E. C. (1978). Towards an objective nephanalysis. *J. Appl. Meteorol.* **17**, 1258–1266.

Hawkins, R. S. (1964). Analysis and interpretation of TIROS II infrared radiation measurements. *J. Appl. Meteorol.* **3**, 564–572.

Henderson-Sellers, A. (1979). Clouds and the long term stability of the earth's atmosphere and climate. *Nature* **289**, 786.

Henry, B. (ed.) (1980). *Arid Zone Hydrology*, FAO, Rome.

Hershfield, D. M. (1961). *Rainfall Frequency Atlas of the United States, for Durations from 30 Minutes to 24 Hours and Return Periods from 1 to 100 Years*. U.S. Weather Bureau Tech. Rep. 40, 115 pp.

Hielkema, J. U. and Howard, J. A. (1976). *Application of Remote Sensing Techniques for Improving Desert Locust Survey and Control*. Report No. M/100112-X, FAO, Rome, 69 pp.

Hiser, H. W., Senn, H. V. and Davies, P. J. (1963). *Meso-scale Synoptic Analysis of Radar and Satellite Meteorological Data*. Final Report to U.S. Weather Bureau, Contract No. CW6-10242, Institute of Marine Science, University of Miami, Miami, Fla., 45 pp.

Hobbs, P. V., Pruppacher, H. R., Cotton, W. R. and Few, A. A., Jr. (1979). Conference on cloud physics and atmospheric electricity. AMS, 31 July–4 August 1979, Issaquah, Wash. *Bull. Amer. Meteorol. Soc.* **60**, 219–224.

Holland, J. Z. and Rasmussen, E. M. (1973). Measurements of the atmospheric mass, energy and momentum budgets over a 500-kilometer square of tropical ocean. *Mon. Wea. Rev.* **101**, 44–55.

Hollinger, J. P. (1973). *Microwave Properties of a Calm Sea*. Naval Research Laboratory Report No. 7111-2, Washington, D.C., 69 pp.

Howard, J. A. (1977). Concepts of remote sensing applications for food production in developing countries. In *Earth Observation Systems for Resource Management and Environmental Control* (D. J. Clough and L. W. Morley, eds.), Plenum Press, New York and London, 265–277.

Howard, J. A., Barrett, E. C. and Hielkema, J. U. (1978). The application of satellite remote sensing to monitoring of agricultural disasters. *Disasters* **2**, 231–240.

Howard, J. A., Roy, J. L., Hielkema, J. U. and Barrett, E. C. (in press). *Concepts, Methods, and Applications of Remote Sensing Data to Desert Locust Survey*. FAO, Rome, 24 pp.

Hoxit, L., Maddox, R., Chappell, C., Zuckerberg, F., Mogil, H. and Jones, I. (1978).

Meteorological Analysis of the Johnstown Pennsylvania Flash Flood, 19–20 July, 1977. NOAA Technical Report ERL 401-APCL 43, 71 pp.

Hubert, L. F., Timchalk, A. and Fritz, S. (1969). Estimating hurricane wind speeds from satellite pictures. *Mon. Wea. Rev.* **97**, 382–383.

Hudlow, M. D. and Patterson, V. L. (1979). *GATE Radar Rainfall Atlas.* NOAA Special Report, U.S. Department of Commerce, 155 pp., available from Superintendent of Documents, Stock No. 003-019-00046-2.

Hudlow, M. D., Arkell, R., Patterson, V., Pytlowany, P. J., Richards, F. and Geotis, S. (1979). *Calibration and Intercomparison of the GATE C-Band Radars.* NOAA Technical Report EDIS-31, Center for Environmental Assessment Services, NOAA, Washington, D.C., 132 pp.

Ilaco (1975). *Water Resources Development Project, Northern Oman.* Final Report to Government of Sultanate of Oman, Ilaco, The Netherlands, 221 pp.

Ingraham, D., Amorocho, J., Guilarte, M. and Escalona, M. (1977). Preliminary rainfall estimates in Venezuela and Colombia from GOES satellite images. *Preprints, 2nd Conference in Hydrometeorology* (Toronto, 25–27 Oct. 1977), American Meteorological Society, Boston, 316–323.

Irvine, W. M. and Lenoble, J. (1974). Solving multiple scattering problems in planetary atmospheres. In *Proceedings of the UCLA International Conference on Radiation and Remote Probing of the Atmosphere,* 28–30 Aug. 1973 (J. G. Kuriyan, ed.), pp. 1–57.

Jones, J. B. (1961). A western Atlantic vortex seen by TIROS 1. *Mon. Wea. Rev.* **89**, 383–390.

Katzenstein, H. and Sullivan, H. (1960). A new principle for satellite-borne meteorological radar. *Proc. Weather Radar Conference,* American Meteorological Society, Boston, 505–515.

Kaveney, W. J., Feddes, R. G. and Liou, K.-N. (1977). Statistical inference of cloud thickness from NOAA 4 scanning radiometer data. *Mon. Wea. Rev.* **105**, 99–107.

Keigler, J. E. and Krawitz, L. (1960). Weather radar observations from an earth satellite. *J. Geophys. Res.* **65**, 2793–2808.

Kellogg, W. W. (1966). Satellite meteorology and the academic community. In *Satellite Data in Meteorological Research* (H. M. E. van de Boogaard, ed.), National Center for Atmospheric Research TN-11, Boulder, Col., 5–13.

Kessler, E. and Atlas, D. (1959). Model precipitation distributions. *Aerospace Eng.* **18**, 36–40.

Kidder, S. Q. (1976). *Tropical Oceanic Precipitation Frequency from Nimbus 5 Microwave Data.* Atmospheric Science Paper 248, Colorado State University, Fort Collins, 50 pp.

Kidder, S. Q. and Vonder Haar, T. H. (1977). Seasonal oceanic precipitation frequencies from Nimbus 5 microwave data. *J. Geophys. Res.* **82**, 2083–2086.

Kilonsky, B. J. and Ramage, C. S. (1976). A technique for estimating tropical open-ocean rainfall from satellite observations. *J. Appl. Meteorol.* **15**, 972–975.

Kinzer, G. D. and Gunn, R. (1951). The evaporation, temperature and thermal relaxation-time of freely-falling water drops. *J. Meteorol.* **8**, 71–83.

Konrad, T. G. (1978). Statistical models of summer rainshowers derived from fine-scale radar observations. *J. Appl. Meteorol.* 171–188.

Kornfield, J., Hasler, A. F., Hanson, K. J. and Suomi, V. E. (1967). Photographic cloud climatology from ESSA III and V computer-producted mosaics. *Bull. Amer. Meteorol. Soc.* **48**, 878–883.

Kuettner, J. P., Rider, N. E. and Sitnikov, I. G. (1972). *Experimental Design Proposal for the GARP Atlantic Tropical Experiment.* GATE Report No. 1, World Meteorological Organization, Geneva, p. 70.

Kuo, H. L. (1965). On formation and intensification of tropical cyclones through latent heat release by cumulus convection. *J. Atmos. Sci.* **22**, 40–63.

Langmuir, I. (1948). The production of rain by a chain reaction in cumulus clouds at temperatures above freezing. *J. Meteorol.* **5**, 175–192.

La Seur, N. E. and Zipser, E. J. (1964). *Studies of TIROS Data over the Tropical Atlantic Supplemented by Concurrent Aircraft Photography.* Final Report on Grant No. WBG16 to U.S. Weather Bureau, Department of Meteorology, Florida State University, Tallahassee, 100 pp.

LeComte, D. (1977). *Evaluation of Satellite Rainfall Estimates in Northern Haiti.* Unpublished notes, Environmental Data and Information Service, U.S. Dept. of Commerce, Washington, D.C., 8 pp.

Leigh, R. M. (1973). *Tropospheric Relative Humidity and Cloudiness.* Unpublished Ph.D. Thesis, Dept. of Physics, James Cook, University of Northern Queensland, Townsville, 239 pp.

Lethbridge, M. (1967). Precipitation probability and satellite radiation data. *Mon. Wea. Rev.* **95**, 487–490.

Lilly, D. K. (1978). *The Dynamical Structure and Evolution of Thunderstorms and Squall Lines.* Unpublished ms., National Center for Atmospheric Research, Boulder, Col., 58 pp.

Lo, C. S., Barchet, W. R. and Martin, D. W. (1981). Vertical mass transport in cumulonimbus clouds on day 261 of GATE. Submitted to *Mon. Wea. Rev.*

Lorenz, E. N. (1970). The nature of the global circulation of the atmosphere: a present view. In *The Global Circulation of the Atmosphere* (G. A. Corby, ed.), Royal Meteorological Society, London, 3–23.

Lovejoy, S. and Austin, G. L. (1979a). The delineation of rain areas from visible and IR satellite data for GATE and mid-latitudes. *Atmosphere-Ocean* **17**, 77–92.

Lovejoy, S. and Austin, G. L. (1979b). The sources of error in rain amount estimating schemes for GOES visible and IR satellite data. *Mon. Wea. Rev.* **107**, 1048–1054.

Lovejoy, S. and Austin, G. L. (1980). The estimation of rain from satellite-borne radiometers. *Q. J. Roy. Meteorol. Soc.* **106**, 255–276.

MacDonald, R. B. and Hall, F. G. (1977). *LACIE: a Proof-of-Concept Experiment in Global Crop Monitoring.* Houston, Texas, NASA JSC-11703, 25 pp.

MacDonald, R. B., Hall, F. G. and Erb, R. B. (1976). *LACIE: an Assessment after One Year of Operation.* Presented at the Tenth International Symposium on Remote Sensing of the Environment, 6 Oct., 1976, ERIM, Ann Arbor, Michigan, 23 pp.

Maddox, R. A., Hoxit, L. R., Chappell, C. F. and Caracena, F. (1978). Comparison of meteorological aspects of the Big Thompson and Rapid City flash floods. *Mon. Wea. Rev.* **106**, 375–389.

Marshall, T. S. and Palmer, W. McK. (1948). The distribution of raindrops with size. *J. Meteorol.* **5**, 165–166.

Martin, D. W. and Scherer, W. D. (1973). Review of satellite rainfall estimation methods. *Bull. Amer. Meteorol. Soc.* **54**, 661–674.

Martin, D. W. and Suomi, V. E. (1972). A satellite study of cloud clusters over the tropical north Atlantic Ocean. *Bull. Amer. Meteorol. Soc.* **53**, 135–156.

Mason B. J. (1971). *The Physics of Clouds.* Oxford University Press (Clarendon), London and New York, 671 pp.

Mie, G. (1908). Beitrage zur Optik truber Medien, speziell kolloidaler Metallosungen. *Ann. Phys.* **25**, 377–445.

Miller, D. B. (1971). *Global Atlas of Relative Cloud Cover, 1967–70, Based on Data from Meteorological Satellites.* U.S. Dept. of Commerce/U.S. Air Force, Washington, D.C., 237 pp.

Miller, D. H. (1977). *Water at the Surface of the Earth: An Introduction to Ecosystem Hydrodynamics.* Academic Press, New York and London, 557 pp.

Miyakoda, K. (1974). Numerical weather prediction. *Amer. Scientist* **62**, 564–574.

Mook, C. P. and Johnson, D. S. (1959). A proposed weather radar and beacon system for use with meteorological earth satellites. *Proc. Third Nat. Convention on Military Electronics* (Washington, D.C., 29 June–1 July), Professional Group on Military Electronics, Inst. of Radio Engineers, 206–209.

Moore, R. K. (1974). *Design Data for Radars Based on 13·9 GHz Skylab σ° Measurements.* University of Kansas Center for Research, Inc.

Mosher, F. R. (1979). *Visible Flux across Finite Clouds.* Ph.D. Thesis. The University of Wisconsin, Madison, 109 pp.

Mower, R. N., Austin, G. L., Betts, A. K., Gautier, C., Grossman, R., Kelley, J., Marks, F. and Martin, D. W. (1979). A case study of GATE convective activity. *Atmosphere-Ocean* **17**, 46–59.

McCullough, D. G. (1968). *The Johnstown Flood.* Simon and Schuster, New York, 302 pp.

McDonald, W. F. (1938). *Atlas of Climatic Charts of the Oceans.* U.S. Weather Bureau, Dept. of Agriculture, Washington, D.C.

McGinnis, D. F., Scofield, R. A., Schneider, S. R. and Berg, C. P. (1979). *Satellites as an Aid to Water Resource Managers.* Preprint 3486, American Society of Civil Engineers Convention and Exposition, Boston, 2–6 Apr., 21 pp.

McKee, T. B. and Cox, S. K. (1974). Scattering of visible radiation by finite clouds. *J. Atmos. Sci.* **31**, 1885–1892.

McKee, T. B. and Cox, S. K. (1976). Simulated radiance patterns for finite cubic clouds. *J. Atmos. Sci.* **33**, 2014–2020.

McKee, T. B. and Klehr, J. T. (1978). Effects of cloud shape on scattered solar radiation. *Mon. Wea. Rev.* **106**, 399–404.

McNab, A. L. and Betts, A. K. (1978). A mesoscale budget study of cumulus convection. *Mon. Wea. Rev.* **106**, 1317–1331.

Nagle, R. E. and Serebreny, S. M. (1962). Radar precipitation echo and satellite cloud observations of a maritime cyclone. *J. Appl. Meteorol.* **1**, 279–295.

National Aeronautics and Space Administration (1977). *Proposed NASA Contribution to the Climate Program.* NASA, Greenbelt, Md., 222 pp.

National Oceanic and Atmospheric Administration (1976). *Big Thompson Canyon Flash Flood of July 31–August 1, 1976,* National Disaster Survey Report 76-1, U.S. Dept. of Commerce, Rockville, Md., 41 pp.

National Oceanic and Atmospheric Administration (1977). *Johnstown, Pennsylvania, Flash Flood of July 19–20, 1977.* Natural Disaster Survey Report 77-1, U.S. Dept. of Commerce, Rockville, Md., 60 pp.

National Oceanic and Atmospheric Administration (1979). *A Study of the Caribbean Basin Drought/Food Production Problem.* U.S. Agency for International Development, Final Report, Project No. 940-11-999. Washington, D.C., 112 pp.

Neiburger, M. (1949). Reflection, absorption and transmission of insolation by stratus cloud. *J. Meteorol.* **6**, 98–104.

Nimomiya, K. (1974). Bulk properties of cumulus convections in the small area over the Kuroshio region. *J. Meteorol. Soc. Japan* **52**, 188–202.

Nitta, T. (1977). Response of cumulus updraft and downdraft to GATE A/B-scale motion systems. *J. Atmos. Sci.* **34**, 1163–1186.

Nordberg, W., Bandeen, W. R., Conrath, B. J., Kunde, V. and Persano, I. (1962). Preliminary results of radiation measurements from the TIROS III meteorological satellite. *J. Atmos. Sci.* **19**, 20–30.

Nordberg, W., Conaway, J. and Thaddeus, P. (1969). Microwave observations of sea state from aircraft. *Q.J.R. Meteorol. Soc.* **95**, 408–413.

Nordberg, W., Conaway, J., Ross, D. B. and Wilheit, T. T. (1971). Measurements of microwave emission from a foam covered, wind driven sea. *J. Atmos. Sci.* **28**, 429–435.

Oliver, V. J. and Scofield, R. A. (1979). Estimation of rainfall from satellite imagery. *Preprints, Sixth Conference on Weather Forecasting and Analysis* (Albany, 10–14 May 1976), American Meteorological Society, Boston, 242–245.

Palmen, E. and Newton, C. W. (1969). *Atmospheric Circulation Systems.* Academic Press, New York and London, 603 pp.

Paris, J. F. (1971). *Transfer of Thermal Microwaves in the Atmosphere.* Dept. of Meteorology, Texas A and M University, College Station, Texas, 257 pp.

Park, A. B. (1975). *Programme Plan for Developing the Capability of Forecasting Crop Production.* Consultant's Report WS/F8302, FAO, Rome, 39 pp.

Park, A. B. (1977). A crop production alarm system. In *Earth Observation Systems for Resource Management and Environmental Control* (D. J. Clough and L. W. Morley, eds.), Plenum Press, New York and London, 331–352.

Pedgley, D. E. (1974). Use of satellites and radar in locust control. In *Environmental Remote Sensing: Applications and Achievements* (E. C. Barrett and L. F. Curtis, eds.), Edward Arnold, London, 143–152.

Pike, A. C. (1971). Intertropical Convengence Zone studied with an interacting atmosphere and ocean model. *Mon. Wea. Rev.* **99**, 469–477.

Plank, V. G. (1969). The size distribution of cumulus clouds in representative Florida populations. *J. Appl. Meteorol.* **8**, 46–67.

Plank, V. G., Atlas, D. and Paulsen, W. H. (1955). The nature and detectability of clouds and precipitation as determined by 1·25 centimeter radar. *J. Meteorol.* **12**, 358–378.

Probert-Jones, J. R. (1962). The radar equation in meteorology. *Q.J.R. Meteorol. Soc.* **88**, 485–495.

Purdom, J. F. W. (1973). Satellite imagery and the mesoscale convective forecast problem. *Preprints of the Eighth Conference on Severe Local Storms* (Denver, 15–17 Oct., 1973), American Meteorological Society, Boston, 244–251.

Purdom, J. F. W. and Gurka, J. A. (1974). The effect of early morning cloud on afternoon thunderstorm development. *Preprints of the Fifth Conference on Weather Forecasting and Analysis* (St. Louis, Mar., 1974), American Meteorological Society, Boston, 58–60.

Radok, U. (1966). *An Appraisal of TIROS III Radiation Data from Southeast Asia.* Atmospheric Science Paper No. 102, Colorado State University, Fort Collins, 37 pp.

Rainbird, A. F. (1969). Some potential applications of meteorological satellites in flood forecasting. In *Hydrological Forecasting, Proceedings of the WMO-Unesco Symposium on Hydrological Forecasting, Australia,* World Meteorological Organization, Technical Note No. 92, WMO No. 228, TP. 122, 73–80.

Ramage, C. S. (1975). Preliminary discussion of the meteorology of the 1972–73 El Niño. *Bull. Amer. Meteorol. Soc.* **56**, 234–242.

Rao, M. S. V. and Theon, J. S. (1977). New features of global climatology revealed by satellite-derived oceanic rainfall maps. *Bull. Amer. Meteorol. Soc.* **58**, 1285–1288.

Rao, M. S. V., Abbott, W. V. and Theon, J. S. (1976). *Satellite-Derived Global Oceanic Rainfall Atlas (1973 and 1974)*, NASA SP-410, Washington, D.C., 31 pp. plus appendices.

Reed, R. J. and Recker, E. E. (1971). Structure and properties of synoptic-scale wave disturbances in the equatorial western Pacific. *J. Atmos. Sci.* **28**, 1117–1133.

Renardet Saute I.C.C. (1975). *Water Resources Survey in North-east Oman.* Interim Report to the Government of the Sultanate of Oman, 116 pp.

Reynolds, D. W. and Smith, E. A. (1979). Detailed analysis of composited digital radar and satellite data. *Bull. Amer. Meteorol. Soc.* **60**, 1024–1037.

Reynolds, D. and Vonder Haar, T. H. (1973). A comparison of radar determined cloud height and reflected solar radiance measured from the geosynchronous satellite ATS-3. *J. Appl. Meteorol.* **12**, 1082–1085.

Reynolds, D. W., McKee, T. B. and Danielson, K. S. (1978). Effects of cloud size and cloud particles on satellite-observed reflected brightness. *J. Atmos. Sci.* **35**, 160–164.

Reynolds, D. W., Vonder Haar, T. H. and Grant, L. O. (1978). Meteorological satellites in support of weather modification. *Bull. Amer. Meteorol. Soc.* **59**, 269–281.

Richards, F. and Arkin, P. (1979). *Spatial and Temporal Variations in the Relationship between Satellite Cloud Coverage and Precipitation.* Unpublished report, Center for Environmental Assessment Services, Environmental Data and Information Services, Washington, D.C.

Richards, F. and Arkin, P. (1981). On the relationship between satellite-observed cloud cover and precipitation. *Mon. Wea. Rev.* **109**.

Riehl, H. (1954). *Tropical Meteorology.* McGraw-Hill Book Co., New York, 392 pp.

Riehl, H. and Malkus, J. S. (1958). On the heat balance in the equatorial trough zone. *Geophysica* **6**, 503–538.

Rodgers, E., Siddalingaiah, H., Chang, A. T. C. and Wilheit, T. (1979). A statistical technique for determining rainfall over land employing Nimbus 6 ESMR measurements. *J. Appl. Meteorol.* **18**, 978–991.

Rodda, J. C. (1971). The precipitation measurement paradox—the instrument accuracy problem. *Reports on WMO/IHD Projects No. 16,* WMO No. 316, 55 pp.

Roffey, J. (1975). *Programme Plan for Improving Desert Locust Survey and Control using Satellite Application Techniques.* Consultant's Report, FAO, Rome, 24 pp.

Rosenkranz, P. W. (1978). General theoretical aspects of passive microwave remote sensing. Appendix to *High Resolution Passive Microwave Satellites* (D. H. Staelin and P. W. Rosenkranz, eds.), Massachusetts Institute of Technology, Cambridge.

Sadler, J. C., Oda, L. and Kilonsky, B. J. (1976). *Pacific Ocean Cloudiness from Satellite Observations.* Dept. of Meteorology, University of Hawaii, 137 pp.

Savage, R. C. (1976). *The Transfer of Thermal Microwaves through Hydrometeors.* Ph.D. thesis, Dept. of Meteorology, University of Wisconsin, Madison, 147 pp.

Savage, R. C. (1978). The radiative properties of hydrometeors at microwave frequencies. *J. Appl. Meteorol.* **17**, 904–911.

Savage, R. C. and Weinman, J. A. (1975). Preliminary calculations of the upwelling radiance from rainclouds at 37·0 and 19·35 GHz. *Bull. Amer. Meteorol. Soc.* **56**, 1272–1274.

Scherer, W. D. and Hudlow, M. D. (1971). A technique for assessing probable distributions of tropical precipitation echo lengths for X-band radar from Nimbus-3 HRIR data. *BOMEX Bull. No. 10,* 63–68.

Schwiesow, R. L. (1972). Atomic, molecular, particulate, and collective generalized scattering. In *Remote Sensing of the Troposphere.* Environmental Research Laboratories, NOAA, Boulder, Col., Chapter 10.

Scofield, R. A. (1978a). Using satellite imagery to estimate rainfall during the Johnstown rainstorm. *Preprints, Conference on Flash Floods: Hydrometeorological Aspects,* (Los Angeles, 2–5 May 1978), American Meteorological Society, Boston, 181–189.

Scofield, R. A. (1978b). Using satellite imagery to detect and estimate rainfall from flash-flood producing thunderstorms. *Preprints, Conference on Weather Forecasting and Analysis and Aviation Meteorology,* (Silver Spring, Md., 16–19 Oct., 1978), American Meteorological Society, Boston, 132–141.

Scofield, R. A. (1978c). Using satellite imagery to estimate convective rainfall. Appendix C in *Meteorological Analysis of the Johnstown, Pennsylvania Flash Flood, 19–20 July, 1977,* (L. Hoxit, *et al.,* NOAA Technical Report ERL 401-APCL 43), 65–71.

Scofield, R. A. and Oliver, V. J. (1977a). *A Scheme for Estimating Convective Rainfall from Satellite Imagery.* NOAA Technical Memorandum NESS 86, Washington, D.C., 47 pp.

Scofield, R. A. and Oliver, V. J. (1977b). Using satellite imagery to estimate rainfall from two types of convective systems. *Papers, Eleventh Technical Conference on Hurricanes and Tropical Meteorology,* (13–16 Dec. 1977, Miami Beach), American Meteorological Society, Boston, 204–211.

Seevers, P. M., Drew, J. V. and Carlson, M. P. (1975). Estimating vegetative biomass from Landsat-1 imagery for range management. In *Proceedings of the Earth Resources Survey Symposium,* Houston, Texas, NASA TMX-58168, 1–8.

Sekhon, R. S. and Srivastava, R. C. (1971). Doppler radar observations of drop-size distribution in a thunderstorm. *J. Atmos. Sci.* **28**, 983–994.

Shenk, W. E., Holub, R. J. and Neff, R. A. (1976). A multi-spectral cloud type identification method developed for tropical ocean areas with Nimbus 3 MRIR measurements. *Mon. Wea. Rev.* **104**, 284–291.

Shifrin, K. S. and Chernyak, M. M. (1968). Microwave absorption and scattering by precipitation. In *Transfer of Microwave Radiation in the Atmosphere,* Israel Program for Scientific Translataions, pp. 69–78.

Shifrin, K. S. and Ionina, S. N. (1968). Thermal radiation and reflection of microwaves from swelling sea surface. In *Transfer of Microwave Radiation in the Atmosphere,* Israel Program for Scientific Translations, pp. 19–43.

Shifrin, K. S., Rabinovich, Yu. I. and Shchukin, G. G. (1968). Microwave radiation field in the atmosphere. In *Transfer of Microwave Radiation in the Atmosphere,* Israel Program for Scientific Translations, 1–14.

Shtern, M. I. (1972). *Investigations of the Upper Atmosphere and Outer Space Conducted in 1970 in the USSR.* NASA TT-F-666, Washington, D.C.

Sikdar, D. N. and Suomi, V. E. (1971). Time variation of tropical energetics as viewed from a geostationary altitude. *J. Atmos. Sci.* **78**, 170–180.

Sikdar, D. N. (1972). ATS-3 observed cloud brightness field related to a meso-to-subsynoptic scale rainfall pattern. *Tellus* **24**, 400–413.

Sikdar, D. N., Suomi, V. E. and Anderson, C. E. (1970). Convective transport of mass and energy in severe storms over the United States—an estimate from a geostationary altitude. *Tellus* **22**, 521–532.

Silver, S. (1951). *Microwave Antenna Theory and Design.* MIT Radiation Laboratory Series, 12, McGraw-Hill, New York.

Simpson, J. (1975). Precipitation augmentation from cumulus clouds and systems: scientific and technological foundations, 1975. *Adv. Geophys.* **1**, 1–72.

Simpson, J. and Wiggert, V. (1969). Models of precipitating cumulus towers. *Mon. Wea. Rev.* **97**, 471–489.

Smith, E. A. and Kidder, S. Q. (1978). *A Multispectral Satellite Approach to Rainfall Estimates.* Unpublished manuscript, Colorado State University, Fort Collins, 26 pp. plus tables and figures.

Smith, W. L., Woolf, H. M., Hayden, C. M., Wark, D. Q. and McMillin, L. M. (1979). The TIROS-N operational vertical sounder. *Bull. Amer. Meteorol. Soc.* **60**, 1177–1187.

Snider, J. B. and Westwater, E. R. (1972). Radiometry. In *Remote Sensing of the Atmosphere,* Environmental Research Laboratories, NOAA, Boulder, Colorado, Ch. 15.

Spillane, K. T. and Yamaguchi, K. (1962). Mechanisms of rain producing systems in southern Australia. *Aust. Meteorol. Mag.* **38**, 1–19.

Squires, P. (1958a). The microstructure and colloidal stability of warm clouds. Part I—The relation between structure and stability. *Tellus* **10**, 256–261.

Squires, P. (1958b). The microstructure and colloidal stability of warm clouds. Part II—The causes of the variations in microstructure. *Tellus* **10**, 262–271.

Staelin, D. H. (1969). Passive remote sensing at microwave wavelengths. *Proc. IEEE* **57**, 427–439.

Staelin, D. H. and Rosenkranz, P. W. (eds.) (1978). *High Resolution Passive Microwave Satellites.* Research Laboratory of Electronics, Massachusetts Institute of Technology, Cambridge.

Staelin, D. H., Gustincic, J. J. and Rosenkranz, P. W. (1978). Systems. In *High Resolution Passive Microwave Satellites* (D. H. Staelin and P. W. Rosenkranz, eds.), Massachusetts Institute of Technology, Cambridge, Chapter 8.

Stepanenko, V. D. (1973). *Radiolokatsiya y Meteorologii,* Second ed., Gidrometeoizdat (Leningrad).

Stepanenko, V. D. (1968a). Contrasts of radio brightness temperatures in clouds and precipitation. In *Transfer of Microwave Radiation in the Atmosphere* (K. S. Shifrin, ed.), Israel Program for Scientific Translations (1970).

Stepanenko, V. D. (1968b). Certain geometric features of clouds and precipitation from radar observations. In *Transfer of Microwave Radiaton in the Atmosphere* (K. S. Shifrin, ed.), Israel Program for Scientific Translations (1970).

Stoldt, N. W. and Havanac, P. J. (1973). *Compendium of Meteorological Satellites and Instrumentation.* NSSDC 73-02, NASA Goddard Space Flight Center, Greenbelt, MD., 455 pp. plus appendix and indices.

Stout, J., Martin, D. W. and Sikdar, D. N. (1978). *Rainfall Estimation from Geostationary Satellite Images over the GATE Area.* Final Report of NOAA grant 04-5-158-47, Space Science and Engineering Center, University of Wisconsin, Madison, 26 pp. plus tables and maps.

Stout, J. E., Martin, D. W. and Sikdar, D. N. (1979). Estimating GATE rainfall with geosynchronous satellite images. *Mon. Wea. Rev.* **107**, 585–598.

Struzer, L. R., Nechayev, I. N. and Bogdanova, E. G. (1965). Systematic errors of measurements of atmospheric precipitations. *Meteorol. Gidrol.* **10**, 50–4 (in Russian); also *Sov. Hydrol. Selected Papers,* 500–504 (in English).

Tanaka, H. (1979). Attenuation of millimeter waves through intense rain cells using the ETS-II (Kiku-2) satellite. *Proceedings of the Symposium on the Use of Satellite Data in Meteorological Research* (Tokyo, Nov. 3–5, 1978) pp. 19–23.

Theon, J. (1973). A multispectral view of the Gulf of Mexico from Nimbus 5. *Bull. Amer. Meteorol. Soc.* **54**, 934–937.

Timchalk, A. and Hubert, L. F. (1961). Satellite pictures and meteorological analyses of a developing low in the central United States. *Mon. Wea. Rev.* **89**, 429–445.

Tsang, L., Kong, J. A., Njoku, E., Staelin, D. H. and Waters, J. W. (1977). Theory for microwave thermal emission from a layer of cloud or rain. *IEEE Trans. Antennas Propag.* **AP-25**, 650–657.

Tucker, G. B. (1961). Precipitation over the North Atlantic Ocean. *Q. J. R. Meteorol. Soc.* **87**, 147–158.

Twomey, S. (1959). The nuclei of natural cloud formation, Pt. 2. The supersaturation in natural clouds and the variation of cloud droplet concentration. *Geofis. Pura. Appl.* 243–249.

Twomey, S., Jacobowitz, H. and Howell, H. B. (1967). Light scattering by cloud layers. *J. Atmos. Sci.* **24**, 70–79.

US Weather Bureau (1961). *Requirements and Design Concepts for an Initial Satellite Weather Radar.* Manuscript of the Meteorological Satellite Laboratory, US Dept. of Commerce, MSL Report No. 4, 66 pp.

Vaeth, J. G. (1965). *Weather Eyes in the Sky.* Ronald Press Co., N.Y., 124 pp.

Van de Hulst, H. C. (1957). *Light Scattering by Small Particles.* John Wiley, New York, 470 pp.

Vette, J. I. and Vostreys, R. W. (1977). *Report on Active and Planned Spacecraft and Experiments.* NSSDC/WDC-A-R and S 77-03, NASA, Greenbelt, Md., 305 pp.

Volchok, B. A. and Chernyak, M. M. (1969). Transfer of microwave radiation in clouds and precipitation. *Transfer of Microwave Radiation in the Atmosphere,* NASA TT F-590, 90–97.

Vonbun, F. O. and Sherman, J. W. (1975). Active microwave remote sensing of oceans. In *Active Microwave Workshop Report* (R. E. Matthews, ed.). NASA SP-376, 157–286.

Vonder Haar, T. H. (1969). Meteorological applications of reflected radiance measurements from ATS 1 and ATS 3. *J. Geophys. Res.* **74**, 5404–5412.

Waldteufel, P. (1973). Attenuation des ondes hyperfrequences par la pluie: une mise au point. *Ann. Téléc.* **28**, 255–272.

Waloff, Z. (1976). *Some Temporal Characteristics of Desert Locust Plagues.* Anti-Locust Memoir 13, Centre for Overseas Pest Research, London, 36 pp.

Wark, D. Q., Yamamoto, G. and Lienesch, J. H. (1962). Methods of estimating infrared flux and surface temperature from meteorological satellites. *J. Atmos. Sci.* **19**, 369–384.

Wasserman, S. E. (1977). The availability and use of satellite pictures in recognizing hazardous weather. In *Earth Observation Systems for Resource Management and Environmental Control* (D. J. Clough and L. W. Morley, eds.), Plenum Press, New York and London, 419–435.

Waters, J. W. (1976). Absorption and emission by atmospheric gases. In *Methods of Experimental Physics, Vol. 12: Astrophysics, Part B* (M. L. Meeks, ed.), Academic Press, New York and London.

Weickman, H., Long, A. B. and Hoxit, R. (1977). Some examples of rapidly growing oceanic cumulonimbus clouds. *Mon. Wea. Rev.* **105**, 469–476.
Weinman, J. A. and Davies R. (1978). Thermal microwave radiances from horizontally finite clouds of hydrometeors. *J. Geophys. Res.* 3099–3107.
Weinman, J. A. and Guetter, P. J. (1977). Determination of rainfall distributions from microwave radiation measured by the NIMBUS 6 ESMR. *J. Appl. Meteorol.* **16**, 437–442.
Weinman, J. A. and Swarztrauber, P. N. (1968). Albedo of a striated medium of isotropically scattering particles. *J. Atmos. Sci.* **25**, 497–501.
Weinstein, M. and Suomi, V. E. (1961). Analysis of satellite infrared radiation measurements on a synoptic scale. *Mon. Wea. Rev.* **89**, 419–428.
Weiss, L. L. and Wilson, W. T. (1950). Precipitation gage shields. *Int. Assoc. Sci. Hydrol. Publ.* **43**, 462–484.
Wendling, P. (1977). Albedo and reflected radiance of horizontally inhomogeneous clouds. *J. Atmos. Sci.* **34**, 642–650.
Wexler, H. (1954). Observing the weather from a satellite vehicle. *J. Brit. Interplanetary Soc.* **13**, 269–276.
Wexler, H. (1957). The satellite and meteorology. *J. Astronaut* **4**, 1–6.
Wexler, R. and Atlas, D. (1963). Radar reflectivity and attenuation of rain. *J. Appl. Meteorol.* **2**, 276–280.
Whitney, L. F., Jr. and Fritz, S. (1961). A tornado producing cloud pattern seen from TIROS I. *Bull. Amer. Meteorol. Soc.* **42**, 603–614.
Widger, W. K. (1964). A synthesis of interpretations of extratropical vortex patterns as seen by TIROS. *Mon. Wea. Rev.* **92**, 263–282.
Widger, W. K., Jr. and Touart, C. N. (1957). Utilization of satellite observations in weather analysis and forecasting. *Bull. Amer. Meteorol. Soc.* **38**, 521–533.
Wiesner, C. J. (1970). *Hydrometeorology.* Chapman and Hall, London, 232 pp.
Wigton, W. H. (1976). *Use of Landsat Technology by Statistical Reporting Service.* U.S. Dept. of Agriculture, Washington, D.C.
Wilheit, T. (1972). The electrically scanning microwave radiometer (ESMR) experiment. In *The Nimbus 5 User's Guide* (R. R. Sabatini, ed.), Goddard Space Flight Center, Greenbelt, Md., 59–105.
Wilheit, T. H. (1975). The electrically scanning microwave radiometer (ESMR) experiment. In *Nimbus 6 User's Guide,* NASA Goddard Space Flight Center, 87–108.
Wilheit, T., Theon, J., Shenk, W. and Allison, L. (1973). *Meteorological Interpreations of the Images from Nimbus 5 Electrically Scanned Microwave Radiometer.* NASA X-651-73-189, Goddard Space Flight Center, 21 pp.
Wilheit, T. T., Theon, J. S., Shenk, W. E., Allison, L. and Rodgers, E. B. (1976). Meteorological interpretations of the images from the Nimbus 5 electrically scanned microwave radiometer. *J. Appl. Meteorol.* **15**, 166–172.
Wilheit, T. T., Chang, A. T. C., Rao, M. S. V., Rodgers, E. B. and Theon, J. S. (1977). A satellite technique for quantitatively mapping rainfall rates over the oceans. *J. Appl. Meteorol.* **16**, 551–560.
Woodley, W. L. and Sancho, B. (1971). A first step toward rainfall estimation from satellite cloud photographs. *Weather* 279–289.
Woodley, W. L., Olsen, A., Herndon, A. and Wiggert, V. (1974). *Optimizing the Measurement of Convective Rainfall in Florida.* NOAA Technical Memorandum ERL WMPO-18, Boulder, Col., 99 pp.
Woodley, W. L., Griffith, C. G., Griffin, J. and Augustine, J. (1978). Satellite rain

estimation in the Big Thompson and Johnstown flash floods. *Preprints, Conference on Flash Floods: Hydro-Meteorological Aspects* (Los Angeles, 2–5 May), American Meteorological Society, Boston, 44–51.

Woodley, W., Griffith, C. G., Griffin, J. S. and Stromatt, S. C. (1980). The inference of GATE convective rainfall from SMS-1 imagery. *J. Appl. Meteorol.* **19**, 388–408.

World Meteorological Organization (1967). *The Role of Meteorological Satellites in the World Weather Watch.* WWW Planning Report 17, WMO, Geneva, 76 pp.

World Meteorological Organization (1974). *Guide to Hydrological Practices.* WMO-No. 168, WMO, Geneva.

World Meteorological Organization (1975). *The Physical Basis of Climate and Climate Modelling.* GARP Publications Series No. 16, WMO, Geneva, 265 pp.

Wylie, D. P. (1979). An application of a geostationary satellite rain estimation technique to an extratropical area. *J. Appl. Meteorol.* **18**, 1640–1648.

Wyrtki, K. (1975). El Niño—the dynamic response of the equatorial Pacific Ocean to atmospheric forcing. *J. Phys. Oceanogr.* **5**, 572–584.

Yanai, M., Esbensen, S. and Chu, J. (1973). Determination of bulk properties of tropical cloud clusters from large scale heat and moisture budgets. *J. Atmos. Sci.* **30**, 611–627.

Zipser, E. J. and La Seur, N. E. (1965). *The Distribution and Depth of Convective Clouds over the Tropical Atlantic Ocean, Determined from Meteorological Satellite and Other Data.* Final Report to National Environmental Satellite Center, Grant WBG 36, Department of Meteorology, Florida State University, Tallahassee, 96 pp.

Index

A

Agristars (Agriculture and Resources
    Inventory Surveys through
    Aerospace Remote Sensing),
    264
Agroclimatology, 251, 252
Agrometeorology, 251, 252
    crop production modelling, 260
Albedo, 23
    and cloud geometry, 115
    and cloud texture, 112
    and cloud type, 35
    and optical thickness, 118
Atmosphere
    homogeneity, 137
    zenith opacity, 137, 138
Atmospheric gases and radiation, 126
Atmospheric window, 103

B

Barbados Oceanographic
    Meteorological Experiment,
    59, 60
    echo length and precipitation, 60
Best drop size distribution, 141
Biomass estimation, 256–8
    and grazing potential, 258
Bi-spectral monitoring schemes, 82–92
    assessment, 94
    infrared and visible images, 82

joint frequency distribution, 88, 89
optimum boundary and rain area,
    94
statistical population discrimination
    87–94
thresholding example, 83–5
Brightness temperature, 25, 140,
    143
    curves, 154, 156
    ESMR-6, 174, 175
    rain correlation, 152, 187
    and rainrate, 158, 160, 176
    upwelling, 165, 167
    vertical polarization, 168
    water and emissivity, 167
    and water roughness, 168

C

Catchment monitoring, 212–16
    north-west Montana, 212, 213
    river flow volumes, 215
    satellite reliability, 214
Center for Environmental Assessment
    Service (CEAS)
    China, 254
    Haiti, 255
    operational nature, 253
    weather reports, 239–41
Cloud
    appearance and variability of
        precipitation, 50

Cloud (*cont.*)
    area and rainfall rate, 69
    characteristics in satellite visible
        images, 24
    classification, 33
    cold centres and dark, 86
    composition and liquid water, 111,
        112
    convective, 35–7, 65, 76
    cuboidal building block, flux, 119,
        120
    cuboidal stack, 118
    droplet size, 10, 107
    evaporation, 100, 101
    extratropical, organization, 12
    finite, 141–3
    formation, 9, 10
    geometry, 115–18
    global cover, 9
    grey, 103
    growth, 77
    life-history estimation methods,
        65–81
    light scattering and geometry, 111,
        123
    merging, 77
    motion model, 56
    Nimbus-6 and ESMR-6 views, 166
    plane-parallel, 112, 137–41
    precipitating, distinction of, 25
    precipitation, criteria for, 35
    precipitation probability and
        temperature and brightness,
        40–41
    and rain, diurnal variability, 55
    reflectivity, 33, 62, 112
    superimposed types, 104
    synoptic systems, 33–5
    temperature, 76, 81
    thickness and precipitation, 106–8
    top temperature and rainfall, 39, 77,
        104
    towers, 76
    transmittance, 112
    turret effect on model, 118, 119
    type and brightness assessment, 25,
        26
    type recognition, basis of automatic,
        27
    types, 12

Cloud bands, intra- and
        trans-tropical, 195
Cloud brightness
    and droplet size, 115, 117
    effect of turrets, 118, 119, 121
    factors affecting, 110–21
    and geometry, 115–19
    light scattering, 110
    relation to thickness, 102, 108, 109
    scattering direction, 110, 111
    single scattering theory, 111
    and sun angle, 115, 117
    temperature, contours, 140, 142,
        143, 154
    upwelling, temperature and
        radioactive variables, 139
Cloud brightness and rainfall, 31–9,
        75, 102, 105–9
    bi-spectral measurements, 86
    and convective activity, 35–7
    and convective regime, 37
    factors governing, 123
    and radar echoes, 38
    and sea level pressure chart, 32
    snapshot views, 31–7
    temperature effect, 41, 75, 88, 89
    time-lapse views, 36–9
Cloud emissivity, 103
Cloud growth, 77
    and rainrate, empirical functions, 81
Cloud-indexing methods, 43–64, 311
    Applications Group, National
        Environmental Satellite
        Service, US, 43, *see also*
        Cloud-indexing methods, NESS
    Applied Climatology Laboratory,
        Bristol, UK, 43–51, *see also*
        Cloud-indexing methods,
        Bristol
    common root, 43
    life-history techniques, 55
    and locust monitoring, 246
    man-machine mix, 57
    Oman, 208
    reflective cloud in tropics, 62
    river monitoring in Mekong, 219
    spring wheat, 261–3
    Stanford Research Institute, 60, 61
Cloud-indexing methods, Bristol,
        43–51

Cloud-indexing methods, Bristol (*cont.*)
  accumulated rainfall, 44
  climatological, 52
  computer assistance, 51
  data sources, 48
  desert locust, NW Africa, 50
  flow diagram, 48, 49
  gauge and satellite cells, 48
  precipitation observation regression,
    46, 47
  preparatory stages, 45, 46
  rainfall probability and satellite
    observed sky, 45
  reliability of 12h images, 48, 49
  short period rainfall estimates,
    Sumatra, 47
  short term forecast improvement,
    47, 48
  sites, 48
  summary, 50
  tropical far east, 44–6
  verification attempts, 50, 51
Cloud-indexing methods, NESS, 52–9,
  235
  area-brightness relationship, 57
  basin rainfall, 214
  basis of first rainfall estimates, 52
  cloud-weighting, 54
  diurnal variability of cloud and rain,
    55
  estimate variability, 53
  extratropical rainfall, 57
  fixed weight, early graphs, 52, 53
  flood forecast in Zambia, 52
  and low altitude convectional cloud,
    54, 55
  maximum areal estimates, 53, 54
  overestimates, 54
  quick, cheap general assessment, 55,
    56
  rain-cloud movement and imaging
    times, 58, 59
Cloudiness categories and seasonal
    precipitation rates, 60, 61
Cloud model methods, 95–101
Cloud system indexing, 59
Cloud thickness and rainfall, 106–8
  and brightness, 108, 109
  echoes, 106
  and temperature, 107, 109

Computer
  grid point rain rates, 78
  rain estimate and flood forecast, 81
Continents
  macroscale rainfall mapping, 199,
    200
  need for improvement in rainfall
    mapping, 200
Convective model methods, 95–101
  parameterization, 95
Cropcast, 57, 261
Crops
  assessment, *Landsat*, 256, 257
  climate relationships, 256
  estimation by reflectance, 257
  factors affecting, 252–6
  growth and production, 251–65
Crop prediction programmes, 258–65
  agroclimatological, 259, 260
  agrometeorological, 260
  cloud-indexing of spring wheat,
    261–4
  integrated systems, 264, 265
  satellite assistance, 258, 259
  statistical, 259, 260
Crop production
  area × yield, 259
  model groups, 259
Cumulonimbus
  area doubling, 105
  area and echo area, 69
  convection, 95
  evolution of ensemble and rainrate,
    66
  explosive growth, 37, 38
  upward mass transport, 100
Cyclones, 11
  enhancement and topography, 11
  flood disasters, 236
  occluded maritime, 33

D

Data analysis, multispectral, 51
Disasters
  rural and agricultural impact, 222
  weather-related, 222
Drop
  microwave scattering and
    absorption, 129

Drop (*cont.*)
  Mie efficiency factors, 129
  radius and Mie efficiency factors, 132
  size distributions, 141
  small volume, scattering and
      absorption, 131–3
Drought, 256

E

Earth radar reflectivity, 273–5
EarthSat, 1975 exercise, 57
East African Army Worm, 250
El Niño phenomenon, 185
Emissivity of surfaces, 127, 171
ESMR-5
  background contrast, 147, 148
  brightness temperature and rainfall,
      152, 187
  characteristics, 147
  grid print map analysis, 149
  melting snow absorption and scatter,
      164
  non-beam filling, 159, 161
  Pacific Ocean rainfall, 194
  precipitation anomalies and foam,
      187, 192
  precipitation frequency, 192, 193
  and radar image, 150
  and radar rainrates, 160, 162, 163
  rain mapping, 147
  rainrate errors, 161
  smoothing and field of view, 159
  storm evolution, 153
  tropical storms, 152, 153
  Wilheit model, radar comparison,
      156, 157
ESMR-6
  brightness temperature, 175, 176
  differences from ESMR-5, 164, 165
  instrument improvement, 177, 178
  model evolution, 165–77
  moisture category categorization,
      168, 169
  rain map, 177
  SE US rainfall patterns, 169, 172,
      173
  variable surface emissivity, 171
  vertical and horizontal polarization,
      170

ESMR rainfall atlas
  information yielded, 195–200
  interannual variations, 199
  oceanic rainfall area, 195–7
  substructure, Indian Ocean, 197–9

F

Flash floods
  accuracy of earth location, 234
  anvil sizes, 231
  Big Thompson River, 1976, 223–8
  and image sequences, 234
  Johnstown, 1889, 228, 229
  Johnstown, 1977, 229–33
  Kansas City, 1977, 231–5
  rainfall quantity, 226–8, 233
  rainfall type, 225
  satellite cover, 229
Floods
  cost in US, 75
  forecasting, 52
  intense rains, 223–38
  lack of information, 75
  tropical storm, 236–8
Foam and surface wind, 187, 192
Food Information and Early Warning
    System of FAO, 264
Forests, destruction, 4, 251

G

GARP, *see* Global Atmosphere
    Research Programme
GATE, *see* GARP Atlantic Tropical
    Experiment
Gaut-Reifenstein model for tropical
    storms, 152, 153
General circulation, 6
Geostationary Operational
    Environmental Satellite
    (GOES)
  Colorado-Wyoming-Nebraska
      floods, 224, 225
  thunderstorms, 76, 78
Global Atmosphere Research Project
    (GARP)
  defined objectives, 297
  first (FGGE), 296, 297

Global Atmosphere Research Project (GARP) (*cont.*)
  global observation specifications, 298
Global Atmosphere Research Project Atlantic Tropical Experiment (GATE) 5, 62–4
  bi-spectral thresholding, 83–5
  convective scale rainfall maps, 65
  frequency of readings, 71
  Griffith-Woodley rainfall estimation, 71–3
  infrared rainfall mapping, 187–91
  longitudinal profiles, 192
  master array, 203
  rainfall mapping, 186, 187
  rainfall profiles, 192
Global Telecommunication System (GTS), 15
  decline in data, 15
GOES, *see* Geostationary Operational Environmental Satellite

H

High resolution picture taking system, 18
Hurricanes, 236, 237
  Anita, satellite monitoring, 236, 237
  bright cloud, 71
  cloud precipitation cross-section, 36, 37
  cloud shield, 37
  satellite estimate rainrate, 238
Hydrology
  component assessment, 308
  problems for satellites, 308
  satellite payloads, 310
Hydrology, applied 206–20, *see also* Catchments, Rivers, Water resources
Hydrometeor growth, 108
  and brightness temperature, 143

I

Ice, 130
  and brightness temperature, 154
  Marshall-Palmer distribution, 134
  refraction index, 130

Infrared images, 102–23
  composition, 103
  wavelengths, 103
Infrared measurements, value, 103
Infrared sensors and cloud temperature, 82

L

Langmuir chain reaction, 11
Latent heat release, cyclone, 204
Legendre polynomials, 134
Life-history methods, 65–81, 93, 311
  cloud area-echo area, 69, 70
  cloud area and location, 69
  dry regions, 80
  GATE, 71
  Griffith-Woodley, 69–75, 94, 236, 237, 311
  Scofield and Oliver, 75–81, 94, 225, 237, 311
    modified parameters, 79
  sequential pictures, 67
  Stout, Martin and Sikdar, 65–81, 94
  tropical storms, 236
Locusts (*Schistocerca gregaria*), 241
  breeding sites, 243
  Desert Locust Satellite Application Project (FAO), 243–5
  ground inspection, 245
  hazard maps, 247–9
  monitoring development, 244
  monitoring, flow diagram, 245
  population expansion factors, 243
  rainfall-dependent hazards, 246, 247
  vegetation mapping, 244

M

Marshall-Palmer drop size distribution, 133, 140, 141
  measurement, 153
  and temperature, 140
Microwave radiation
  active systems, 269–92, 309
  atmospheric absorption, 127, 128
  baseline microwave observations, 178
  beam pattern, 159
  brightness temperature, 125

Microwave radiation (*cont.*)
  characteristics, 125
  cloud particle absorption and
      scattering, 129–34
  drop scattering and absorption,
      129–31
  finite clouds, 141–3
  frequency and single-scattering
      albedo, 136
  ice, 130
  observation parameters, 144, 145
  passive methods, 144–78
  plane parallel clouds, 137–41
  polarization and object image, 126
  polarization plane, 125, 126
  properties in atmosphere, 124–43
  radioactive transfer equation,
      134–43
  scattering in clouds, 135
  sensitivity, 144
  sources and sinks, 127–36
  surface and materials, 127
  visible, infrared data use, 163
  water emissivity, 126
Microwave, passive systems, 144–78,
      301, *see also* ESMR-5
  history, 145
  in Nimbus-5, 145
  sensor characteristics, 146
Mie equations, 129–33
Model,
  anvil expansion, 38
  climate, fine tuning, 299, 301
  cloud geometry, 113–20, 141
  cloud motion, 56
  cloud in nature, 121, 122
  convective, 95–101
  crop production, 259
  cyclone, 34
  future methods, 177
  one dimensional cloud, 97–101
  precipitation, 2 km bubble, 98
  radiation transfer, Gaut-Reifenstein,
      152, 153
  Savage rain cloud, 165
  three layer cumulo-nimbus, 100
  Wilheit (1972), 147
  Wilheit (1977), 153, 192
Moisture budgets, 201, 202
  satellite rainfall techniques and, 202–5

Moisture soundings, accuracy, 23
Mosquito breeding grounds, Lanssat
      monitoring, 243
Multispectral scanners, 18

N

National Disaster Survey Report
  Big Thompson flood, 225
  Johnstown, 1889, 228–31

O

Oceans
  ESMR rainfall atlas, 195, 196
  macroscale rainfall mapping, 183–99
  mesoscale rainfall mapping, 201–5
  reflectivity backscatter, 273, 274
Oman
  altitude and rainfall, 208
  cloud indexing technique, 208
  rainfall, 208
  water resource evaluation, 207–12
Oxygen absorption spectrum, 128

P

Pests, weather sensitivity, 241
Polarization
  and discrimination, 168
  temperatures, 169
  vertical and horizontal, 170
Precipitation, *see also* Rain and
      Rainfall
  convectional cellular, 11
  stable upglide, 11
Precipitation index, 186
Precipitation process, 9, 10
President's Council on Environmental
      Quality, 4

Q

Qatar, natural water reserves, 208,
      209

R

Radar, 5, 7, 301
  antenna, directional efficiency, 270

Radar (*cont.*)
  area rain amounts, 93
  attenuation, 275–9
  attenuation and rainrate, 278
  calibration, 14
  cloud brightness, 38
  cloud precipitation relationships,
    33–5
  cumulonimbus echoes, 38, 39
  Earth reflectivity, 273–5
  echo delay, 283, 285
  fixed beam angle disadvantage,
    284–6
  gauge comparison, 74, 75, 226
  image, hand rectified, 150
  land water rainfall, 182
  meteorology, 269
  microwave comparison, 157, 160
  model of cyclone, 34
  not used on satellites, 269
  performance factors, 290, 291
  polarization, 280
  power received, factors affecting,
    280
  precipitation, design requirements,
    286
  rainfall physics, 270–81
  reflectivity, 271, 272
  satellite,
    geometry, 283
    present needs, 287, 288
    proposals, 281–92
  and satellite rain estimates, 67, 68,
    71–4, 83–5, 202, 226
  scan angle and coverage, 282, 283
  scanning pencil beam antenna, 288
  space shuttle, 288
  vertical discrimination and nadir, 284
  wavelength and rainrates, 290, 291
  weather, problems, 13, 14
Radiometer
  Electrically Scanning Microwave
    (ESMR), 47, 147–64
  multichannel, 25, 145, 178
  scanning, 22
  spin-scan, 22, 23
  temperature, humidity, infrared
    grid-point analysis, 151
  vertical profile, 23
  wavelengths, 22

Rain
  amount and echo area, 92, 93
  incidence, 9
Raincloud polarization, 167
Raindrops, *see also* Droplets
  polarization, 280
  reflectivity, 270–3
Rainfall
  anomalies and extremes, 238–41
  characteristics, 4–12
  climatic average, frequency, 4, 5
  and cloud area, 66
  and cloud top temperature, 39
  distribution over land and water,
    182
  duration-intensity curves, 6
  event duration, 5
  global average, 6, 7
  interannual variation, 199
  key to detection, 41
  lack in USSR, 242, 243
  major areas, sub-structure, 197–9
  maximum point, 5
  organization and distribution
    patterns, 6
  satellite radar, comparison, 67, 68,
    226
  tropical and cloud type, 69
  tropical Far East, satellite
    estimation, 184
  variability, 4–7
Rainfall data
  Asian, crop-growing demand, 59
  increased demand for, 16
  uses of, 16
Rainfall estimation
  accuracy and echo area, 60
  accuracy and intensity, 57
  catchments, 212–16
  and cloud area, 69
  and cloudiness, 60, 61
  comparisons, 62, 63
  decision tree equations 76, 77
  factors governing cloud brightness,
    123
  floods, 233
  flood, Scofield and Oliver *vs*
    raingauge, 77
  and observed, stability adjustment,
    99

Rainfall estimation (*cont.*)
  and radar estimations, 63
  research, 306
  satellite-gauge estimations, 70, 71,
      75, 77–9
  satellite overestimate, 212–16
  satellite-radar comparison, 67, 68,
      71–4, 82–4, 90, 91
  visible and infrared images, 102–24
Rainfall forecasting, development, 307
Rainfall inventories, 181–205
  macroscale, 181, 182
Rainfall mapping
  agriculture, 252–65
  gaps in continental, 200
  GARP, 186
  GATE, 186–91
  hazard maps for locusts, 241, 247–9
  macroscale over continents, 199,
      200
  macroscale over oceans, 183–99
      tropical far east, 183
  mesoscale over oceans, 201–5
  and moisture budgets, 201, 202
  Oman, 209–11
  reflective clouds, 183, 185
Rainfall measurement
  conventional, 12–14
  forest, 13
  radar, satellite, 281–92
  radiometer, 178
  from ships, 13
  sum of area and cloud area change,
      65
  verification of forecast, 51
Rainfall monitoring, satellite assisted
  active, 27
  agricultural areas, China and US,
      253
  cloud brightness, 31–9
  costs, 48
  delayed response, 27
  elements, 28
  existing systems, 22–6
  future of, 306–11
  future strategy, optimum approach,
      303, 305
  multi-satellite, 302–6
  multi-sensor approach, 302
  network deterioration, causes, 15, 16

Oman, 208–12
  paradigm, 307
  passive, 25–7
  and pests, 241–9
  potential users, 294–301
  precipitation network, 14, 15
  preliminary studies, 31–42
  present choice, 301–11
  satellite platforms 302–6
  single satellite, 302
  single sensor, 301
  status of, NASA view, 299, 300
  verification standardization, 306,
      307
Rainfall probabilities and satellite
      observations, 45
Rainfall processes, 9–11
  dynamics of, 10
  and evaporation, 10
  and temperature, 10
  and water droplet size, 10
Rainfall studies, significance, 3
Rainfall system, workmap for China,
      58, 59
Rainfall types, 11, 12
  conventional, cellular, 11
  intermediate, 12
  orographic, 12
  stable upglide, 11
Raingauges, 12, 13, 77
  accuracy, 47
  errors, 13
  International Reference
      Precipitation Gauge, 293
  problems, 13, 293
  -satellite data in Oman, 209, 210
  types, 293
Rainrate
  brightness and radiance, 86
  brightness temperature, 153
      background emissivity, 141
  and cloud area, effect of
      temperature, 98
  and echo area, 92
  and echo height, 140
  intensity and cloud size, 176, 177
  map, instantaneous, 7, 8
  and storm centre distance, 205
  Wilheit's maps, 192, 193
Rayleigh regime, 131–3

Reflectance, 33, 62, 112
    and absorption, 112
    cuboidal clouds, 115
    finite and semi-finite cubic clouds,
        114
    and thickness and clouds, 114
Reflective surfaces, 23
Remote sensing systems, *see also*
        Radiometer
    data analysis, 23
    disasters, 221–3
    radar, satellite definition, 17
River monitoring, 216–20
    gauging methods, 217
    Mekong, 219
    Sumatra, 218
    Upper Nile, 214, 215
    Upper Sinu, flow volume estimation,
        215–17

S

Satellite
    Applications Technology, 19, 37
    contributions to meteorology, 20, 21
    earth resource, 18
    Essa, 19
    geosynchronous orbit, 19
    hydrological, payloads, 310, 311
    hydrologic, payloads, 308
    -improved rainfall maps, 49
    Landsat, 18, 256, 257
    limitations of data, 21, 22
    Meteor, 19
    Molniya, 19
    Nimbus, 19
    NOAA, 18
    orbital period, 18
    polar orbit, 18, 19, 22, 47, 302, 309
    rainfall estimation and observations,
        57
    raingauge competition or
        complementation, 44
    rain map, 90, 92
    restricted and general purpose, 309
    synchronous meteorological, 19
    Tiros, 19, 31–3
    water assessment, 206–20
Satellite data
    infrared, 25, 39
    microwave, 25
    multispectral, 25
    objectives, 1971, 296
    user community needs, 309
    users, 295, 296
    visible, 23–5
Satellite-derived Global Oceanic
        Rainfall Atlas, 192
Satellite, environmental, 18, 19
    orbits, 18, 19
    utility, 20–2
Satellites, geostationary, 18, 22, 47,
        49, 51, 55, 65–81, 302, 303,
        309
    NASA and radar, 287
    network deficiencies, 216
    radar, 289
    sequential pictures and rain
        estimate, 67
Satellite monitoring
    complete areal indications, 47
    disasters, 221–50
    existing systems, 22–6
    first for weather, 19
    floods, 223–38
    future of, 306–11
    hourly estimates and radar, 67
    need for 14–16
    pests and plagues, 241–50
    present choice, 302–6
    rainfall applications, 205
    rainfall extremes, 238–41
    rainrates and radar, 203
    snapshot views, 31–7
    time-lapse views, 37–9
    types used, 18–20
    use of, 17, 18
Scatterometer, 275
Screwworm Eradication Data System,
        243
Sea level pressure chart, superimposed
        radar echoes, 32, 33
Sensors
    bi-spectral, 82
    microwave on satellites, 146
    scanning radiometers, 22, 23
    wavelengths used, 22
Snow, 164
Snowfall monitoring, 4
Soil emissivity, 127

Storms
    area–depth relations, 7–9
    evolutionary view, 153
    forecasting, 80
    satellite radar, 289
    tropical, Carolina images, 78
Strategy for global rainfall monitoring,
        303–305
    flow diagram, 304
    optimum approach, 303
    problems, 305
    sponsors, 305
Stream flow monitoring problems, 60
Subsynoptic-scale cloud masses, 11
Surinam River, Afobaka Dam areal
        precipitation, 214

T

Thresholding, 83–5, 91
Thunderstorms
    bi-spectral monitoring, 83, 86
    cloud brightness, 108
    estimation errors, 59
    evening and morning cloudiness, 55
    GOES images, 76
    Griffith and Woodley life-history
        scheme, 70, 71
    mass and latent heat transport, 38
    radar and raingauge data, 39, 40
Tropical islands, cumuliform
        convection "hot spots", 182
Tropics
    air mass convection, 97
    Ava, radiaton transfer, 152–5
    cloudiness, analysis, 195
    cumulonimbus, section, 38
    cyclone, latent heat release, 204
    estimation method, 64
    fewer rainfall stations, 15
    marine convection, 36
    organized weather systems, 45, 62
    rainfall intensity, 6
    rainfall type, 57, 62
    rain layer heights, 157

U

United States Climate Programme,
        297–301
    climate spectrum, 297, 299

    goals, 297
    models, 299, 301
    soil moisture and root zone, 299
Upper Nile, water resource
        development, 214, 215

V

Vegetation
    emissivity, 127
    monitoring and locusts, 244
    transformed index (TVI), 257
Vegetation flush, arid areas, 51
Visible images, 105–9

W

Water
    absorption spectrum of vapour, 128
    demand, 3, 4
    polarization, 167
    rationing, 207
    refraction index, 131
    roughness, 167
    shortages, 206, 207
    sources and properties, 3
Water droplets, size and rainfall
        dynamics, 10
Water emissivity, 126, 127
Water management, 308
Water resource evaluation, 206–12, 308
    Middle East, Oman, 207–21
    representative basins, 207
Weather
    CEAS reports, 239–41
    hazard types, 222
Weather forecasting
    Bristol, successes, 47
    computer use and cloud
        temperature, 81
    storm monitoring, 80
Wegener-Bergeron process, 10, 11, 108
Wheat harvesting predictions, 57
World Meteorological Organization, 14
    gauge networks, 14
    minimum network density, 14
World Weather Watch, 15
    satellite objectives, 295

Z

Zero buoyancy, 96, 97